Variability and Consistency in Early Language Learning

Variability and Consistency in Early Language Learning

The Wordbank Project

Michael C. Frank, Mika Braginsky, Daniel Yurovsky,
and Virginia A. Marchman

The MIT Press
Cambridge, Massachusetts
London, England

This book was set in Source Serif Pro and Source Sans Pro by Westchester Publishing Services. Printed and bound in the United States of America.

Library of Congress Cataloging-in-Publication Data

Names: Frank, Michael C. (Professor of human biology), author. | Braginsky, Mika, author. | Yurovsky, Daniel, author. | Marchman, Virginia A., author.
Title: Variability and consistency in early language learning : the wordbank project / Michael C. Frank, Mika Braginsky, Daniel Yurovsky, and Virginia A. Marchman
Description: Cambridge : The MIT Press, 2021. | Includes bibliographical references and index.
Identifiers: LCCN 2020025467 | ISBN 9780262045100 (hardcover)
Subjects: LCSH: Children—Language—Testing—Data processing. | Language acquisition—Research—Data processing.
Classification: LCC LB1139.L3 F66 2021 | DDC 401/.93—dc23
LC record available at https://lccn.loc.gov/2020025467

10 9 8 7 6 5 4 3 2 1

Contents

Preface

Overview

The emergence of children's early language is one of the most miraculous parts of human development. The ability to communicate using language arrives with incredible rapidity—most parents judge that their child is producing words with the intent to communicate before their first birthday (Schneider et al., 2015), and the onset of comprehension is even earlier (e.g., Bergelson and Swingley, 2012; Tincoff and Jusczyk, 1999).

New words enter children's expressive vocabularies slowly at first, but this process accelerates over the second year such that children reach an average of 300 words by 24 months and more than 60,000 by the time they graduate from high school (Fenson et al., 2007). At the same time, there are significant individual differences in the rate and timing of language acquisition. For example, although some 18-month-olds already produce 50 to 75 words, others produce no words at all and will not do so until they are two years or older (e.g., Brown, 1973; Bloom, 2002; Clark, 2016).

How do children learn their first language? To what extent do different children and children learning different languages follow the same path into language? Are these paths similar or idiosyncratic? These questions about the consistency and variability of early language lead directly into the central question of language acquisition: What are the mechanisms that lead to the emergence of human language?

Answering this question is complicated immensely by the fact that there is no one single event that constitutes language acquisition. Early language learning involves the accumulation of thousands of words, a host of grammatical rules and constructions, and the pragmatics of language use. This process takes place over the course of years of growth and millions of separate interactions. This problem of time scales makes measurement a tremendous challenge. In addition, during the period in which language emerges, language ability varies wildly from child to child, and most children are at best reluctant experimental participants. Accurate measurement of language development across individuals is thus a major challenge.

Parent report is one powerful method for addressing these challenges. The MacArthur-Bates Communicative Development Inventory (CDI) is a simple survey instrument that measures early language outcomes and that was designed to address these issues.[1] The CDI is a checklist for parents to fill out to report on their child's progress in language. In different versions of the form, parents mark whether their child "understands" or "understands and says" particular words out of a list of several hundred. Separate sections for gestures, word forms, and grammar are also present in some versions. Despite their simplicity, over the past 25 years of use, CDI forms have been shown to be reliable and valid measures of children's early vocabulary size as well as being a valuable tool for measuring other aspects of early language. In addition, CDI forms have been adapted to more than 100 different languages around the world. Research based on the CDI has contributed tremendously to our understanding of the growth of language in early childhood.

In this book, we examine the question of variability and consistency in early language through the lens of the CDI. Our book is an outgrowth of the Wordbank project (Frank et al., 2016a), which has as its goal to archive CDI data in a structured format so that the data can be explored and analyzed in the service of describing early language. The database currently contains data from more than 82,055 CDI form administrations across 29 languages or dialects. Wordbank is also continuously growing as new researchers contribute data. We believe this database is the largest and most diverse set of data on early language acquisition currently in existence.

The first contribution of this work is unification. Over the course of our work with Wordbank, we have developed a consistent framework for representing and analyzing CDI data. This framework allows us to unify a variety of influential previous analyses of CDI data, from sex differences in learning to the relation between grammar and the lexicon and the representation of nouns in early vocabulary. Research in early language learning often is based on a fragmentary empirical picture where many important theoretical conclusions are based on analyses of transcripts from a small number of children or on analyses of experimental or parent-report data from English learners only. We hope that bringing together

1. For purposes of clarity and ease throughout we refer to CDIs (the family of instruments) rather than the MB-CDI (the particular English forms). We will use this term throughout even though technically some of our data come from "checklists" containing only vocabulary items rather than true Communicative Development Inventory forms.

a large set of analyses of vocabulary data and implementing them consistently, openly, and reproducibly on the same dataset will help to create an empirical starting point for future work.

The second contribution of this work is generalization. By creating a systematic framework for evaluating prior analyses of CDI data, we are in a position to notice parallels among analyses or paths left untaken by previous analysts. For example, after systematizing our treatment of noun biases in chapter 11, we apply the same tools to analysis of semantic categories in chapter 12.

The third contribution of this work is to develop theory that treats the question of consistency and variability in early language. We introduce the notion of a continuum describing the relative consistency of phenomena across languages and cultures. At one end of this continuum are phenomena that exhibit substantial consistency and hence can provide clues to "process universals"—the aspects of the process of early language learning that may be shared across learners in very different circumstances. This notion contrasts with notions of "content" or "structural" universals in which specific principles regarding the structure of languages are innately given. The general idea of universals of process has a long history in the field (e.g., Bates et al., 1989; Clark, 1977), but the Wordbank project provides an opportunity to apply new empirical data and analytic power to these ideas.

This set of contributions reflects an important guiding principle of our work here. Studies that at first glance seem like "mere replication"—in which a particular analysis is replicated on a larger dataset—are in fact important opportunities for theoretical development. Replicating with a larger dataset typically leads to a more precise estimate of the phenomenon of interest that can be used not only for confirmation but also for quantitative comparisons and computational modeling. But this increased precision then allows for the examination of variability and consistency across meaningful units like children, words, instruments, or languages. Such estimates in turn are—as we argue throughout—relevant to foundational theoretical questions. Thus, there is never "mere" replication. More precise measurements sit hand and hand with better theory (Greenwald, 2012).

Outline

The first chapters of this book provide an overview of the practical and theoretical issues that we cover. Chapter 1 gives a broad theoretical overview of our claims and sets up some of the empirical themes that we return to throughout—especially the notion of using the consistency and variability of phenomena to make generalizations about the process of acquisition. Chapter 2 then discusses the practicalities of the CDI and the Wordbank project. Chapter 3 gives a methodological and

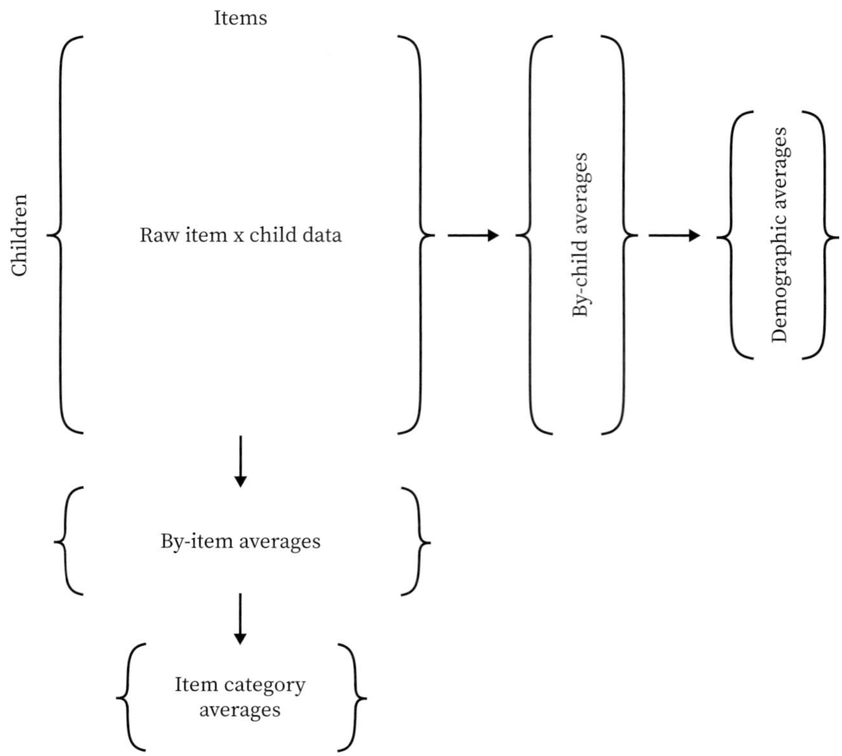

Figure P.1
A graphical outline of the book.

descriptive overview of the dataset we analyze. And chapter 4 discusses the psychometric properties of parent report data, addressing questions about the strengths and limitations of this method.

Chapters 5 through 15 then form the empirical heart of the book. Each applies a particular analysis of interest to our dataset. Although our goal is a full exploration of the phenomena of child language acquisition, the analyses we report are constrained by the structure of the data in Wordbank. At its heart, the individual instrument datasets stored in Wordbank are matrices of item by child data (see figure P.1).

Considering the data in this way leads to a number of obvious data analytic strategies, many of which correspond directly to previous approaches to CDI data. For example, averaging across items leads to *by-child averages* where each child receives a comprehension or production score. Chapter 5 considers this view of the data, examining developmental change and variability in such data. Then in chapter 6, we report the ways that vocabulary growth varies across child sex, birth order, and maternal education (a rough but cross-culturally valid proxy for socioeconomic status).

Averaging across the other margin of the data leads to *by-item averages*. These can be examined in a number of different ways. Gesture items and their growth trajectory—and relationship to overall vocabulary size—are examined in chapter 7. Chapter 8 then considers the growth trajectories for individual words, focusing especially on early vocabulary. Chapters 9 and 10 follow this approach and attempts to predict the trajectories of individual words based on both environmental and conceptual features of these words. This last approach calls for the incorporation of other resources, and so we use a variety of English and cross-linguistic resources to supplement Wordbank data in this chapter.

Next, we investigate the grouping of items into categories (both syntactic and semantic). Chapter 11 considers the categorical composition of early vocabulary, giving special consideration to the "noun bias" that is found in many—but not all—of the languages in Wordbank. Chapter 12 adopts the same approach for semantic, rather than syntactic, categories. This approach leads us to consider other aspects of morphosyntax that are reflected in the CDI forms. Chapter 13 explores the relationship between vocabulary growth and the growth of grammar. Chapter 14 examines morphological generalization. Finally, chapter 15 returns to the question of individual variation using tools built up in previous chapters to quantify differences in the style and trajectory of learning across children.

The book ends with three synthetic chapters. Chapter 16 synthesizes observations across languages for the preceding chapters to quantify variability and consistency directly across phenomena. Chapter 17 considers the question of what the process of language acquisition looks like from the bird's-eye view afforded by our data. Lastly, chapter 18 considers methodological morals from our work here and the outlook on measuring children's early language beyond the CDI.

Finally, the appendices describe and cite the individual datasets reported in Wordbank and provide supplemental analyses validating particular analytic practices that we adopt.

How to Read This Book

You can read this book as a narrative monograph. If you intend to do so, we recommend you read chapters 1, 2, 3, and 4 before moving on to substantive chapters of interest.

You can also read this book as a reference. If you take this approach, just dive into any chapter that is of interest, knowing that you may need to use some of the terminology defined in chapter 3 to interpret the constructs and analyses that are used. Further, you may have concerns about the reliability and validity of CDI-type instruments; these are addressed in chapter 4.

A number of the analyses reported here were first described in earlier conference proceedings or publications (e.g., Frank et al., 2016a; Braginsky et al., 2015, 2016, 2019). Rather than reprinting these verbatim, this manuscript updates them using the unified analytic approach and larger dataset described in chapter 3. The version of these analyses represented in this manuscript should be considered more definitive than any previously published or presented version.

The dataset in Wordbank is constantly growing and changing as we add new features, new data, and new languages. In addition, as users of the data occasionally identify issues and errors, we make corrections to the database. In writing this manuscript, we have attempted to find a middle ground between a completely dynamic document that responds to any change in the database, on the one hand, and a traditional, static book, on the other. A static manuscript would be a shame given the potential for dynamic extension and updating with new data. However, if the data were completely dynamic, any claim we made in prose would risk being out of date almost as soon as we wrote it.

For this reason, we work using snapshots of the underlying database. Every so often, we will return to the manuscript and recompile the online edition and then check references to the data that might have changed.

Acknowledgments

Our sincere thanks go to all of the generous researchers—too many to name here, but listed on the Wordbank contributors page (http://wordbank.stanford.edu/contributors) and in appendix A—who contributed their data to the database. Even during our time working on this project, norms of data sharing have shifted; some substantial portion of this shift is due to the generosity of those researchers who shared their data early on in the process. Thanks especially to Rune Nørdgard Jørgensen for sharing the full Cross-Linguistic Lexical Norms (CLEX) CDI dataset (Jørgensen et al., 2010) and for his kind assistance in the transition from the CLEX-CDI website to the Wordbank website.

Major thanks are also due to the MacArthur-Bates CDI Advisory Board, especially Philip Dale and Larry Fenson, for their continued intellectual, financial, logistical, and personal support of the Wordbank project.

Thanks to Danielle Kellier for substantial work importing data and maintaining the Wordbank website as well as updating the universal lemma mappings. Initial programming work was done by Ranjay Krishna, and some database imports were performed by Elise Sugarman.

Thanks to NSF Award #1528526, "Wordbank: An Open Repository for Developmental Vocabulary Data," for financial support of the Wordbank project as well as to the Stanford Psychology Department for a small seed funding award that supported the initial development of the site.

This book benefited from thoughtful comments by Philip Dale and Eve Clark as well as comments on chapter 11 by Dedre Gentner.

1 Theoretical Foundations

One of the defining human characteristics is the ability to use language in its lexical and combinatorial richness. The study of language acquisition has been a traditional locus of our search to understand the nature of this ability. What allows human children to acquire a language has been the subject of one of the historical "great debates," in which highly diverse proposals about the architecture of the human mind and the nature of human uniqueness have been formulated. Does language arise from domain-specific adaptations for syntactic structure (Chomsky, 1981, 2014)? Or does it arise from a combination of environmental input and sophisticated, general-purpose learning mechanisms (Elman et al., 1996)? Two poles have traditionally emerged in this discussion: from domain-general empiricist proposals to domain-specific nativist proposals, as illustrated in figure 1.1.

In this chapter, we begin by presenting the perspective from these poles but contend that they are rarely helpful in the practice of understanding the scope and course of early language learning. Instead, we argue for the development of theories that describe the scope and course of language learning as a whole as well as its quantitative variation across children, languages, and cultures. These concerns lead us to frame our own study in terms of a set of distinct theoretical issues: capturing consistency and variability; drawing connections across time scales of learning; and the notion of process, rather than content, universals. We end by discussing the relation of our theoretical stance to others and the roles of replication and larger datasets in building quantitative theories.

1.1 The Picture to Date

Nativist proposals regarding language acquisition have typically been tightly focused on syntactic development, assuming that children' phonology, morphology, and lexicon are learned from their input. Within the domain of syntax, the emphasis is on the complexity of the grammatical structures that young children

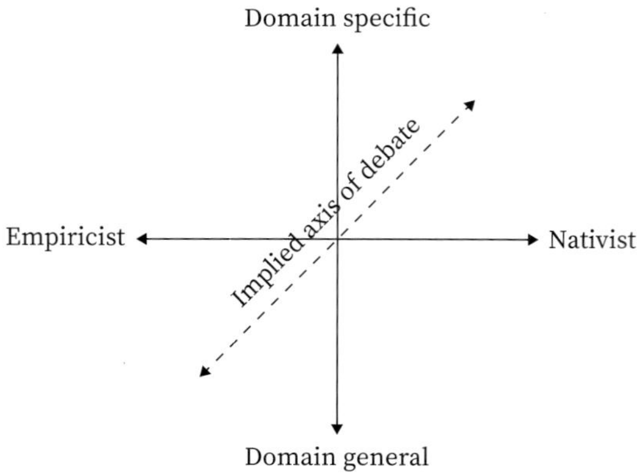

Figure 1.1
Schematic of the space of theoretical debates around language acquisition.

appear to show mastery of (e.g., Crain and Thornton, 2000). To the extent that these are rare in children's environment, this rarity could suggest an innate foundation for children's knowledge (e.g., Legate and Yang, 2002). The type of proposals that are on offer within this space often include "principles and parameters"–type theories, in which languages share a set of syntactic principles that govern syntactic combination but vary on a relatively small set of parameters that control how these structures vary (Baker, 2005; Yang, 2002).

These proposals are challenged by cross-linguistic variation in syntactic abstractions (Evans and Levinson, 2009), by the character and scope of children's syntactic generalizations (which are often tied to specific lexical structures; Tomasello, 2000), and by evidence of early input-sensitive learning and generalization both within specific domains (e.g., Meylan et al., 2017) and in artificial language tasks (Gómez and Gerken, 1999). Such proposals also tend to downplay the inherent variability that characterizes language learning across individuals (Bates et al., 1988, 1994), focusing instead on the universals or commonalities that exist across children. While nativist proposals acknowledge that individual variation exists, variation is not typically used as a lever into the process of acquisition, existing in a "theoretical vacuum" (Bloom, 2002, p. 52). Far greater emphasis in this tradition is placed on how language acquisition works in the general case.

In contrast, empiricist proposals for language learning emphasize the ability of children to learn structure across domains, the richness of the distributional input that children are exposed to, and their ability to create appropriate abstractions from structured input. One important component of these proposals is children's general statistical learning abilities (Saffran et al., 1996; Fiser and Aslin, 2002). Work in this tradition demontrates that even quite general statistical tools can recover some aspects of linguistic structure across a variety of domains when they are applied to sufficient data (Redington et al., 1998; Frank et al., 2012; Elman et al., 1996). Further, general statistical learning abilities are likely extremely useful for learning sound structures (Feldman et al., 2009) and vocabulary (Smith and Yu, 2008). Thus, one strength of more empiricist proposals is their ability to deal with a broader set of acquisition phenomena than syntax alone.

Empiricist proposals regarding linguistic structure are sometimes referred to as "emergentist" or "constructivist" accounts. A primary challenge of such accounts is that, even at an early age and with limited input, children sometimes show evidence of abstractions that encode key aspects of linguistic competence. For example, young children show some systematicity in the word order of their productions even in the absence of structured input (Goldin-Meadow and Mylander, 1983; cf. Bowerman, 1973). Further, at least by some time in their third year, typically developing children appear to interpret the arguments of novel verbs (Gertner et al., 2006) and to show evidence of category structure for syntactic categories such as determiners (Yang, 2013) and structures such as the dative (Conwell and Demuth, 2007). These results are sometimes intepreted as suggesting that while general learning mechanisms are necessary for language acquisition, they are not sufficient—especially in the domain of syntactic structure.

The "great debate" between these viewpoints is philosophically appealing but has also led to a polarization of the field of language acquisition. Typically, researchers work in their own siloed traditions (empiricist or nativist) and focus on individual phenomena that do not make contact with one another—a classic version of Kuhn's (1970) "paradigms." Research in the nativist tradition often focuses on particular syntactic phenomena that are largely neglected in the empiricist tradition (e.g., the "optional infinitive," Wexler, 1998; but cf. Freudenthal et al., 2010). In contrast, research in the empiricist tradition has often used artificial language learning tasks that are argued not to reflect the underlying structural properties that are claimed to be innate (e.g., Lany and Saffran, 2010; cf. Yang, 2004). Empiricist theories are also more likely to recognize the causes and consequences of individual variation in rates of learning, including potential sources of that variation that arise from variations in the circumstances in which children learn and the opportunities they have for engaging with language in a meaningful way (e.g., Huttenlocher et al., 1991; Hoff, 2006; Weisleder and Fernald, 2013). By focusing on different paradigms and phenomena and theorizing using distinct vocabularies, these traditions make limited contact with one another. One prominent language development conference has traditionally featured two distinct plenary talks, one from each of these two different traditions—a clear sign of polarization.

To be tested, proposals must make specific predictions. But these theories are frameworks, and few proposals within these frameworks can be said to generate testable and clearly competing predictions, even within a specific language and linguistic domain. Any individual observation typically cannot be said to be inconsistent with any but the absolute strongest nativist or empiricist position. Computational models have been an important method for allowing proposals to be instantiated to the degree that they can make testable predictions. In practice, however, models typically end up less differentiated than framework rhetoric suggests. To get off the ground in performing a particular empirical task, theories must often help themselves to generous amounts of both innate structure in the form of structured inputs from social, cognitive, or perceptual domains and statistical learning abilities (Roy and Pentland, 2002; Alishahi and Stevenson, 2008; Frank et al., 2009, 2010; Yang, 2004).[1]

Further, when nativist and empiricist viewpoints do make differing predictions, they are often in phenomena that—from a bird's eye, or even a parents' eye, view—are relatively trivial in the general course of language development. This observation is especially true in the domain of syntax. Abstraction debates have played out in the acquisition of the definite determiner "the" (Valian, 1986; Pine and Lieven, 1997; Yang, 2013; Meylan et al., 2017), auxiliary inversion (Pullum and Scholz, 2002; Legate and Yang, 2002), and the use of anaphoric "one" (Akhtar et al., 2004; Regier and Gahl, 2004; Lidz et al., 2003), for example. These highly specific and complex phenomena are occasionally observable in the children of linguistically trained parents, but even the closest observer would be forgiven for being more compelled by watching the increasingly creative and complex ways that children interpret, use, and play with language rather than the occasional syntactic slip. Further, all of these phenomena reflect a broader historical argument that syntactic structure is the heart of the uniquely human language faculty (Chomsky, 1957) and that other aspects of language tend to be shared with other species (Hauser et al., 2002) and are therefore somehow less interesting and less critical as objects of study.

From an evolutionary perspective, however, syntax is far from the only unique or notable feature of human communication (Pinker and Jackendoff, 2005; Tomasello, 2010). The nature and range of children's communicative gestures, the variety of sounds used in human languages, and the diversity of lexical items available to speakers are all relatively unprecedented, especially within the primate lineage. And these observable aspects of language—as well as the emergence of syntactic structure more broadly—are some of what makes the broad course of language

1. The research on the nature of inflectional morphology—the "past tense debate"—is one place where computational models have played a foundational role in instantiating theoretical claims about innateness and representational structure (e.g., Rumelhart and McClelland, 1986; Pinker and Prince, 1988; Plunkett and Marchman, 1991, 1993, 1996; Marcus, 1995; Marchman, 1997).

acquisition striking from the perspective of a researcher, clinician, or a parent. As observers, we notice the first communicative signals, the emergence and rapid growth of vocabulary, the beginning of the productive combination of words, increases in the length and complexity of utterances, and the patterns of error and overgeneralization that remain in early childhood. Moreover, there is considerable evidence for continuity across these domains (e.g., Bornstein and Putnick, 2012; Tsao et al., 2004; Bates et al., 1988; Cristia et al., 2014). Children's earliest gestures and sounds relate to their oral language comprehension and production. And these in turn relate to later skill in using language as a tool for learning, through both auditory (or visual, in the case of sign language) and written modalities.

These broader patterns of language learning are the natural focus of investigations like ours that use parent report to learn about children's language. Parents are attentive and accurate observers of communicative gesture, vocabulary, and word combination. But without linguistic training they may not even notice subtleties like nonproductive determiner use, auxiliary inversion, or anaphoric "one." Further, investigations of the broader dynamics of acquisition using parent report can, in many cases, make productive contact with the rich literatures on early communication, speech perception (e.g., Kuhl, 2004), word learning (e.g., Bloom, 2002; Snedeker, 2009), and grammatical productivity through verb structure (Fisher et al., 2010). While debates over the nature of syntactic knowledge and abstraction have raged, these subfields of language acquisition that focus on more directly observable phenomena have prospered. Such research makes at most limited contact with broad questions of nativism and empiricism, in part because these fields deal with phenomena that are language-specific—sounds, lexical items, grammatical constructions—and hence, *must* be learned from the input. The question is then about the mechanisms and constraints that guide this process of learning rather than about any posited universal or innate content (even at the level of abstractions).

1.2 Making Progress

A unifying framework is necessary to move beyond great debates. What should a unifying theoretical framework for language learning look like? In spite of the critiques above, we still believe in the importance of the search for core, universal aspects of language learning that elucidate the process by which children acquire this uniquely human ability. Yet we also believe that the sort of theory that describes such universals will likely look radically different from its historical antecedents. We will now sketch some aspects of what such a theory might look like and how this vision connects to our present investigation.

We start with the observation that any universal of language acquisition is likely to be a statistical or quantitative universal; we refer to these as "consistencies." The variation across the world's languages is such that very few properties will be truly invariant (Evans and Levinson, 2009). Further, we are unlikely to be able to access the kinds of samples that would allow us to make claims of universality (one estimate suggests that strong statistical support for claims of universality requires evidence from more than 100 languages from diverse families; Piantadosi and Gibson, 2014). Thus, we will talk about the relative consistency and variability of particular phenomena rather than any sorts of absolutes, and we expect any empirically grounded theory will do so as well.

In addition, language learning takes place at the time scale of years. A parent's responses on a Communicative Development Inventory (CDI) form provide a global snapshot of a child's language at a particular point in time rather than demonstrating the operation of a mechanism or principle. Substantial reconstruction is necessary to understand how processes operating over seconds—for example, online statistical learning (Aslin and Newport, 2012) or pragmatic inference (Bohn and Frank, 2019)—would result in particular structures accreting over time in the vocabulary. Thus, consistencies we observe are at best the basis for abductive inferences about underlying mechanisms.

These consistencies are consistencies in the learning of vocabulary and constructions rather than syntactic rules. Words and constructions *must* be learned from data. Thus any putative universals identified in our investigation must not be "content universals" that specify particular grammatical rules or linkages. They must be "process universals" in the sense that they specify mechanisms or processes that unfold over time and operate over children's interactional input in ways that produce the observed consistencies.

Before we begin spelling out these ideas in more depth, we note that there are several important precedents for this line of argument. First, harkening back to Bates et al. (1988) and Elman et al. (1996), our proposal is a shift from a focus on consistencies in representational content to a focus on consistent learning mechanisms. Second, Clark (1977) distinguishes between "process" and "product." In the language of this distinction, our work in this book aims at uncovering *processes* that lead to the *products* that parents observe and report on the CDI. Finally, these "process universals" are likely to be similar to the "operating principles" discussed by Slobin (1973; 1985), although Slobin's initial idea was that these would be elements of the Chomskian Language Acquisition Device and hence likely language-specific. In contrast, as we will see in this chapter and throughout our narrative, in our analysis many apparent process universals seem to be driven by nonlinguistic factors such as children's general statistical learning abilities as well as the ways they allocate their attention to the world around them.

1.3 Variability and Consistency

A starting point for our analysis is the observation by Bates et al. (1988) that what "hangs together" in development can provide clues about the architecture of the underlying language learning mechanisms. We think of these correlations—the statistical instantiation of "hanging together"— as providing targets for theorists. A successful theory can gain support by providing accounts for these correlational observations as its explananda. Crudely put, if a theory posits that some aspect of language acquisition is universal, it should be relatively more consistent in our data.

What are the units over which we should compute variability and consistency? We refer to these as "signatures"—loosely, measurements that can vary across populations. In practice, a signature can be the output of any analysis, with the simplest being vocabulary size or variability itself (as in chapter 5). Signatures are linked to particular theoretical goals by arguments about the validity of an analysis—for example, the argument of Bates et al. (1994) that the over representation of nouns in early vocabulary is a meaningful dimension of variation among individuals. A signature for our purposes is thus an analysis that yields a set of numbers. In nearly every chapter of this book, we define one or several signatures whose variability we can measure.

Different sources of variability provide different sorts of evidence. One sense of the notion of "universal" that dates to early generative syntax is the notion of typological invariance—invariance across languages (Greenberg, 1963). Following this general idea, we focus on variability in some signature across languages in the majority of the book. The implied inference is that consistency across languages in some signature implies that the signature results from a mechanism that is independent of the language being learned and the context in which it is learned. (We will discuss more on inferences about mechanism below.)

When we assess the variability of some signature across languages, we are also examining variability across datasets; hence, many things vary that are not the target of our inference. Different datasets are constructed by different researchers with different goals. They use different instruments with different items, different lengths, and different categorical compositions. These instruments are administered to different samples with different sampling strategies. The nature of the administration of the instructions—whether the form is given online or on paper—is sometimes different as well.

Thus, when a particular response appears to be consistent, we can say a fortiori that none of these sources of variation appear to have affected the consistency of the response. (Or at very least that, if they have, they have canceled each other out in a highly nonrandom way.) When variability does occur, we cannot make the opposite inference. Thus, variability has many explanations, but consistency tends to point us toward a single inference.

We focus on cross-*dataset* variability as the primary source of variability in this book. We refer to this variability throughout as cross-*linguistic* variability, though in fact there are a number of caveats that must be stated. First, many things vary among datasets far beyond language (as noted previously). And second, some datasets represent the same language in different dialects (e.g., Australian and British English). Some even reflect the same language and dialect and are measured using two different instruments (e.g., the two Beijing Mandarin datasets). In some cases, we can even leverage these parallels to help us rule out alternative explanations. There are three reasons we focus on cross-dataset variability.

First, datasets vary so much, even assuming this variation is somewhat random, that claims of consistency are that much stronger when they emerge from this sort of data. Imagine, counterfactually, that all of the instruments we used had exactly the same structure and item set and that all were administered identically. Certainly, this lack of variation would make our life easier in a number of ways when making quantitative comparisons among datasets! But it also then means that these consistencies would be confounded in our data—particular item sets (plausibly) or administration instructions (somewhat less plausibly), or both, could be sources of an observed consistency in the data. In contrast, while the messiness and inconsistency of the data in the Wordbank dataset make many aspects of our analysis much harder, it actually increases the strength of the inferences we can draw when, despite this, we see some phenomena emerge with striking consistency.[2]

Second, the genesis of the investigations documented in this book was in part the observation that several phenomena that we examined were strikingly consistent across languages. For example, the gender effects shown in chapter 6 were much more consistent than any of us thought since we were at the time ignorant about the previous literature in this particular area (Eriksson et al., 2012). Empirically, when we browsed the Wordbank dataset, we found a lot to look at that was both surprising and interpretable when we examined well-known signatures across languages. Thus, our motivation was in part circular: the emergent success of the cross-dataset approach made us curious about what comparisons were possible.

The final reason we consider cross-dataset variability as our primary lever is a negative one. The obvious competitor as a source of variability is variation across individuals. We examine this variability to some extent in chapters 4 and 5 and even more extensively in chapter 15. While we document substantial and

2. Of course, we also consider the confounds that do remain. In particular, confounding related to the parent report structure of the CDI is a major risk that we discuss at length in chapter 4 and also in chapter 18.

stable variability across individuals (echoing Bates et al., 1994), this variability empirically proves to be less of a lever into theoretical issues of interest than we would hope. One reason for this is data-related—we have far more cross-sectional than longitudinal data in the Wordbank dataset—and hence, we cannot track stability and change over time as easily or powerfully as we would like (but cf. chapter 13). Further, we have very few additional measures on most children in the dataset (beyond the broad demographic features, such as maternal education, that we report on in chapters 6 and 9). In addition, as we show in chapter 15, though there is some reliable stylistic variation among children, much of the apparent variability in children's style of language learning can be traced to variation in learning rate. Thus—and in contrast to the exciting emergent conclusions from cross-linguistic variation—for us, individual variation appears to be a less powerful theoretical lever.

Nevertheless, individual variation across individuals does exist, and it is robust. Indeed, in chapter 5, we show remarkable consistency across languages in the extent to which there is variability across individuals. We remain optimistic that continuing to explore the extent and consistency of this variability will provide a window into the universal processes that guide learning and help to define the upper- and lower-limits on typical development. Our pessimism about the promise of theorizing around individual variability is one that is limited only to the current dataset. In the longer term we are optimistic about the lessons that can be learned from individual variation in language learning.

In chapter 16, we bring together estimates of variability of individual signatures from each of the earlier constituent chapters. We combine these into a single, data-driven continuum from consistency to variability and use this continuum to drive speculations about the sorts of mechanisms that would produce the observed pattern of data. We next consider some theoretical constraints for these speculations.

1.4 Process Universals

1.4.1 Preconditions

Imagine we were to uncover an aspect of language development that was completely consistent across languages. (Surprisingly, as we will see in chapter 17, there are some!). What could we then infer from this observation? Not much, it turns out. The observed regularity could be due to different sources in different datasets or it could be uninteresting from a theoretical perspective.

First, even if the consistency were interesting, any inference from it will always be abductive—an inference backward from observation to cause. These abductive inferences will always be underconstrained and tentative; thus they will always be at best empirically grounded speculations that should be brought together with other data to make a test. In some sense, this is the fundamental caveat governing our entire enterprise. Our research design is correlational, and so definitive causal inferences are not possible.

Inferences can go wrong even within this more limited correlational paradigm. For example, we might observe that across languages, a particular word (or more precisely, the set of word forms corresponding to the same concept) was always produced earliest. It could be the case that the word happened to be learned earliest in some languages because it was short and easy to pronounce while in other languages, it was learned early due to a high frequency of usage in the input. This example illustrates the difficulties of reverse inference from consistency. Similarly, we could observe that a certain distributional form always described children's vocabulary estimates across languages. This regularity could be due to the operation of the central limit theorem rather than any interesting or substantive mechanism that we might be interested in as psychologists.

These problems mean that we need to have two (somewhat informal) conditions on the consistencies that we posit. First, we need to consider the possibility of multiple routes to the same observed consistency. To the extent that observed regularities are specific and surprising, it will be less likely that there are multiple routes across different languages to observing the same thing. Second, for any potential causal story that we posit, we need to be able to hypothesize a plausible or interesting causal story that does *not* generate the observed regularity. The tightness of this comparison with a counterfactual governs the strength of the inference.

1.4.2 The Nature of the Processes

Suppose a consistency we identify meets the conditions we describe above. It is sufficiently surprising that we do not see a parsimonious story for how the data for different languages could have been generated by different processes; there are close counterfactual circumstances in which this consistency would not emerge. Further, in our example, suppose we have a sufficiently large and diverse set of languages and cultures represented in our dataset such that we can justify using the title "universal" rather than the more descriptive and limited term "consistency." We can then imagine trying to constrain the nature of the sort of universal that could give rise to this type of consistency.

What can we say about such putative universals? First, to the extent that they arise from reports of children's vocabulary, they cannot be universals of *content*. Words are learned from a set of specific interactional circumstances in a child's history (e.g., a trip to the zoo where a giraffe was seen for the first time). Thus, there is no viable sense in which possible universals for learning of this sort could be content universals; the English word "giraffe" is not innately given. For this reason, we describe these putative universals as "process" universals: they relate to the process by which each individual extracts generalizations from their own idiosyncratic experiences.

Further, since these processes are fundamentally learning processes, they operate at the time scale of moment-to-moment interactions (Frank et al., 2009; McMurray et al., 2012). Using the CDI, we observe the accretion of vocabulary and linguistic competence over the course of millions of these interactions (see, e.g., Dupoux, [2018], for estimates). This mismatch in time scales makes inferences about the nature of the processes even trickier. We are studying "macroeconomic" indicators about the global outcomes of the learning system, hoping to connect these to the microeconomic dynamics of individual "markets" for specific words.

Yet from the perspective of the language learning literature, there are still obvious candidates for the sort of process universals we are talking about, the general idea of "statistical learning" is one (Saffran et al., 1996; Aslin and Newport, 2012; Saffran and Kirkham, 2018). From a very early age, children are sensitive to regularities in their sensory environments and track statistical associations among elements in these environments. Concrete examples of these mechanisms in action include the tracking of syllable-to-syllable conditional probabilities (Saffran et al., 1996) and the tracking of word-referent correspondences (Smith and Yu, 2008), but in principle these mechanisms are likely operating over every level of representation present in early language (Shukla et al., 2011; Dupoux, 2018).

This viewpoint is quite consistent with the general perspective of "language as skill learning" advocated by Chater and Christiansen (2018). Since all aspects of language learning are subject to the general dynamics of repetition and practice, we should expect to see effects of word frequency pervasively throughout our data—and indeed we do. These effects manifest themselves directly in chapter 10 as well as perhaps more indirectly in the demographic variation we study in chapter 6.

Processes of statistical learning operate in a social and pragmatic context (Bohn and Frank, 2019). Statistical learning processes operate over social input that includes information from social partners (Yu and Ballard, 2007). In addition, it is likely that statistical learners take into account the nature of the social context in the inferences that they make (Frank et al., 2009; Shafto et al., 2012; Frank et al., 2013). Such social inferences are the focus of much research (including our own). But because they are among the most ephemeral aspects of in-the-moment interactions, they will probably be visible only very indirectly in our data (for example, as they are affected by sociodemographic variation, e.g., in chapter 6).

Processes of generalization are also strong candidates for process universals. Because the CDI does not query the details of lexical meaning, we cannot detect conceptual generalization principle—though these are likely at work (Xu and Tenenbaum, 2007)—but we can look for some evidence of morphosyntactic generalization mechanisms. The nature of these generalization mesomechanisms is highly controversial (for all the reasons discussed above), but every account of learning requires some type of generalization from specific lexical items to syntactic constructions or morphological rules (Tomasello, 2003; Yang, 2016). We discuss constraints on these generalization processes in chapter 13.

In addition, a number of process*ing* factors might lead to *processes* that are universal. At the same time as children's language abilities are growing, a variety of core aspects of cognition are undergoing developmental change as well. Children's general speed of processing is changing (Kail, 1991; Frank et al., 2016b)—including changes in memory (Ross-sheehy et al., 2003; Rovee-Collier, 1997), attention (Colombo, 2001), and executive function (Davidson et al., 2006). Processing factors also influence the speed and efficiency with which children can comprehend words in real time, linking both to developmental change and individual differences that are stable and meaningful and that extend beyond language (Fernald and Marchman, 2012; Marchman and Fernald, 2008; Fernald et al., 2006). Although much of the developmental literature on these cognitive constructs focuses either on infancy or the preschool years—because one- and two-year-olds are hard to measure with standard cognitive psychology tasks—the assumption is that these processes are developing continuously throughout the period we focus on. These developmental changes mean that the processes we describe are not static but—inasmuch as they draw on these capacities—are themselves a moving target.

Finally, it is important to note that process universals need not be internal to the child.[3] Instead, they could be universals of interaction between children and their caregivers. Specific signatures could emerge from these processes in just the same way as they could from processes internal to the learner. To take a concrete example, the timing of turn taking in conversation is one such proposed interactional universal (Stivers et al., 2009). It is entirely possible that some of the specific consistencies in the content of children's early vocabulary (chapter 8) emerge from consistencies in children's early environments, including universal features of what children and their caregivers talk about and why.

1.4.3 Alternatives

In chapter 16, we will examine the empirical support for the claim of consistent signatures in language learning across languages. Building on these data, our further claim is that these consistencies are supported by universal processes. It is

3. Our proposal diverges here from earlier accounts like Slobin (1973)'s operating principles, which were assumed to be learner-internal.

helpful to consider what the alternative is to this position. Perhaps the most important alternative view is that the process of language acquisition is specific and particular, rather than universal. Two prominent sets of particulars ground this alternative.

The first is the vast semantic and syntactic variation across the world's languages. For example, as illustrated by Slobin (1996) and others, languages vary dramatically in the ways that they assign semantic content to verbs. If the semantic partition of verbs led to large-scale differences in the time line or mechanism of acquisition, we might see systematic differences in the predictors of age of acquisition for verbs in these languages—yet we do not. Further, languages are more or less morphologically complex, and while the most highly morphologically complex, polysynthetic languages are not represented in Wordbank, we do have data from relatively less complex (Mandarin) and more complex (Russian) languages. If morphosyntax were relatively more or relatively less important in the acquisition of particular languages, we might expect radical differences in the noun bias, the grammar-lexicon correlation, or the predictors of age of acquisition across languages. Yet we do not observe these. Of course, there is always room for finer-grained predictions—with more detailed predictive models and better typological coverage, perhaps we will discover such variable morphosyntactic signatures. Our point here is merely that, to the extent that we observe consistencies, neither morphosyntactic nor semantic variability across languages dominates the process of vocabulary acquisition.

The second set of language-specific particulars that might lead to variance across languages is the vast cultural variability across the communities represented in the Wordbank data. Our data contain both "individualist" and "collectivist" cultures (Markus and Kitayama, 1991; Nisbett et al., 2001) as well as both "loose fit" and "tight fit" cultures (Gelfand et al., 2011). To the extent that parenting differs across these cultures—and there is good evidence that it does (e.g., Bornstein, 2013)—we should see variance in the trajectory of language learning. For example, it would be quite reasonable to predict that the female advantage in vocabulary acquisition might vary as a function of cross-national gender biases (e.g., as measured in Nosek et al., 2009). Yet it is strikingly consistent overall (presaging the conclusions in chapter 6), again arguing that—at the broadest level, at least—variable cultural factors do not dominate other processes in vocabulary acquisition.

1.5 Replication and Theory Building: Conclusions

In this chapter, we have sketched a bit of what we see to be the unique theoretical contributions of work with a much larger dataset than is usual in developmental language acquisition research. In a nutshell, doing this work at scale allows for the identification of sources of variability in the "signatures" of language learning.

What these signatures are is a matter for further development. Each chapter will describe and motivate the particular signatures that it includes. Further, the consistency of these signatures can provide a motivation for positing process universals that underlie the emergence of these signatures (pending the caveats stated above). We return to the general picture of language learning that emerges from our study in chapter 17.

One final note about the theory that emerges from the work we do here. One set of concepts that is subsumed in our interest in consistency is that theory be supported by observations that are reproducible, replicable, and robust (Munafò et al., 2017). A theory of consistencies is, again, a fortiori all of these. If a particular characteristic can be shown again and again across individuals, samples, and languages, it is replicable. Indeed, one view of our enterprise is that its impact is fundamentally in the consolidation of knowledge through unifying—replicating—previous work.

Crudely put, we have compiled all of the CDI data that we could and all of the CDI analyses and then executed the cross of analyses and datasets. The project of this book is thus in some sense a cross-linguistic replication study. And so, when we state that a particular phenomenon is consistent across languages, it is by definition replicated—but it is additionally robust to a number of different procedural decisions (such as the design or administration of the CDI form) that end up varying widely in our data.

Finally, this work is also fully computationally reproducible—the analytic conclusions we draw here are based on a set of open data and code that can be rerun to create the figures and tables in the manuscript. This characteristic alone does not guarantee their correctness, but at a minimum their provenance is known.

Moving onward from the high-level theoretical framing we have given here, our next chapter introduces some of the practical foundations of our study, including the reliance on parent report and the nature of the Wordbank project.

2 Practical Foundations

In general, a major issue of experimental psychology is that the constructs may depart from the ecological task that is of principal interest (Cronbach and Meehl, 1955). Early language is a rare case in that these problems are minimized. Measures of early language comprehension and production tend to be face valid and tightly linked to the construct of interest. And early language measures are often very closely related to the ecological task—linguistic communication—that is the theoretical target for explanation. Thus, early language is the rare case where, in principle, consistency and variability can both be explored in a single set of measurements (Bates et al., 1994).

Unfortunately, researchers interested in consistency have often measured theoretically important, carefully chosen phenomena using small convenience samples that suffice to show a proof-of-concept but do not provide information about variability. In contrast, work on variability among individuals has often focused on larger samples with more reliable tasks that—perhaps as a consequence of their reliability—are less tightly linked to a particular theoretical construct of interest. As we will discuss, the CDI is a rare case where the overall assessment of language is reliable enough to index variability yet where the individual items are detailed enough to be used to study consistency.

We begin by introducing the Communicative Development Inventory (CDI) and contrasting it with other methods of measuring early language. We then discuss how the cross-linguistic use of parent report methods creates both challenges and opportunities, ending with our development of Wordbank as a way of archiving data on cross-linguistic parent report.

Authors' Note: Some material in this chapter is adapted from Frank et al. (2016a) and Marchman and Dale (2017).

2.1 Measuring Early Language

2.1.1 The Logic of Parent Report and Its Strengths

How do you measure young children's language? Parent report survey instruments like the MacArthur-Bates CDI (Fenson et al., 1994, 2007) and the Language Development Survey (LDS) (Rescorla, 1989) provide an inexpensive method for researchers to get a global picture of children's language. The CDIs in particular were developed across a period of more than 40 years. Originally designed for use in a research study (Bates, 1976), the instruments have evolved from a structured face-to-face interview to a paper-and-pencil format and are now increasingly administered online (e.g., the web-cdi project; Kristoffersen et al., 2013, for Norwegian; laboratorium.detskarec.sk for Slovak). While other assessment tools exist for slightly older children, to our knowledge, no other measure allows cost-effective global language assessment for children in the critical age ranges between the emergence of language and the period when children become more able to engage in structured, face-to-face activities (around 30 months).

Parent-report instruments, like the CDI and LDS, take advantage of the fact that parents (or other primary caregivers) are experts in the behavior of their own child. Parent reports are based on experiences with the child that are not only more extensive than any researcher or clinician can obtain, but are also more representative of the child's ability. Parents have experience with their child at play, at meals, at bath and bedtime, at tantrums—in short, with the full range of the child's life and therefore with the full range of language structures used in these contexts. Parents also have opportunities to hear their child interact with other people: other caregivers, grandparents, siblings, and friends. Because responses on these instruments represent an aggregation over much time and many situations, they are less influenced by factors that can mask a child's true ability in the laboratory or clinic—such as shyness or compliance—or that can impact the validity of naturalistic sampling, such as word frequency. As Bates et al. (1988) point out, "parental report is likely to reflect what a child knows, whereas [a sample of] free speech reflects those forms that she is more likely to use" (p. 57).

Because of its format, parent report enables the collection of data from far larger samples of children than would be possible with standardized tests or naturalistic observation. Information from more adequate samples, especially in the form of norms, can benefit both clinical practice and research. Fenson et al. (1994), for example, used the norming data from English versions of the CDIs—a sample of 2,550 children aged 8 to 30 months—to address questions about variability in communicative development. Large samples are especially needed to provide an accurate statistical description of extreme scores; for example, what score corresponds to the 10th percentile? What does the most advanced child (e.g., > 90th percentile) look like at a given age?

Research on questions such as environmental influences on language development benefits substantially from large samples. Correlational research in general is hampered by the problem of multicollinearity. Predictor variables—such as parental education, the number of books in the home, family size, use of questions versus imperatives—are likely to be intercorrelated, making it difficult to separate the effects of each of them individually. Large samples that have a substantial amount of nonoverlapping variance are essential for addressing these questions.

The core of the CDIs is the vocabulary checklist. This list is essentially a "bag of words" representing the set of words that best capture variation in lexical development across the full spectrum of child ages and abilities. Parents choose the words they believe their child can currently "understand" (comprehension—measured for younger children) or "understand and say" (production—measured for both younger and older children). A child's score on a vocabulary checklist represents their comprehension or production "vocabulary size," which indexes that child's relative status against other children assessed with the same list. In their initial English, Italian, and Spanish instantiations, the vocabulary checklists were developed in two versions: Words and Gestures (WG, 8–18 months), which contains about 400 words, and the Words and Sentences (WS, 16–30 months), which contains about 700 words. This structure has often been replicated across cross-linguistic adaptations, though there is some variation in form construction (see chapter 3) and some forms include substantially different numbers of words or include/exclude other measures.

The vocabulary checklists contain words from many different semantic (e.g., animal names, household items) and syntactic (e.g., action words, connectives) categories, resulting in broader samples of lexical knowledge than are available from other methods. Importantly, however, these words are not chosen to create a complete list of all words understood or produced by a child. Instead, CDI word lists are constructed to include a set of words that most children will know as well as a sampling of intermediate and more difficult words that will be useful in assessing variability among children.[1]

An additional advantage of the parent-report method is that parents can also report on many different subcomponents and correlates of early vocabulary development (limited, of course, by what parents can observe and can reliably report). In particular, the CDI instruments ask about use of communicative gestures,

1. While there have been some efforts to estimate the child's total vocabulary from CDI scores by Mayor and Plunkett (2011), the resulting estimator is calibrated based on a small handful of diary studies and cannot easily be extended across forms or languages.

grammar, and symbolic play in addition to vocabulary comprehension and production. Information about what early vocabulary development correlates with, and what it does not, can yield important theoretical information about the common mechanisms underlying learning.

Of course, parent report has substantial limitations that can lead to both measurement error and bias. These are addressed to some extent by design features of the CDI and are further addressed by evidence for the reliability and validity of the instrument. Because these concerns are so central to our enterprise here, we discuss these issues at length in chapter 4 from both theoretical and analytic points of view.

2.1.2 Other Methods of Measuring Early Vocabulary

As we discuss throughout, parent report is an imperfect method. But often critics of parent report forms like the CDI fail to consider the weaknesses of the alternatives. Here we briefly consider two alternatives: naturalistic observation and experimental testing.

Since naturalistic observations are highly face valid and have the potential for tremendous ecological richness, they are the other leading candidate for measurement of early language. Unfortunately, such observations are extremely costly and time-consuming to transcribe and annotate. These difficulties lead to a tradeoff where most observation-based studies include either dense data about a small number of children or smaller amounts of data with a larger sample size. Dense datasets currently provide the best method for in-depth study of the interaction between learning mechanisms and language input in individuals (e.g., Lieven et al., 2009; Roy et al., 2015), but the generalizability of these studies is necessarily limited by their small sample sizes. And sample sizes for such studies are in turn limited by the costs and practicality of gathering and transcribing such data (see, e.g., Bergelson and Aslin, 2017, for the state of the art). At the other end of the spectrum, assessment of many individual language samples can yield information about individual variability (e.g., Dickinson and Tabors, 2001; Cartmill et al., 2013; Weisleder and Fernald, 2013), but at a cost in terms of depth.

Further, standardization and avoidance of confounds in naturalistic observation studies is challenging. Although parent report seems at first glance to be much more subject to the biases of individual parents, in fact many of the same confounds arise in other paradigms. For example, should an observation session be during play with the parent or an experimenter? Given that parents vary in their talkativeness during a play session, play with a parent is bound to measure parents' ability to elicit language as well as variation in children's ability or knowledge. But for toddlers, temperamental variation is extreme, so an experimenter play session may simply be impractical for some children (and language use may be limited by shyness rather than a lack of ability). Another difficulty is that words will not occur in a laboratory play session in the same distributions as they would occur across other more naturalistic and varied contexts: a word's frequency of occurance will be biased by the particular activities and objects used in the play session. Although these difficulties can be navigated through careful procedural and statistical control, the point of this example is that no observational method offers a perfect solution.

Estimates of production vocabulary from naturalistic observation are highly correlated with the CDI within studies (e.g., Bornstein and Haynes, 1998) but are likely to be affected substantially by length of the session, context, and interlocutor when comparing across studies (see, e.g., Hidaka, 2015, for discussion). And although there exist methods to extract insights about global vocabulary from naturalistic observation, these statistical extrapolations are relatively new and have not been validated extensively (Hidaka, 2015). Finally, naturalistic observations do not measure children's language comprehension, a variable of interest for many early language researchers.

Experimental testing, in contrast, is an excellent method for measuring individual aspects of children's linguistic abilities—for example, their comprehension of a handful of words or their speed of processing (e.g., Bergelson and Swingley, 2012; Fernald et al., 2006). These methods are much less subject to the confounding of observational methods. But an infant or toddler can only provide a limited number of trials during a single measurement session, even in tasks using eye movements. Thus, the ability to measure a particular child's global language ability is limited. Further, the specific words to measure for children of different ages vary—those words that are appropriate for measuring a 14-month-old's competence are trivially easy for a 24-month-old. And attrition can be quite high for a long measurement session, requiring repeated testing for many participants. Other comprehension vocabulary measures are also available across some range of languages (e.g., the Peabody Picture Vocabulary Test 4, Dunn and Dunn, 2007; the Computerized Comprehension Task [CCT] Friend and Keplinger, 2008), but only some of these assessments are tailored for children younger than 2 ½ years.

In sum, despite some clear weaknesses, for breadth, depth, and ease of data gathering, parent report is unmatched. In chapter 4, we provide a more extensive discussion of issues surrounding the reliability and validity of parent report.

2.2 Cross-Linguistic Use of the CDI

2.2.1 Adaptation, Not Translation!

The CDI was originally designed for English and Italian and quickly adapted to Spanish. Since these initial efforts, parallel CDI instruments have now been adapted or are underway for more than 100 languages (https://mb-cdi.stanford.edu

/adaptations.html), with data from 29 languages or dialects represented in this book. The ethic behind the development of these parallel instruments is "adaptation, not translation"—in other words, create forms with the same spirit as the English form—but not simply by translating the items (Dale, 2015). Instead, developers have been strongly encouraged to craft instruments that reflect the linguistic and cultural contexts that influence the early acquisition of vocabulary and other aspects of language in that particular linguistic and cultural context.

The resulting forms vary widely, including differences in length and intended age range. Some forms include hundreds of items more than the original 680 words on the English Words and Sentences form; others are so-called short forms and include only a hundred or a few hundred carefully selected words. Some are designed to capture development from the emergence of language through ages 3–4 years, while others are focused on very early development (like the English Words and Gestures form, designed for ages 8–18 months). These differences also reflect differences in goals for the developers of adaptations—for example, some focus on research assessment while others are designed for clinical screening.

While many words on the English-language checklist may easily translate to other languages, some may not be present in another language, and still others will be present but be less relevant within the same developmental time frame—for example, *cheese* in Japanese or *snow* in Arabic. Conversely, additional words may be needed in the new language that were not included on the English-language vocabulary checklist—for example, *tortilla* in Mexican Spanish. In all languages, though, the vocabulary checklists include a range of words that appear earlier and later in normal development as well as a roughly similar proportion of words from different lexical classes—for example, nouns, verbs, adjectives, and so on. Taken individually then, each adapted instrument captures key trends in vocabulary development when scores are aggregated across all items.

Due to variation in language structure, the interests of the developers of CDI adaptations, and the target age ranges of the forms, the CDI instruments vary in structure across languages. Most adaptations of the WG generally include gestures as well as vocabulary comprehension and production; however, it is not always the case. Further, while adaptations of the WS always include vocabulary production, not all instruments also contain some measures of grammar—for example, early use of closed-class morphology or combinatorial syntax. Cross-linguistic differences also render the structure and format of many parts of the grammar sections to be very different and hence not always amenable to comparisons across languages. Finally, a few instruments included in our dataset are pure vocabulary checklists, with no other sections included.

In sum, CDIs are a useful tool for many languages, but the forms differ substantially among languages. When these differences obviously confound our analysis, we present the relevant control or comparison analyses (e.g., as in chapter 11).

2.2.2 Our Approach to Comparison

The wide cross-linguistic adoption of the CDI provides an opportunity for cross-linguistic comparison, but it also creates many challenges that are not present in datasets that are designed from the start for such comparisons. Differences in instruments and items as well as differences in samples and administration conditions all make it potentially quite problematic to compare scores and score distributions across forms. We discuss differences in instruments and items here and defer discussion of differences in samples and administration to chapter 3.

Differences in length among CDI forms mean that comparisons of raw scores across instruments are inappropriate. Dividing raw scores by the total number of items on a form results in proportions, which are somewhat more comparable but still potentially misleading. A more comprehensive form with more items on it will yield lower proportions for children with the same vocabulary size. Despite this weakness, we typically use proportions for visualizing differences across forms as it is cumbersome to compare raw scores with different totals. More discussion of absolute and relative vocabulary size differences among instruments can be found in chapter 5.

Like other psychometric instruments, CDI instruments can also be normed, and many of the most popular forms are. In the standard norming process, the form is administered to a large, typically developing sample so that percentile ranks can be computed. Following norming, the percentile of a particular raw score for both children in the norming sample and new administrations can be computed and used in place of the proportions or raw scores. These percentile ranks can be useful for clinical purposes, but they also complicate comparison across instruments because of potential differences in the norming populations. In addition, the Wordbank dataset includes normed and unnormed forms, and for the normed forms, we sometimes have access to both the norming dataset and other data but sometimes only have access to the norming data. For these reasons, we do not employ normative percentile ranks in our analyses.

As the preceding discussion shows, there are serious difficulties that crop up immediately in comparisons across instruments. We will grapple with these difficulties throughout the book, but we generally adopt two approaches that help us navigate this complexity.

Our first approach to cross-linguistic and cross-instrument data is to provide standardized analyses within each instrument and language without assuming equivalence across words, instruments, or populations. Thus, we will typically investigate a particular phenomenon (say, the "noun bias" or the "female advantage") independently and in parallel for each of the instruments available to us.[2] We can then—still with caution—analyze and compare the magnitude of this phenomenon across languages, having abstracted away from the specifics of each particular instrument. We sometimes colloquially refer to this approach as "every form an island," meaning that each instrument is analyzed separately and only the analytic results at the highest level are compared.

Our second approach recognizes the necessity to make cross-linguistic and cross-form comparisons at the level of particular words. Cross-linguistic conceptual comparisons are fraught, both philosophically (e.g., Quine, 1960) and practically. We refer to the practical issue as the *tortilla* problem: in American English, we have the word *bread,* which translates to *pan* in Mexican Spanish. But the Spanish word *tortilla* takes some of the cultural role of *bread* in English; thus, *bread* has two reasonable translations.[3] Thus, to facilitate (cautious) cross-linguistic comparison, we developed a set of rough-and-ready translation equivalents. We call these "unilemmas" (short for "universal lemmas"). A "lemma" is a canonical form of a word typically used for gathering frequency counts across different morphological variants (e.g., *walk* is the lemma for *walks, walked,* and *walking*). Unilemmas are used for mapping distinct lexical forms across languages. Unilemmas enable a number of desirable analyses, but more practically, they also provide consistent glosses that make it easier for researchers to work in languages with which they are not familiar. For convenience, our unilemmas are written in English, but they could of course have been written in any other language as well. Further details on the unilemmas are given in chapter 3.

Even with the care we used here to construct a robust set translation equivalents, individual items are likely to only be roughly equivalent cross-linguistically and may have significantly different referential scopes for children learning the different languages. That is, if a parent indicates that a child can produce the word *dog* in English and another parent indicates the translation equivalent in, for example, Spanish (*perro*), it may nevertheless be the case that these words are produced with different frequencies and in different contexts by children speaking the two languages.

2. An exception to this approach is that we do sometimes interpolate words' trajectories across matched instruments for the same language—for example, the proportion of children who say the word *cat* on both Words and Gestures and Words and Sentences forms for American English; see appendix C.

3. These translation issues go the other way as well: *reloj* translates to two distinct words (*clock* and *watch*) in English.

2.3 Wordbank

2.3.1 History

To take advantage of the opportunity posed by the broad use of CDI instruments in the child language community—and in particular the widespread cross-linguistic adaptation of the CDI—in 2014 we began constructing Wordbank, an open repository for CDI data. Our inspiration for Wordbank came from two successful projects for sharing data on children's language acquisition. The first is the Child Language Data Exchange System (CHILDES) (MacWhinney, 2000). A database of transcripts of children's speech and speech to children, CHILDES has grown into a robust and important tool for the community, with many contributors and affiliated projects. The second is the Cross-Linguistic Lexical Norms (CLEX) site (Jørgensen et al., 2010), which is closer in content to Wordbank and is effectively our precursor. CLEX archived normative data from a range of CDI adaptations across languages, allowing browsing of acquisition trajectories for individual items or age groups.

Wordbank initially built on CLEX, offering the same functionality but allowing flexible and interactive visualization and analysis as well as direct database access and data download. In addition, Wordbank's goal was always to extend beyond normative data by dynamically incorporating data from many different researchers and projects of varying sizes and scopes. While the resulting datasets in Wordbank are much more heterogeneous than if they were just based on norming samples alone, they are also larger and more representative than the individual norming datasets (in some cases) and available in languages where no norms exist (for others).

We began the Wordbank project—our first large-scale data-aggregation project—with a relatively naive attitude. We thought "if you build it, they will come"—that contributors would flock to the opportunity to share their data with the world. We were unprepared for the challenges of contacting academics around the world, asking them to volunteer their time and hard-won data to an unknown cause, and then understanding the myriad formats and conventions represented in the data we eventually received. For the first couple of years, our data were largely co-extensive with those gathered by CLEX.

Fortunately, in the years since the Wordbank project began, attitudes towards data sharing have been shifting rapidly (in part as a result of work on replication and reproducibility, e.g., Open Science Collaboration, 2015). In addition, the credibility of the Wordbank project has gradually grown, in part due to the support of the MacArthur-Bates CDI Advisory Board. And, as we successively received more data, our expertise in dealing with heterogeneous datasets has grown. Thus, the dataset has grown quickly in recent years. We hope that in the future, authors see their contribution to Wordbank as an aspirational endpoint for future studies using CDI instruments.

2.3.2 Gaps

From the perspective of the study of child language, there are a number of notable omissions from the datasets represented in Wordbank and the analyses reported in this book. We discuss three of these below: our focus on typical development, monolinguals, and (for the most part) Western, educated, rich, industrialized, and democratic (WEIRD) populations.

First, our analyses here focus exclusively on typical development. The study of atypical language development is an important part of characterizing the mechanisms of acquisition; further, characterizing language development in these circumstances can have important applied benefits—for example, in Studies of language in developmental disorders (Tager-Flusberg et al., 2009; Eigsti et al., 2011), in cases of sensory deficits (Landau et al., 2009), and in cases of abnormal input (Curtiss, 1977), among others. Many of these studies have made use of dense observations of individual children with an approach that is fundamentally different than our large-scale, statistical approach here, however. While CDI-type instruments are increasingly being used with atypical populations (e.g., Charman et al., 2003; Luyster et al., 2007), in practice these datasets still tend to be smaller, concentrated in on English-speaking children, and difficult to access publicly. Thus, Wordbank does not currently archive sufficient data from atypical populations to justify inclusion of these analyses in the current book.

Second, the Wordbank dataset focuses on monolingual acquisition. CDI instruments were initially developed to provide normative measurements of variation within a single language. Since then, however, they have increasingly been used for comparison between monolingual and bilingual groups based on the administration of CDIs in both languages (e.g., Pearson et al., 1993; Hoff et al., 2012). These studies initially focused on specific bilingual populations (e.g., Spanish/English bilinguals). Recent studies have moved beyond this strategy and have begun to examine general trends across multiple bilingual pairings (e.g., Bilson et al., 2015; Floccia et al., 2018). Questions of bilingual acquisition are fascinating and important from both a theoretical and practical perspective. But there are practical obstacles to applying our approach to bilingual data, and this means that our book does not consider the bilingual acquisition situation. Practically speaking, because most of the largest CDI datasets were generated from monolingual norming studies, the vast majority of our data are not bilingual. Further, the combinatorics of bilingualism mean that data on nearly all language pairs will be non-existent. For these reasons, our book focuses on monolingual acquisition, though we recognize this as a limitation that must be addressed by future work.

Finally, the sample of languages we include is limited by our access to data. We have made efforts to include any large CDI datasets whose existence we are aware of—including extensive outreach to CDI authors, especially by professional networking through the CDI Advisory Board. Despite these efforts, our dataset is limited both by the sample of languages in which such studies have been conducted and by international attitudes toward data sharing. Thus, although we do cover many languages around the world (see chapter 3 for a map), these languages are skewed toward Europe and the United States as well as toward Western, educated, rich, industrialized, and democratic (WEIRD) populations (Henrich et al., 2010).

In sum, while there are inherent limitations in comparing different instruments across languages—limitations that we return to again and again throughout the book—our dataset is the first that allows the exploration of data on variability and consistency in early language within and across such a large and diverse set of languages. As such, the availability of cross-linguistic CDI adaptations remains at the core of the analyses that we offer within Wordbank.

3 Methods and Data

We begin by introducing the structure of our dataset and the database that contains it. In the second section, we give some descriptive information on the datasets included in the database.

3.1 Database

Why use a database to store vocabulary data? Consider the standard format of raw Communicative Development Inventory (CDI) data, illustrated in figure 3.1, for a small slice of the original CDI norming data (Fenson et al., 1994, 2007).

Each row is a child, each column gives a variable—either a demographic variable or the response of the parent on a particular item. Although this format is useful for homogeneous administrations of a single instrument, it cannot accommodate multiple instruments, multiple languages, or datasets with different sources or kinds of demographic information. Consolidating data across different instruments is very difficult, and tracking data on children with multiple longitudinal administrations of a single instrument must also be done in an ad-hoc manner. The move to a database format allows far more flexible and programmatic handling of heterogeneous data structures from different sources.

Further, as information about particular entities becomes available—for example, cross-linguistic mappings of lexical items—this information can be added in a way that preserves previous analyses. In a tabular format, such functionality is not guaranteed, and changes to the structure of the dataset will necessarily break code for previous analyses. A database, especially when supplemented with an appropriate application programming interface (API, more on this below), can solve this problem elegantly.

Authors' Note: Some material in this chapter is adapted from Frank et al. (2016a).

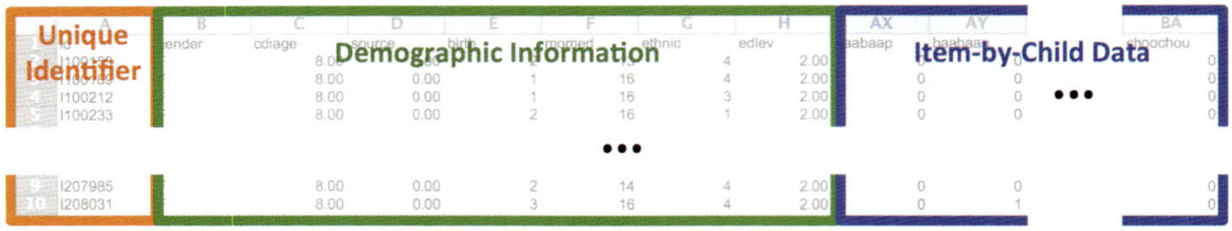

Figure 3.1
Example data from the CDI norming sample (Fenson et al., 2007). Each row has a unique child identifier, demographics, and word-by-word checklist data.

3.1.1 Database Architecture

A relational database such as Wordbank is at its heart an ontology: a set of entities that are described in a series of tables linked by unique identifiers. The primary entities in the Wordbank database are the following:

• **Instrument**: A specific parent-report survey or questionnaire with a particular set of items. For example, the American English Words and Sentences form is an individual instrument.

• **Item**: A particular question on an instrument. A specific word like *dog* is our canonical CDI item, but other items include questions about gestures, morphological and syntactic complexity, and other aspects of early language or behavior.

• **Administration**: A particular instance of an instrument being given to a child, with an associated child age and source (the contributing lab).

• **Child**: A unique individual, with associated demographics.

• **Language**: A particular language or language community for which a CDI instrument has been adapted. Note that this definition of language distinguishes between American English and British English; in the text of the book we use the phrase "language or dialect," but this label is too clunky for a variable name.

These entities are related by two primary groups of tables in Wordbank. The *common* tables store data that is shared among CDI instruments, including information about *administrations* (individual instances of a form being filled out for a child), and *items* (words and other questions on a form). Then the *instrument* tables store the item-by-item response data for particular CDI instruments. We currently include all items on CDI instruments, including questions about communication, gesture, morphology, and grammar (though in quite a few of the datasets that we archive, these nonvocabulary questions have not been digitized, so data on these are sparse at present; see, e.g., chapters 7 and 13).

Wordbank is designed so that it can accommodate data from a wide variety of instruments, both within and across languages. Indeed, at the time of rendering, the site includes data from 82,055 administrations of the CDI across 29 different languages or dialects and 56 different instruments (much though not all of this larger number comes from having available both younger- and older-child focused forms in each language/dialect).

3.1.2 Implementation

Wordbank is constructed using free, open-source tools. The database is a standard `MySQL` database, managed using `Python` and `Django`. All code for Wordbank is hosted in GitHub repositories, with the primary site repository containing data and database code, the `R` package repository containing code for the API, and the book repository containing the code and text for this manuscript.

All data uploaded to Wordbank are open and freely available for download, both through the site itself (https://wordbank.stanford.edu) and through the GitHub repository. The site includes only de-identified data that cannot be linked to individual parents and children under US Department of Health and Human Services' "Safe Harbor" standard. Because of these features, the Stanford Institutional Review Board determined that the Wordbank project does not constitute regulated human subjects research.

3.1.3 The Web Interface

Our website (https://wordbank.stanford.edu) provides a set of interactive visualizations so that interested readers can explore the data in the database. A tutorial introduction to these visualization apps is given in Frank et al. (2016a). We developed these apps to allow quick access to basic generalizations about the distribution of vocabulary totals (similar to analyses in chapters 5 and 6) and to information about individual items (similar to chapter 8).

3.1.4 The `wordbankr` API

An application programming interface (API) is a set of abstractions that allow applications to interact with a resource (e.g., a set of data like Wordbank) through consistent abstractions. Although in principle it is possible to construct raw `SQL` queries to Wordbank, in practice all access is through an `R` API that constructs individual `SQL` calls. This API is distributed to `R` users through the `wordbankr` package, which is available through the Comprehensive R Archive Network (CRAN) (https://CRAN.R-project.org/package=wordbankr).

We developed this package, `wordbankr`, to provide a simple and flexible API for the Wordbank dataset (Frank et al., 2016a), and our current book depends on it heavily. The package provides a consistent set of function calls for retrieving data from the underlying database—for example, `get_instruments` and `get_administrations`—to retrieve all or subsections of these tables, respectively. We do not describe the package in depth here, since it is described in our previous paper and in its online documentation (https://github.com/langcog/online documentation (wordbankr).

3.1.5 Unilemmas: Cross-Linguistic Conceptual Mappings

As described in chapter 2, it is sometimes useful to (cautiously) compare the developmental trajectory for a single concept across multiple languages. To facilitate these comparisons, we created "unilemmas"—cross-linguistic mappings from lexical items to single (English) forms that stand for a particular conceptual abstraction. Some lexical items are represented on only one or a handful of instruments, but there are many that are common across a large number of instruments, leading to an opportunity for cross-linguistic comparison.

Unilemmas were created for particular instruments by following a two-step procedure. First, using a pool of English unilemmas, we proposed candidate mappings for each lexical item on a form. This first step was often accomplished by a non native speaker using translation resources and the context of the form (e.g., that an item occurs in the "animal sounds" section). Second, we recruited a linguistically sophisticated native speaker of the language (often though not always a psychologist or linguist), provided them with the candidate unilemma list, and asked them to review this list item by item and suggest corrections and amendments.[1]

Not every instrument has unilemma mappings, but they are currently available, at least partially, for 46 out of the 56 instruments.

3.1.6 A Note on Age

Developmental psychologists are very fond of using temporal units like months and years as rough guides. Children tend to begin to crawl between 5 and 8 months and say their first word around one year. This practice is fine for rules of thumb, but we also use these units for measurement as though they were precise (e.g., "infants with ages between 7;0 and 8;0") when in fact such infants will vary in the number of days since their birth depending on facts like whether their seven months of life encompassed February or not. A similar problem is true of years as a scientific

1. The specific direction they were given was: "We're looking for the best English translation of these words. These are words that are among the first words that children learn, so your translation should be closest to the meaning of the word as it would be used by a young child (say, under 3 years old). For cases when there are two equally good English words, put both. If you don't think there is a good translation into a reasonable English word that a kid might know, you can leave the alternative translation blank."

unit—because of leap years, the unit of years technically includes 365.2524 days—though the magnitude of the imprecision is smaller.

Despite these issues, months are the currency of language development research, and we often receive contributed datasets with months as the only measure of age. In Wordbank, we define a *standardized month* as 365.2524 / 12 = 30.4377 days. When possible, we compute the number of days from birth to testing and then compute the number of standardized months that the child has lived. If this is not possible, we use months as reported in the dataset. We define an 8 month-old (`age = 8`) as a child who has lived between 8 and 9 standard months: their age is in the range [8–9) standard months. (The alternative definition, from 7;16–8;15, is sometimes used in infancy research but is in our opinion less intuitive.)

3.2 Datasets

This section gives a broad overview of the data we have available. Unlike projects in which data are collected by the organizers, in our work here, we rely on the kindness of others in contributing data that are often years or decades old. Some datasets come via an email containing well-curated tabular data; others were contributed in more idiosyncratic formats or even on paper. One dataset was even retrieved by one of us from a doorstep several hours' drive away in the form of a paper bag full of paper forms. Thus, the amount and type of metadata available for some datasets is limited. For example, we have limited demographic information for some datasets and only vocabulary—not complexity or gesture—items for others. In many cases we do not have full details of instructions and administration for a particular dataset. This section gives an overview of data availability and some demographic comparisons of the samples. Specifics of each dataset—to the extent that they are available—are given in appendix A.

3.2.1 Data Provenance

As mentioned above, datasets come from a variety of sources. In all cases, the preferred citation for each dataset and its contributor is given on the Wordbank contributors page. Several of these datasets were transferred second-hand from a pre-existing database (CLEX-CDI); (Jørgensen et al., 2010), while many of the others were contributed directly via electronic or paper forms. In the case of paper forms, we rekeyed them using double-entry methods (either ourselves or via a commercial contractor).[2]

2. In a check for errors in the rekeying of one Korean dataset, we found that there were 4 incorrect fields in 10 full records for an error rate of approximately 0.06 percent.

Table 3.1
Overview of the available instruments in the Wordbank dataset.

Language	Form	Categories	Items	Words	Age range
American Sign Language	FormA	18	561	536	8–36
American Sign Language	FormBOne	22	699	674	8–36
American Sign Language	FormBTwo	21	562	537	8–36
American Sign Language	FormC	21	563	538	8–36
British Sign Language	WG	22	569	548	8–36
Cantonese	WS	25	815	804	16–30
Croatian	WG	19	396	396	8–16
Croatian	WS	22	717	717	16–30
Czech	WS	21	553	553	16–30
Danish	WG	20	410	410	8–20
Danish	WS	23	858	725	16–36
English (American)	WG	20	492	396	8–18
English (American)	WS	23	797	680	16–30
English (Australian)	WS	16	628	558	12–30
English (British)	Oxford CDI	18	418	418	12–25
English (British)	TEDS Threes	1	113	100	34–47
English (British)	TEDS Twos	1	113	100	20–35
French (French)	WG	24	795	698	8–16
French (French)	WS	23	716	690	16–30
French (Quebecois)	WG	19	408	408	8–16
French (Quebecois)	WS	22	754	664	16–30
German	WS	21	588	588	18–30
Greek (Cypriot)	WS	23	818	818	18–30
Hebrew	WG	20	543	443	11–25
Hebrew	WS	21	638	597	25–36
Italian	WG	20	505	408	7–24
Italian	WS	23	670	670	18–36
Kigiriama	WG	17	293	292	8–15
Kigiriama	WS	22	760	705	16–30
Kiswahili	WG	17	293	292	8–15
Kiswahili	WS	22	760	705	16–30
Korean	WG	19	344	284	8–17
Korean	WS	25	679	641	18–36
Latvian	WG	18	402	402	8–16
Latvian	WS	22	723	723	17–36
Mandarin (Beijing)	IC	14	232	232	12–16
Mandarin (Beijing)	TC	20	710	710	17–30
Mandarin (Beijing)	WS	25	836	799	16–30
Mandarin (Taiwanese)	WG	17	354	354	8–16
Mandarin (Taiwanese)	WS	19	696	696	16–36
Norwegian	WG	21	489	395	8–20
Norwegian	WS	23	868	731	16–36
Portuguese (European)	WG	20	317	317	8–15
Portuguese (European)	WS	22	639	639	16–30
Russian	WG	21	455	427	8–18
Russian	WS	22	728	728	16–36
Slovak	WG	17	456	309	8–16

Table 3.1
continued

Language	Form	Categories	Items	Words	Age range
Slovak	WS	21	691	610	17–36
Spanish (European)	WG	18	303	303	8–15
Spanish (European)	WS	20	588	588	16–30
Spanish (Mexican)	WG	22	525	428	8–18
Spanish (Mexican)	WS	24	747	680	16–30
Swedish	WG	19	385	385	8–16
Swedish	WS	21	710	710	16–28
Turkish	WG	20	418	418	8–16
Turkish	WS	22	711	711	16–36

Each of these datasets is then imported to the database by creating a custom import key that matches individual columns of the dataset to particular database fields (e.g., item types like words or gestures, or standardized demographic fields). These mappings are preserved along with the raw data so that they can be rechecked later.

3.2.2 Overview of the Data

Wordbank currently contains data from 29 language communities. Many of these are from instruments similar to the original Words and Gestures (WG, "infant") / Words and Sentences (WS, "toddler") format, with around 400 items in WG and 700 in the WS. Typically, WG forms are intended for children from 8 to 18 months and WS forms are intended for children 16 to 30 months, but these ranges are flexible. Some WS forms are used up to 36 months or extended as low as 12 months (in cases where a single form is considered desirable by the researchers constructing the adaptation).

Wordbank also includes some other forms that do not fit into this schema. Some of these are "short forms" with no internal category structure and fewer items overall, and these are excluded from many item and category analyses. But others have many structural features of WS and WG forms. For example, the Oxford CDI is a WG-style form with comprehension as well as production estimates that are applied to a larger age range. We group this form with WG forms because it measures comprehension. The Mandarin Infant Checklist (IC) and Toddler Checklist (TC) are checklist forms without grammatical and gesture items but with structured sets of vocabulary items. We include these checklist forms in analyses where WS and WG data are included and group them by their target age range.

Table 3.1 shows an overview of the instruments in Wordbank. The number of administrations available is highly variable across instruments and languages. Figure 3.2 shows the distribution of administrations across forms and

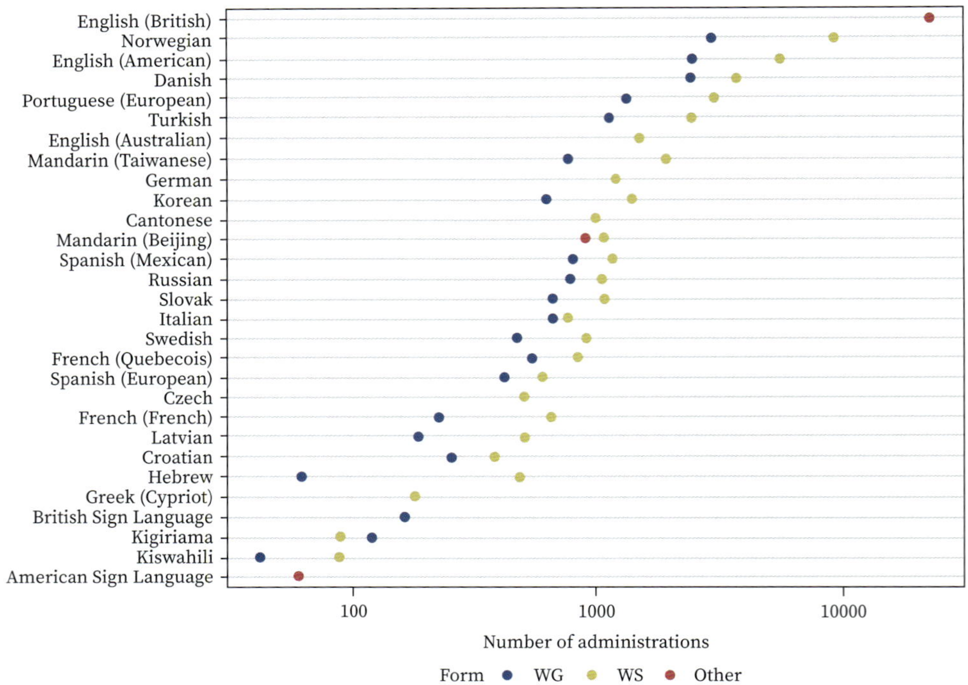

Figure 3.2
Log-scaled number of administrations for each instrument.

languages. We have fewer WG administrations than WS forms for essentially all languages/dialects. These instruments have global reach, although the maximal number cover North America and Western Europe. African, South American, and South/Southeast Asian languages are notably under-/unrepresented. Figure 3.3 indicates which countries have their population represented in the dataset.

3.2.3 Administration Details

Data in the dataset were gathered between the beginning of the first CDI norming study in 1990 and the present, with the majority of datasets gathered within the 10–15 years prior to the writing of this book. The details of administration vary widely from dataset to dataset. Though we have different levels of knowledge regarding the exact details of administration, we know that the three most common circumstances of administration (in no particular order) are.

1. on paper in a lab or other space, with instructions given in person by a researcher (e.g., Jackson-Maldonado et al., 1993);

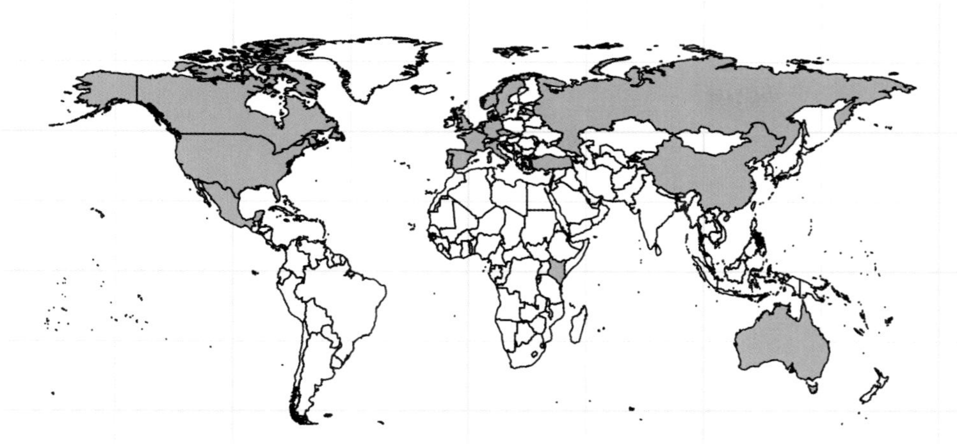

Figure 3.3
World map shaded for those countries whose population is represented in the dataset.

2. on paper, with the form sent by mail with written or telephone instructions from a researcher (e.g., the original American English norming study; Fenson et al., 2007, or the British English Twins Early Development data, which were sent home as part of a packet; Dale et al., 2003); or
3. electronically, with instructions given either electronically or by phone (e.g., Kristoffersen et al., 2013).

We have limited direct evidence about the effects of particular administration details on the overall results, although some early studies found comparable results across mail-in and personal administration (Jackson-Maldonado et al., 1993). Such evidence would require random assignment of parents to administration method rather than, for example, a comparison of administration methods across different populations in which there are obvious sample-related confounds. Nevertheless, the CDI community has amassed a substantial set of anecdotal experiences. For example, improper administration or limited instructions can result in over- or under-reporting, especially with respect to comprehension (see e.g., Feldman et al., 2000).

In one trial we conducted using electronic administration, we found that basic written instructions were misinterpreted by some proportion of parents (as evinced by an atypical number of floor and ceiling responses). This proportion appeared to decrease when we made an attempt to simplify and illustrate the instructions that we gave. Congruent with the general warnings above, such experiences suggest that caution is warranted in interpreting absolute comparisons

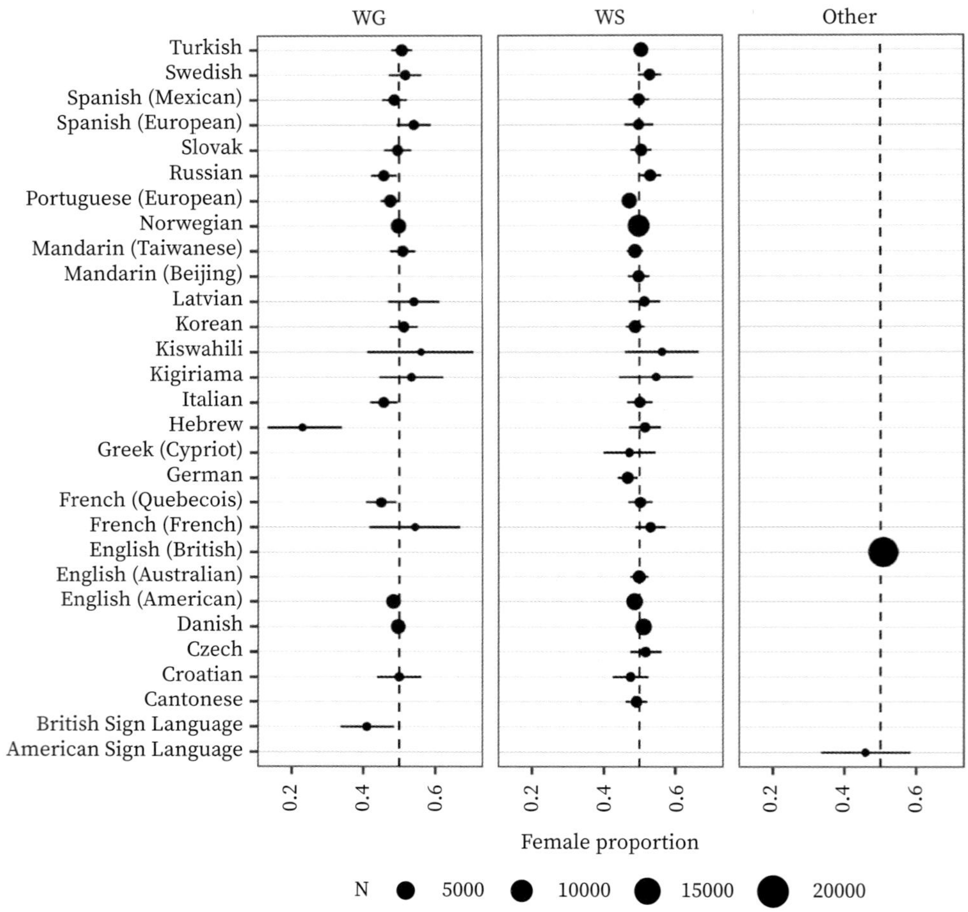

Figure 3.4
Proportion of female-assigned children for each instrument.

among different populations where there are also differences in administration style. We will return to questions about administration in chapter 18.

3.2.4 Demographic Details

In addition to differences in administration and form, samples from different studies also differ in myriad other ways. The most important of these—for example, cultural differences among language communities—are extremely hard to quantify. But we can make a first stab at investigating some similarities and differences among the convenience samples from different studies by comparing demographics where they are available. The demographic makeup of our datasets is shown in figure 3.4 for sex, figure 3.5 for maternal education, and figure 3.6 for birth order.

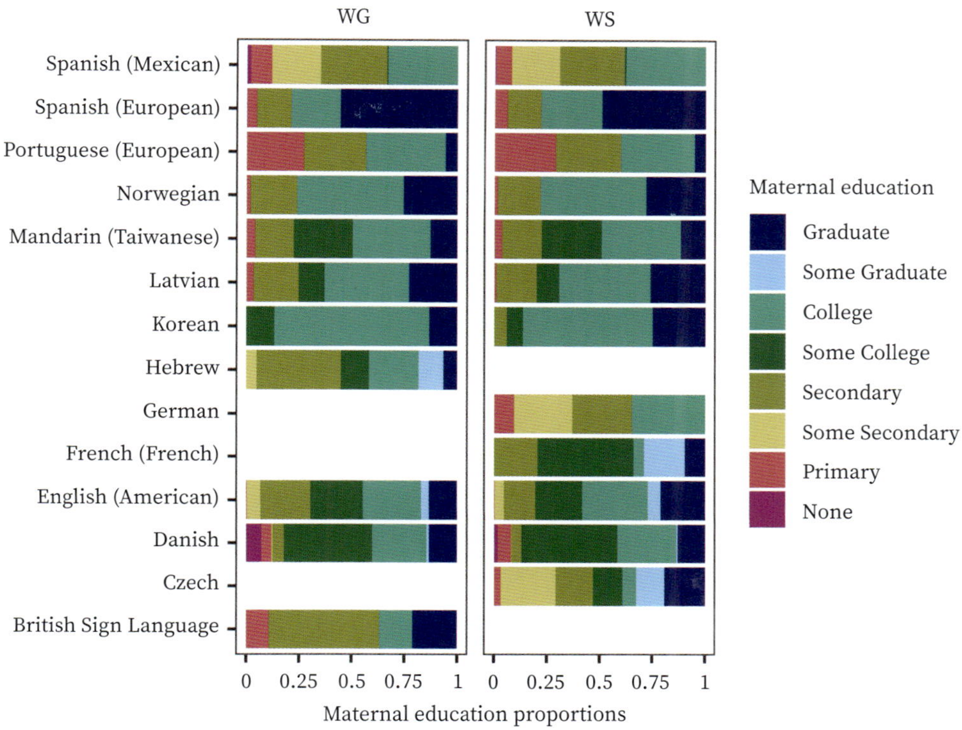

Figure 3.5
Proportions of children with each level of maternal education for each instrument.

Sex proportions tend to be quite close to 0.5, with a few exceptions for small datasets. Several WG datasets (e.g., British Sign Language, Russian, Italian, Quebec French) have more males than might be expected by chance. This pattern is important because (as we will investigate in chapter 6), there are systematic differences in vocabulary size between boys and girls, and so sample differences in sex will lead to absolute differences in mean vocabulary size.

Although we have maternal education data for far fewer datasets, there are also substantial differences among datasets on this variable (we will also return to this issue again in chapter 6). Analyses of this variable are complicated by different reporting formats: for example, the German and Mexican Spanish datasets have no separate categorization for graduate education. That said, even for datasets with the most fine-grained maternal education breakdown, we see substantial differences in distribution.

Finally, when we examine birth order, we also see differences in the proportion of children who are firstborn versus later-born. The majority of the French sample is first-born, while the Czech sample has many more second children, for example.

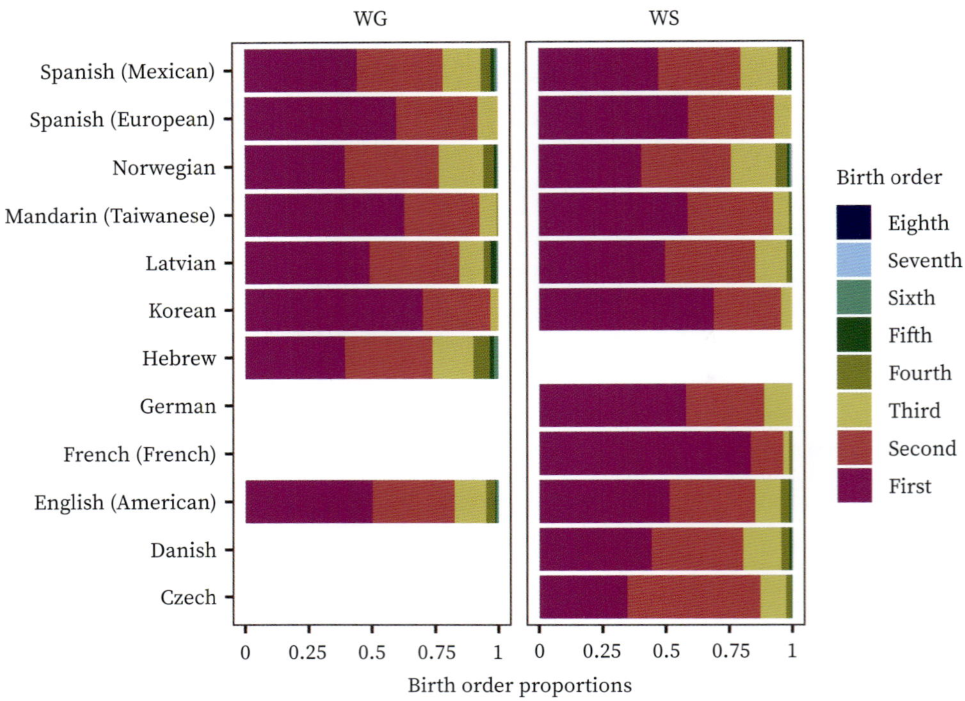

Figure 3.6
Proportions of children with each value of birth order for each instrument.

In summary, our samples differ substantially in their demographic makeup. Presumably, these differences are due both to the composition of the societies being sampled as well as to the sampling procedure employed by the researchers.

3.2.5 Longitudinal Versus Cross-Sectional Data

The strongest developmental inferences can be made by the examination of longitudinal data, in which children's individual development is measured multiple times using the same instrument. Unfortunately, relatively little of our CDI data comes from this type of repeated administration. Figure 3.7 shows the number of administrations for particular languages that come from longitudinal datasets with a particular depth. There is a substantial amount of two-administration longitudinal data for several languages, but only a few have more than two observations for individual children.

In general, this aspect of our data is a consequence of the fact that, for normative datasets, pure cross-sectional data collection is used to ensure statistical independence between datapoints. Thus, we must typically settle for using the large

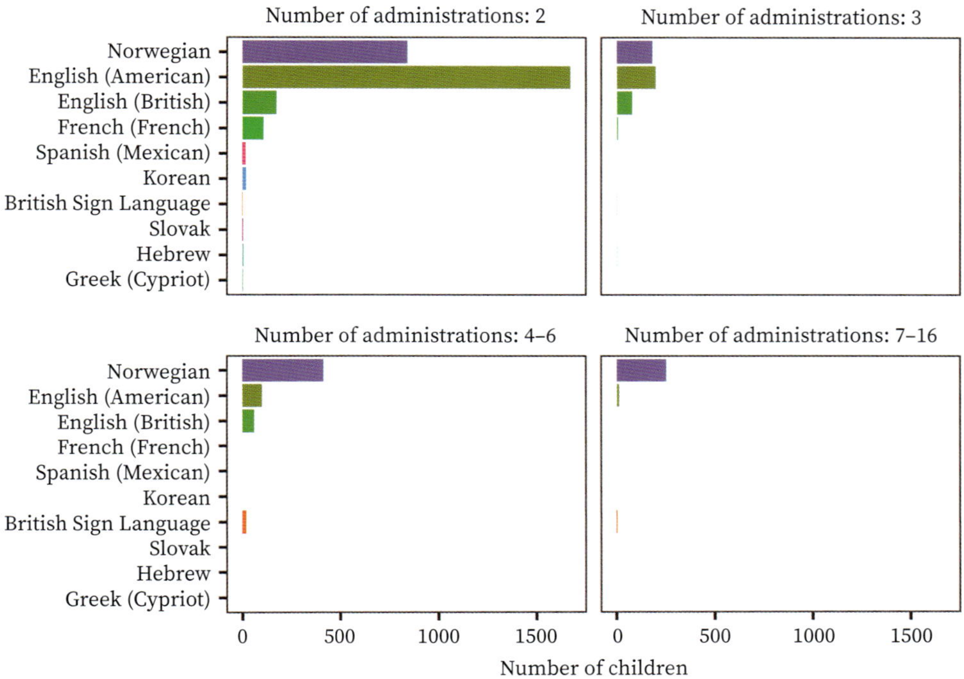

Figure 3.7
Number of children for whom there are multiple administrations per instrument that are split into bins.

amount of available cross-sectional data to average out individual variability. We do use the more extensive Norwegian and English longitudinal data in chapters 13 and 15, however.

3.2.6 Difficult Datasets

One feature of dealing with data from such disparate sources can't be glossed over. There are "difficult" datasets—data that do not make sense with respect to other analyses. This short section documents some of these issues (helping itself more intuitively to a few visualizations that will be developed in more detail in subsequent chapters).

In general, our approach with respect to these data is to embrace the messiness of the data we have. While it is very tempting to remove specific datasets from consideration when they deviate from our expectations, this practice creates a strong circularity in all of our inferences: they will be estimates of variability or consistency stemming from cases where we ourselves have imposed certain consistency standards on our data. While there are some cases where we have a relatively likely

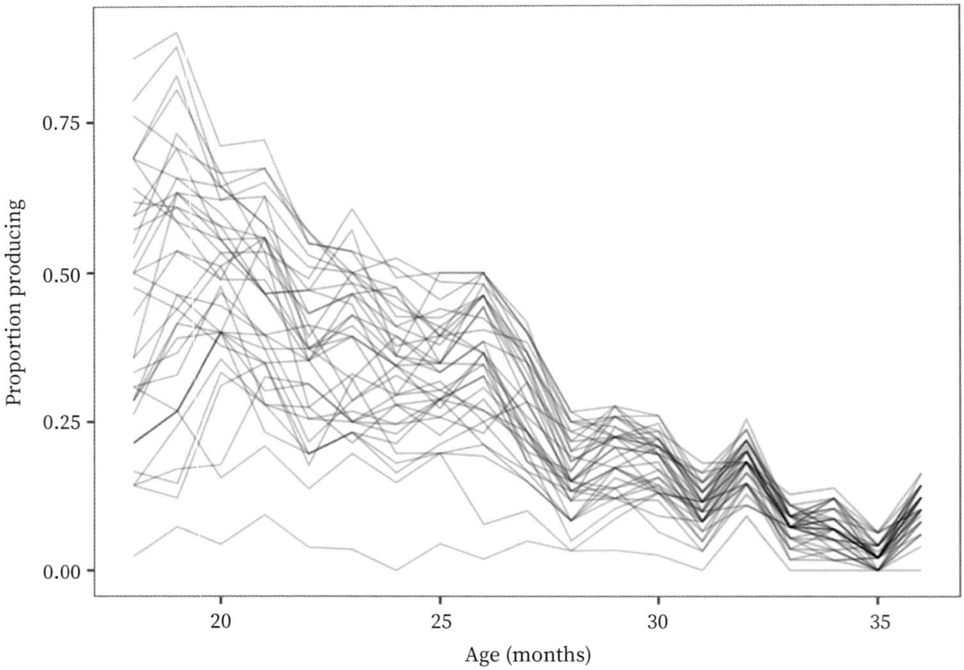

Figure 3.8
Proportion of children reported to produce each item in the Sounds category in Russian WS.

explanation close at hand for the pattern we observe in the data, unless we can confirm this pattern externally, we have chosen not to exclude these data.

One example of this kind of situation comes from the Russian dataset. Although nearly every individual item in every dataset shows a positive developmental slope (indicating learning over time), Russian animal sounds are a distinctive exception, as shown in figure 3.8. Every item in this category *decreases* developmentally in a very consistent and reliable way. What happened? One possibility is that this set of items was reverse coded (and so it should be asymptoting at three years). Another possibility is that Russian parents treat these as “baby words” that a 3-year-old would not or should not produce (e.g., rather than saying *oink* they should say *pig*). We can speculate but we will likely never know.

Extending more broadly (and presaging the discussion in chapter 5), our analyses have revealed two datasets that show large disparities not just in a single category but in the pattern of overall vocabulary sizes: Mandarin (Beijing) WS production and Mandarin (Taiwan) WS comprehension.

In our first example, Mandarin WS, these data are reported by Tardif et al. (2009) in a study of both Mandarin- and Cantonese-learning children. The data reported there show a pronounced Mandarin advantage. As it turns out, this advantage is

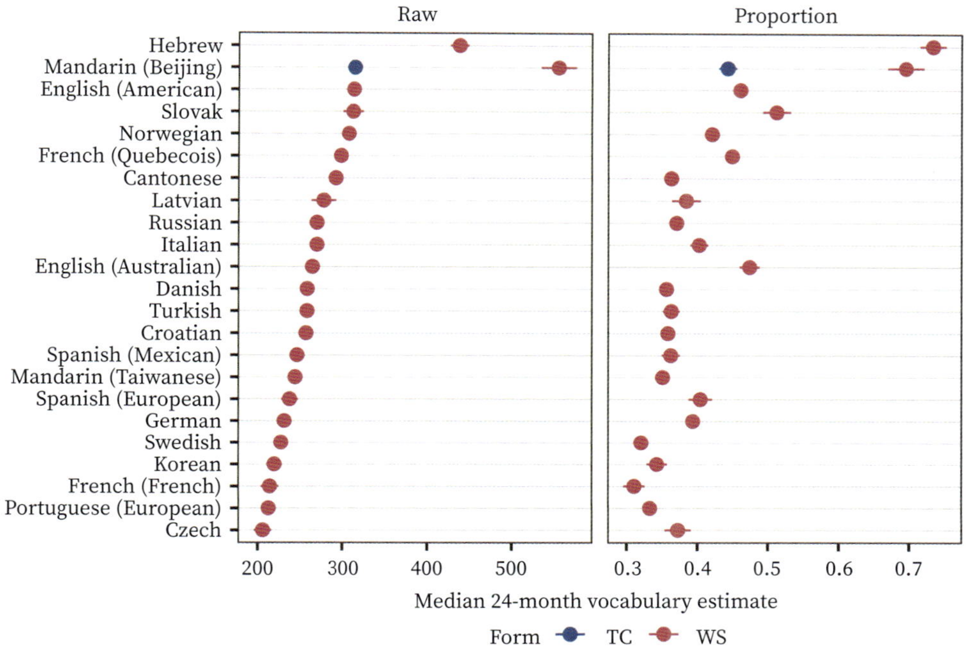

Figure 3.9
Median production vocabulary for 24-month-olds, with total item scores shown in the left panel and proportions on the right. Scores are sorted by total item score. To increase stability, the plotted value is the intercept of a linear model predicting vocabulary as a function of centered age between 18 and 30 months.

almost unprecedented relative to other languages. We plot the median production for 24-month-olds in figure 3.9. This figure reveals how large the Mandarin advantage is and the high level of vocabulary reported for Hebrew speakers as well (this difference is less striking in raw scores because of the relatively smaller number of items on the Hebrew form).

To investigate the Mandarin disparities further, Tardif et al. (2009) discussed a number of possible explanations, given that the administration and sampling procedures were similar in Mandarin and Cantonese. The children in the Mandarin sample are nearly all monolingual, and from one-child families; but these factors did not account for variation between samples. Tardif et al. (2009), therefore, speculate that structural factors regarding Mandarin (e.g., phonological structure relative to Cantonese) might be accounting for the Mandarin advantage.

These speculations seem unlikely in light of the data presented here. First, and perhaps most importantly, the same magnitude response is not shown in the data from the analogous WS Checklist questionnaire of Hao et al. (2008) (blue points). Second, this unusual trajectory is not apparent in the production data from

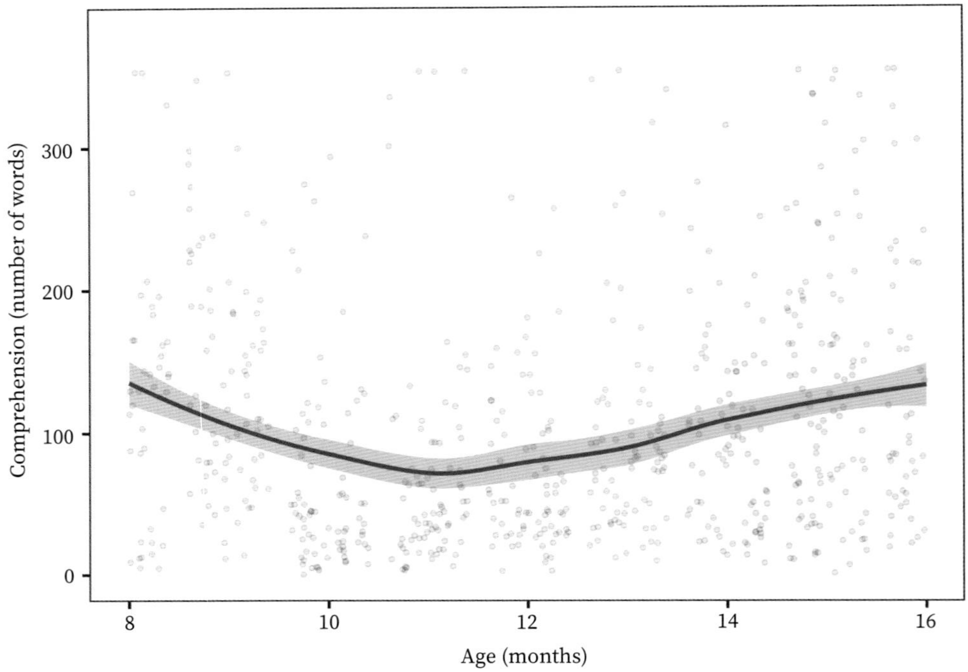

Figure 3.10
Comprehension data from Taiwanese Mandarin.

the Mandarin Beijing WG form. Finally, given the surprising difference between Mandarin and all other languages in the sample, pure phonological factors seem unlikely to account fully for the differences. These differences thus remain somewhat mysterious; perhaps some quirk of administration instructions led to relative over-reporting or perhaps the populations being sampled truly were different. Alongside the Hebrew data, these data serve as an important caution against simple cross-linguistic comparison in raw scores or even percentiles.

Turning to Mandarin (Taiwanese) comprehension scores, we see that they are relatively flat and show very high medians very early in development. Deeper inspection of the full distributional pattern (figure 3.10) suggests that there is relatively little developmental change in comprehension scores on this dataset. In contrast, production appears to follow a more typical pattern. In our experience, this pattern results from parents who do not understand what is being asked on the comprehension section of a form; sometimes they report whether they think a child has heard a particular word or whether they respond to language in more general ways. We have observed a population of "over-responders" of this sort in a number of self-report contexts—often they are parents of very young children who appear loathe to return a form where they have checked essentially no items at all. But such an explanation is only speculation.

We could give other examples. There are quite a number of "difficult datasets" in one way or another in Wordbank. While we have offered some tentative explanations of a few features, these are necessarily post hoc and rely on our assumption that they *should* be relatively similar to other datasets from other cultures and with other forms. Thus, in our further analyses we choose not to omit these data but instead consider them as a caution on making strong inferences from variability rather than from consistency. As we discussed in chapter 1, variability may be caused by a wide variety of sources; it is consistency which is all the more surprising in the face of this sort of variation.

4 Measurement Properties of the CDI

Many researchers are initially shocked to hear that one of the most important methods for studying child language is parent report. Aren't parents extremely biased observers? Yet, as we argued in chapter 2, alternative methods like naturalistic observation or lab experiments can also be biased and are quite costly to deploy at scale. Thus, the goal of this chapter is to revisit the strengths and weaknesses of parent report in depth since the remainder of our book depends on the use of Communicative Development Inventory (CDI) data.

Our goal is to assess the psychometric utility of the CDI. Many studies provide evidence for reliability in the form of concurrent and longitudinal correlations between CDI scores and for validity in the form of correlations between the CDI and other language measures; some of the most prominent of these studies are cited below and a number of others are reviewed in Fenson et al. (2007). We also address some issues that have received a little less attention, however. In the first part of the chapter, we discuss the limitations of the CDI (and the design features that address these limitations); in the second part, we use longitudinal data to examine the test-retest reliability of the CDI; and in the third part, we present evidence for the measurement properties of the CDI (including comprehension questions) from a psychometric perspective.

4.1 Strengths and Limitations of Parent Report

Although the standardization of parent report using the CDI contributes to the availability of large amounts of data in a comparable format, there are significant limitations to the parent-report methodology that are important to understand (Tomasello and Mervis, 1994; Feldman et al., 2000). To begin to do so, it is useful to reflect on what it means when a parent reports that their child "understands" or "understands and says" a word.

In an ideal world, the parent's responses would be an unbiased reflection of their observations of their child's language development. But parent reports are

Figure 4.1
The intuitive structure of parent report.

almost certainly less transparent. Figure 4.1 shows a caricature of the process of parent report. A particular report could depend on direct recall of a particular case when their child actually produced or showed comprehension of the word. For example, when asked if their child produces the word *ball*, a parent is likely recalling situations in which their child has used the word *ball* correctly and then reporting on the success or failure of this process of recollection. Of course, this judgment clearly depends on the parent's ability to accurately judge that the child intended to say the word *ball*, that the child's target word form was *ball*, and that the child has some meaning for the word form *ball* that at least approximates the expected meaning. But, in addition to these factors, parents probably draw on their general assessment of the difficulty of the word and on their overall assessment of the child's linguistic abilities. As even this simple sketch shows, parent-report judgments are based on a fairly complex set of factors. Hence, there are legitimate concerns about the ability of parents to provide detailed and specific knowledge about their children's language. We discuss specific concerns below.

First, parents are imperfect observers. Most parents do not have specialized training in language development and may not be sensitive to subtle aspects of language structure and use. Further, a natural pride in the child and a failure to critically test their impressions may cause parents to overestimate the child's ability; conversely, frustration in the case of delayed language may lead to underestimates. Parent report is most likely to be accurate under three general conditions: (1) when assessment is limited to current behaviors, (2) when assessment is focused on emergent behaviors, and (3) when a recognition format is used. Each of these conditions acts to reduce demands on the respondent's memory. In other words, parents are likely to be better able to report on their child's language at the present time than at times past and better able to report specific items when their child is actively learning those particular items (e.g., reporting on names for animals after a trip to the zoo).

Following point (3), one particular key design principle of the CDI is that parents are better able to choose from a list of items that are likely candidates (recognition) rather than requiring that the parents generate the list themselves (recall). Although this second type of assessment sounds implausibly bad, it is surprising how often it is still used (or, even worse, asking the global question "Does your child know at least 50 words?" that is so commonly used in pediatric assessments).

CDI forms are also designed around commonsense age limitations on parent report. In typically developing samples, the assumption is that parents can track their child's comprehension vocabulary to about 16–18 months, after which the number of words a child can understand is thought to be too large to monitor. For productive vocabulary, the assumption is that specific word productions can be monitored until about 2.5–3 years, after which the number of words a child can say becomes too large. Different instrument developers make different choices about the ceiling of CDI-type forms but relatively few have considered CDI-type parent report for measuring older children's vocabularies (but cf. Libertus et al., 2015).

Second, parent reports likely suffer from a number of biases that interact with sub-portions of the forms and the ages of the target children. For example, it is likely that parents may have more difficulty reporting on children's comprehension or production of function words (e.g., *so, then, if*) perhaps because these words are more abstract and less referential than content words (e.g., *baby, house*). Estimates for function words may then rely more on parent estimates of the words' general difficulty rather than actual observations. We return to this question below in our psychometric analyses.

In addition, asking parents to reflect on their child's language abilities may be particularly difficult for early vocabulary and especially for early comprehension. As Tomasello and Mervis (1994) point out, for the youngest children, especially 8- to 10-month-olds, vocabulary comprehension scores can be surprisingly high, possibly reflecting a lack of clarity in what the term "understands" means for parents of children at this young age (cf. chapter 3, subsection on "difficult data"). On the other hand, more recent evidence has suggested that children in this age range do plausibly have some comprehension skill even if it is somewhat fragmentary (Tincoff and Jusczyk, 1999, 2012; Bergelson and Swingley, 2012, 2013, 2015). Thus, the degree to which very early comprehension reports are artifactual—or were actually ahead of the research literature—is unknown. (Resolving this question will require detailed studies of the correspondence between parent reports and experimental data for individual children). Below we assess some of the measurement properties of comprehension items, but we are unable to resolve the issue fully.

One study that bears on the earliest production data is Schneider et al. (2015), who compiled a number of sources of data on children's first words. Surprisingly, that study found relatively few differences for the age and topic distribution of this very salient milestone across datasets collected via a number of different methods, including for concurrent (CDI) and retrospective reports. The age at which a first word was reported was also relatively similar between CDI data and the concurrent diary reports of a sample of psycholinguists (though some CDI data appeared to be shifted a little bit earlier such that more parents were reporting first words in the 7–9 month period). Thus, there was convergence across different reporting methods in parents' report on first word production. Parent report could be flawed here, but the specific CDI format may not be to blame.

Third, there is some evidence that variability in reporting biases may be moderated by factors such as socioeconomic status (SES; Feldman et al., 2000, 2005; Fenson et al. 2000a). Some studies suggest that parents from some SES groups may be more likely to underestimate child's abilities (Roberts et al., 1999) while others report that parents from lower-SES groups may overestimate children's abilities, especially comprehension at younger ages (Goldfield and Reznick, 1990; Feldman et al., 2000). Later studies, however, have shown that for children over 2 years of age, patterns of validity were consistent in lower and higher-SES groups (Feldman et al., 2005; Reese and Read, 2000). Thus, SES differences could reflect valid delays in children's language development that parallel those obtained with different methods, such as naturalistic observation or standardized tests (e.g., Hammer et al., 2010).

Fourth, as discussed in chapter 2, the items on the original CDI instruments were chosen to be a representative sample of vocabulary for the appropriate age and language (Fenson et al., 1994). The checklists contain some words that most, including the youngest, children are able to understand or produce; some words that are understood or produced by the "average" child; and some words that only children who are relatively more advanced will understand or produce. This structure ensures that the list has the psychometric property of capturing individual differences in vocabulary both across younger and older children and across children of different developmental levels. Validity of the CDIs has been demonstrated in reference to both standardized tests and naturalistic language sampling (see chapter 4 of Fenson et al., 2007).

The checklists were not originally constructed with the assumption that responses on individual items would be reliable and valid, however. (Indeed, as we show below, not all words have ideal psychometric properties—e.g., "mittens"). While item-level responses provide useful information about patterns of words that children are likely to understand or produce, responses on the vocabulary checklist do not necessarily license the conclusion that a child would respond appropriately when asked, "Can you say ——?" by an experimenter in a confrontation naming task. Nonetheless, if parents' observations at the item level reflect any signal—even in the context of significant influence from other factors—then this signal should be observable by aggregating together data from many children. Thus, the item-level analyses we present in chapter 10 (for example) are not predicated on an assumption of high item-level reliability for individual children.

Fifth, while the lengths of the vocabulary checklists on the CDIs may give the impression that they yield an estimate of the child's full vocabulary, in fact, the vocabulary size estimates only reflect a child's relative standing compared with other children assessed with the same list of words. Such estimates should not be misconstrued as a comprehensive estimate of the child's vocabulary knowledge as CDI scores likely substantially understate the size of a child's "true" vocabulary, especially for older children.

Sixth, when a parent reports on a word on the vocabulary checklist, there is no information about the actual form of the word used, and hence, these vocabulary estimates can say little about phonological development (e.g., segmental versus suprasegmental approaches to the analysis of speech). Parents are instructed that they should check that a child can produce a word even if it is pronounced in the child's "special way" and only approximates the adult form. Thus, throughout this book we refrain from analyzing the phonological forms of words reported on CDI instruments (with the exception of chapter 10, in which we use word length in the correct adult form as a predictor of production).

Finally, we also gain no information from parent report about the frequency with which children use a particular word in their spontaneous speech, nor can we know the range of contexts in which individual lexical items are used (e.g., is that word used productively versus in a memorized chunk of speech). Thus, the vocabulary size that is captured by the CDIs reflects the number of different word types that the child is able to understand or produce with little information about nuances in meaning that might be reflected in actual usage.

Despite these limitations, when used appropriately, the CDI instruments yield reliable and valid estimates of total vocabulary size. Because the instruments were designed to minimize bias by targeting current behaviors and asking parents about highly salient features of their child's abilities, they have proven to be an important tool in the field. Dozens of studies demonstrate concurrent and predictive relations with naturalistic and observational measures in both typically developing and at-risk populations (e.g., Dale and Fenson, 1996; Thal et al., 1999; Marchman and Martínez-Sussmann, 2002). In addition, a variety of recent work has shown that individual item-level responses can yield exciting new insights—for example, about the growth patterns of semantic networks when aggregated across children (Hills et al., 2009, 2010). Such analyses have the potential to be even more powerful when applied to larger samples and across languages.

4.2 Longitudinal Stability of CDI Measurements

A classic test of the reliability of a psychometric instrument is its test-retest correlation. Assessing this correlation for CDIs for a single reporter is a bit impractical however, since—unlike, for example, a math test with objective answers and different question forms—this procedure would involve asking a caregiver to fill out the exact same survey twice in a row, and presumably they would remember many of their answers. An alternative possibility would be to measure the same child via multiple caregivers. This procedure was followed by De Houwer et al. (2005), who found that caregivers varied substantially from one another in their responding, but plausibly this variability is due not only to parent bias but also to the different contexts in which caregivers interact with children (e.g., one caregiver takes the child to the zoo more often, another plays kitchen at home).

Avoiding the issues of these procedures, we instead examine correlations in CDI measurements across developmental time. There are only a small number of deeply longitudinal corpora in Wordbank, so we will limit our investigation to two languages: Norwegian and English. Further, the largest group of longitudinal data cover the WS form, so we restrict to these data for simplicity. Within each of these datasets, the modal number of observations is two, but there are some children with more than 10 CDIs available.

In this type of analysis, differences between a particular individual's measurements could vary for two primary reasons: first, measurement error (parent forgetfulness, mistakes, etc.) and second, true developmental change (learning new words). Since vocabulary typically increases over time, we can look at the relative magnitudes of CDI scores via correlations; this is our first analysis. Our second analysis attempts to normalize these absolute differences by extracting percentile ranks and finds that this procedure in fact increases longitudinal

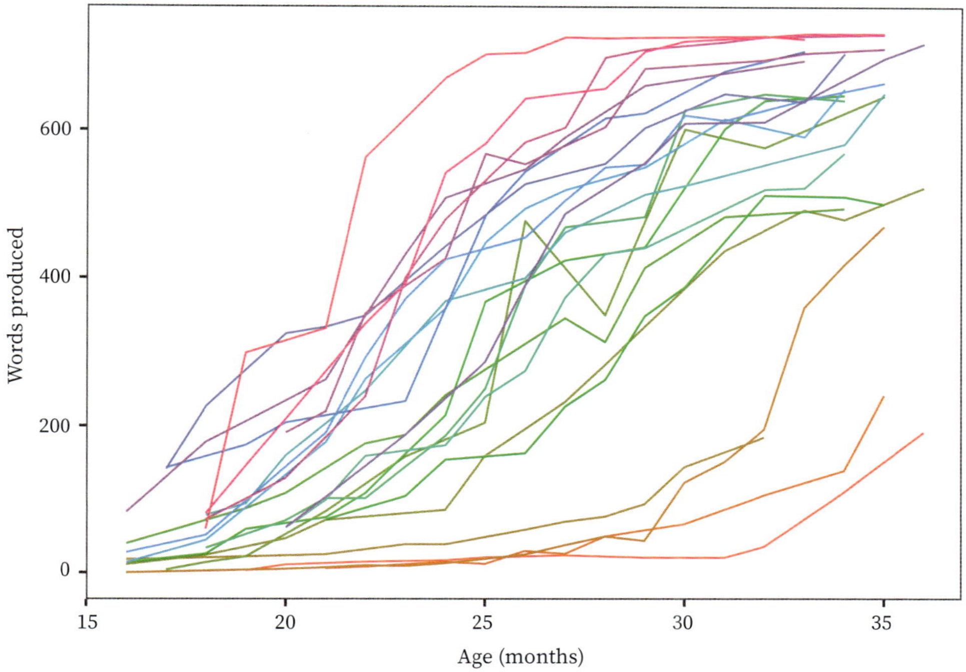

Figure 4.2
Vocabulary size as a function of age for children with more than 10 administrations (color indicates child).

correlations. Because there are two sources of differences between measurements, when correlations are low, we do not have direct evidence for whether (1) children's relative linguistic abilities are shifting with respect to one another or (2) we are observing measurement error. But, when correlations are high, we can assume the converse: measurement error is low *and* developmental stability is relatively high. It turns out that this latter situation is the case. As we will discuss in more detail in chapter 5, there is substantial variability between children in vocabulary size. The current analysis suggests that this variability appears to be quite stable longitudinally.

Figure 4.2 shows the trajectories of children (individual colors) who were measured more than ten times; it includes Norwegian data only due to data sparsity issues in English. These trajectories appear quite stable; the ranking of individuals does not appear to change much over the course of several years. This general conclusion—longitudinal stability of language ability as well as limited measurement error—is ratified by other studies using different datasets; for example, Bornstein and Putnick (2012), who found substantial stability ($r = 0.84$) between

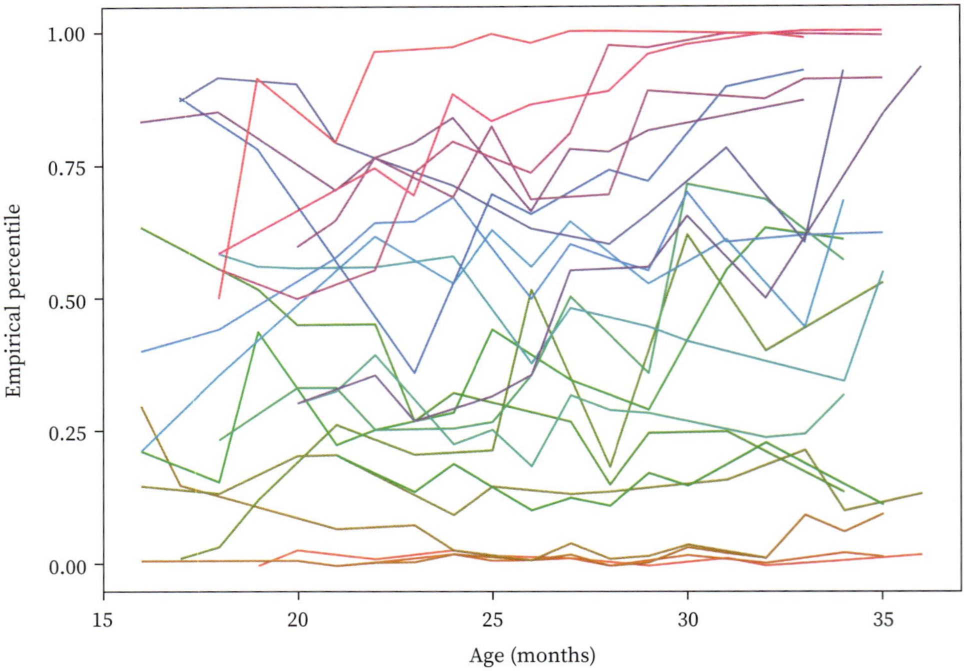

Figure 4.3
Vocabulary percentile as a function of age for children with more than 10 administrations (color indicates child).

latent constructs inferred from early language at 20 months and later language measured at 48 months.

One way to operationalize the question of stability is how children's percentile ranks tend to change over time. We examine this question qualitatively by showing the longitudinal trajectory of individual children's empirical percentile ranks based on the full normative sample for that language.[1] As shown in figure 4.3, these ranks are visually quite stable.

The transformation to percentile ranks allows us to assess the correlation between a child's percentile rank at time 1 and their rank at time 2, depending on the gap between these two. Because of sparsity, we bin children into two-month age bins and eliminate age bins with fewer than 50 children and then calculate between-bin correlations in percentiles. Figure 4.4 shows this analysis, revealing

1. We could use a model-based method (e.g., the `gcrq` method used in the Wordbank app and chapters 5 and 6) but in practice we have enough data in each of these languages so that this method should perform well.

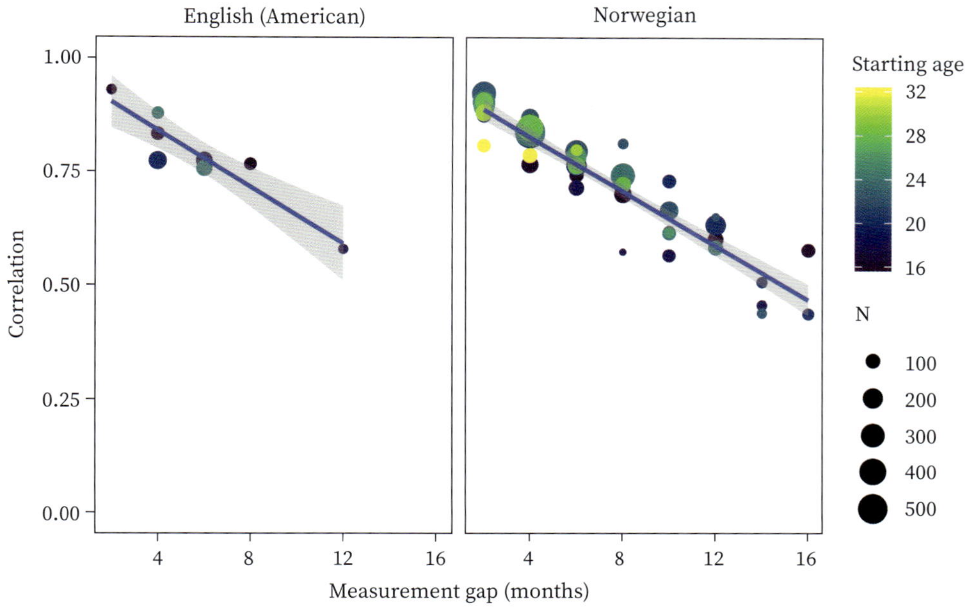

Figure 4.4
Correlations between vocabulary percentiles at multiple age points as a function of the age difference between them.

that percentile ranks are quite stable. Regardless of the age of the children, across a 2–4 month age gap the two percentiles are correlated at better than 0.8.

Longitudinal stability declines to around 0.5 at a maximal remove of 16 months, but this decline should be taken with a grain of salt. First, a 16-month gap amounts to a doubling of the child's age, so stability might be expected to be lower. Second, many children who are measured longitudinally across a 16-month gap will be expected to move from the floor of the form to the ceiling, compromising measurement accuracy. To test this last hypothesis, we evaluated the longitudinal stability of correlations using the same analysis as above but varying whether we used raw scores or percentiles. The percentile method substantially increased correlations.[2]

In sum, the variability between children that we observe in the CDI is quite stable longitudinally. It declines over time, but some of this decline may simply be due to the unavoidable limitations of CDI forms with respect to floor and ceiling effects.

2. We also used latent abilities derived from a four-parameter item response theory (IRT) model as below. While the IRT-derived ability parameters showed a consistent improvement in longitudinal correlations over the use of raw scores, percentiles realized a further gain over the IRT parameters in this case.

4.3 Psychometric Modeling

In this next section, we examine the psychometric properties of the CDI through the lens of psychometric models and item response theory (IRT). In brief, IRT provides a set of models for estimating the measurement properties of tests consisting of multiple items. These models assume that individuals vary on some latent trait and that each item in a test measures this latent trait to some, possibly variable, extent (see Baker, 2001, for detailed introduction). IRT models are a useful tool for constructing and evaluating CDI instruments as they can help to identify items that perform poorly in estimating underlying ability. For example, Weber et al. (2018) used IRT to identify poorly performing items in a new CDI instrument for Wolof (a language spoken in Senegal). IRT can also be used in the construction of computer-adaptive tests; this method has recently been applied to the CDI (Makransky et al., 2016; cf. Mayor and Mani, 2018).

IRT models vary in their parameterization. In the simplest (Rasch) IRT model, each item has a difficulty parameter that controls how likely a test taker with a particular ability will be to get a correct answer. In contrast, in a two-parameter model, each item also has a discrimination parameter that controls how much response probabilities vary with varying abilities. Good tests will tend to have items with high discrimination parameters across a range of difficulties so as to identify test takers at a range of abilities. Three- and four-parameter models add items for estimating lower- and upper-bounds of responding for individual items.

We examine IRT models as a window into the psychometric properties of the CDI. In the first subsection, we explore latent factor scores using the English Words and Sentences (WS) data. In the second subsection, we examine individual items and find generally positive measurement properties, although with some items at ceiling (included via carry-over from the Words and Gestures [WG] form). In the third subsection, we look at differences between comprehension and production in the WG form. In the fourth subsection, we look at the properties of the instrument by word category in both WS and WG.

Overall, the conclusions of our analysis are the following:

- Latent factor scores may have some advantages relative to raw scores in capturing individuals' abilities, but for the purposes of the analyses we perform in the main body of the book, they may carry some risks as well; hence, we do not adopt them more generally.
- In general, CDI WS items tend to perform well, but from a purely psychometric perspective there are a number of items that could be removed from the English WS form because their measurement properties are not ideal.

- Comprehension items, in general, tend to have less discrimination than production, suggesting that they are not as clear indicators of children's underlying abilities.
- Function words tend to have lower discrimination than other items, but the lexical class differences are not huge and do not interact with whether they are measured using production versus comprehension.

These analyses generally ratify the conclusion that the measurement properties of the CDI are good, even for function words and for comprehension measures. These questions may carry slightly less signal about the specifics of a child's vocabulary and load more heavily on a parents' general estimation of the child's linguistic ability, but they do carry some signal that relates to other responses. Further, when the English CDI departs from good measurement practice, it generally does so for completeness (e.g., including *mom* and *dad* words because these are important to parents, even though they do not show good measurement properties or are just different in some other way).

4.3.1 Measurement Properties of Individual WS Items

A first question that we can ask using a fitted IRT model is how well individual items relate to children's overall latent abilities. Practically speaking, in these analyses, we use the `mirt` package (Chalmers, 2012, 2016) to estimate the parameters of a four-parameter IRT model. As described above, the two-parameter model includes difficulty and discrimination parameters for each item. The four-parameter model supplements the standard two-parameter model with two parameters corresponding to floor and ceiling performance for a particular item. Items with high rates of guessing or universal acceptance across test takers would tend to have abnormal values on these bounds.

We fit Rasch and two-, three-, and four-parameter models to the English WS data and performed a set of model comparisons. On all metrics—Akaike information criterion (AIC), Bayesian information criterion (BIC), and direct likelihood comparison—the two-parameter logistic (2PL) model handily outperformed the Rasch model, suggesting that not every item had the same discrimination. Similarly, the 3PL outperformed the 4PL on all metrics, suggesting that adding an upper bound parameter did not increase model fit. On the other hand, the 2PL and 3PL were close in fit, with AIC and log likelihood favoring the 3PL but BIC favoring the 2PL. In the remainder of the analyses below (save for one), we adopt the 2PL for simplicity. In an exploratory analysis, we examine upper and lower bounds from the 4PL because the estimated upper bounds help us reason about those items that are not yet universally known by the older children in our sample.

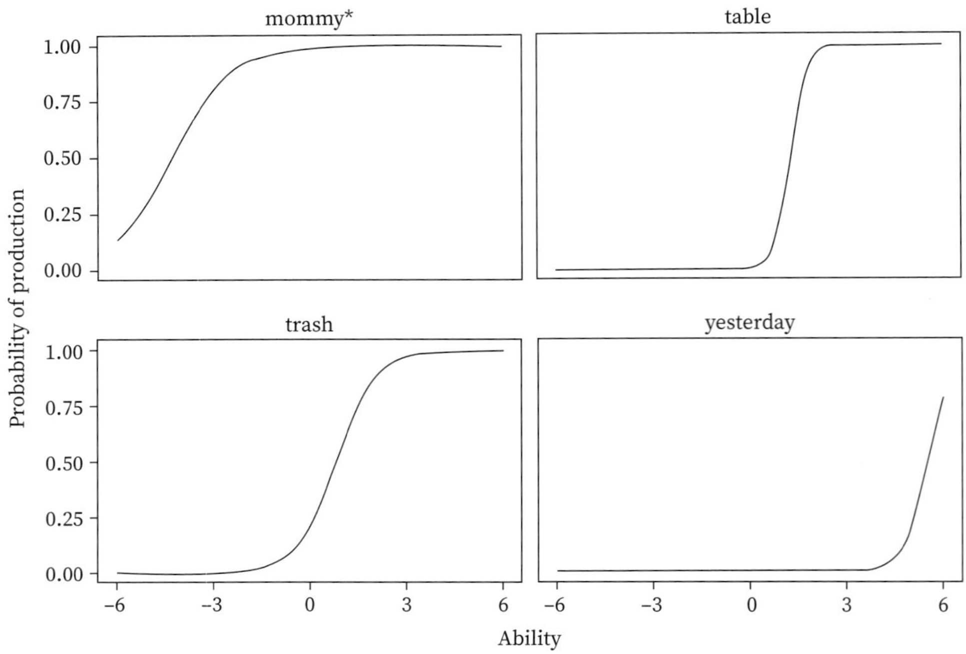

Figure 4.5
Item characteristic curves for a set of individual items from the English WS sample.

We begin by examining some individual item curves from the 2PL fits. Figure 4.5 shows four representative item characteristic curves. Each plots the probability of production by a range of latent ability scores. *Mommy* is produced by children at all abilities and is relatively uninformative about ability. In contrast, *table* and *trash* are both of moderate difficulty, but *table* is more informative because it has a steeper slope. Finally, *yesterday* is more difficult overall—in our sample, many high vocabulary children still did not produce this word. Generally, items with steeper slopes are considered more diagnostic of ability and hence more desirable.

We now examine these properties across the whole instrument. Figure 4.6 shows item discrimination and difficulty across the full set of items with outlying items labeled. Difficulty refers to the latent ability necessary for a child to produce an item, on average. Discrimination refers to how well an item discriminates between children of lower and higher ability (as judged by their performance on other items). For example, the word *table* is spoken by just about half of the children in the sample. Hence, asking whether a child says *table* is a good way to guess whether they are in the top or bottom half of the distribution.

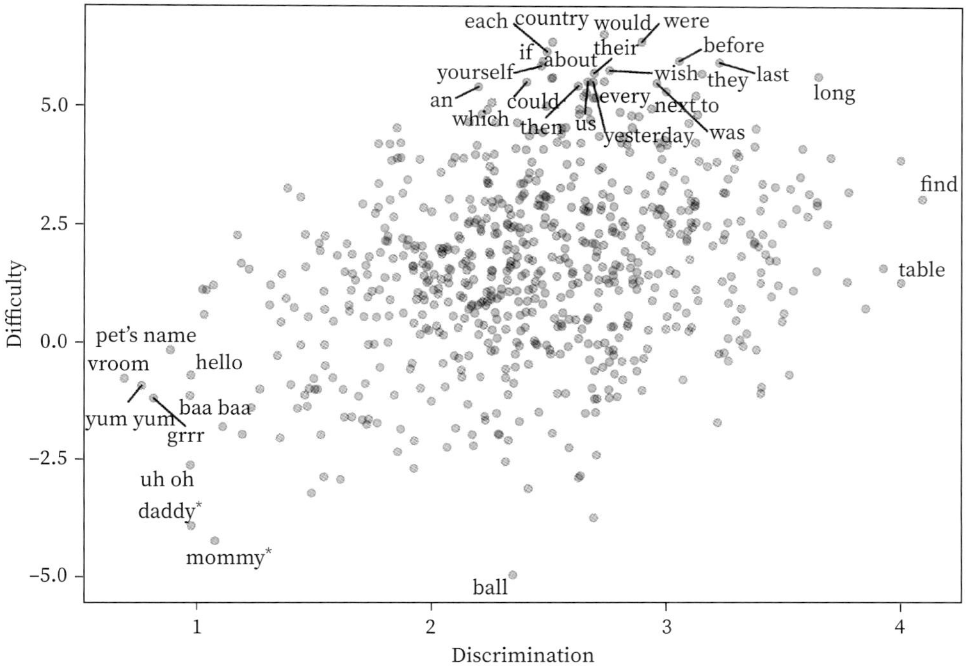

Figure 4.6
Words (points), plotted by their difficulty and discrimination parameters, as recovered by the two-parameter IRT model (see text). Outliers are labeled.

In contrast, visual inspection shows a tail of items with limited discrimination and low difficulty (e.g., *mommy, daddy, uh oh,* etc.). These are clearly items that are produced by nearly all of the children in the sample—they do not discriminate because they are passed by all children in the sample. If the only goal of the instrument were discrimination of different ability levels, they could likely be removed. But, as discussed above, these items tend to be included for completeness. Including these items also helps with compatibility among instruments since the WS instrument is a strict superset of the WG instrument, which is used with younger children and for which there would presumably be more variability in a word like *uh oh.* On the upper part of the plot, we also see a large cluster of words that are quite difficult (e.g., *country, would, were*); these items show some useful discrimination but presumably only for high ability children.

Turning now to an exploratory analysis using the 4PL model, we examine the recovered upper and lower bounds estimated for particular words, as shown in figure 4.7. While overall the 4PL model does not improve fit, these parameters are useful because they show the subset of words that are known by only a small

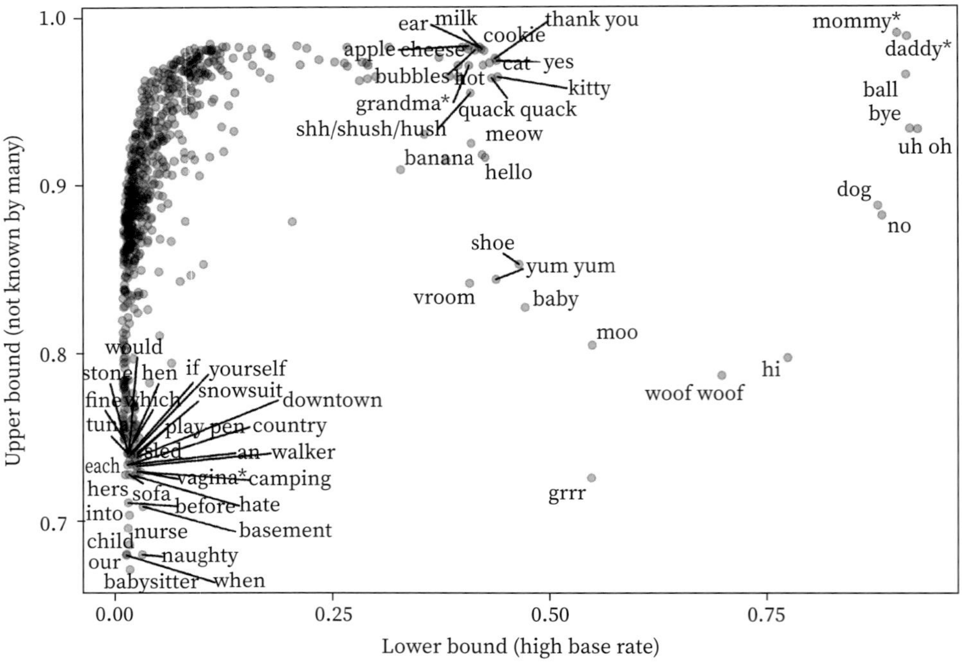

Figure 4.7
Words (points), plotted now by their lower and upper bound parameters from the four-parameter IRT model.

number of children (low ceiling) or are known by almost all children (high floor), respectively. Examining those with a very low ceiling, we see items that are likely to be quite idiosyncratic for a variety of reasons. For example, *babysitter*, *camping*, and *basement* likely vary by children's home experiences (further mediated by access to resources, parenting practices, and circumstances). In contrast, genital items (e.g., *vagina*, or the version of this item used in the child's family) vary by sex (see chapter 9). Examining those items with a very high floor shows early learned words like *mommy*. These words are similar to words with very low discrimination patterns. Because these words are known by essentially all children, the 4PL model may have fit these words as having a high chance level with essentially no discrimination ability.

One way to think about these analyses is that they show that the CDI has not only a large core of words with good measurement properties but also some other words that do not contribute as substantially and add length without adding much signal. If the goal of the CDI were only to provide psychometric estimates of vocabulary size, then these would be good candidates for deletion. But because CDIs

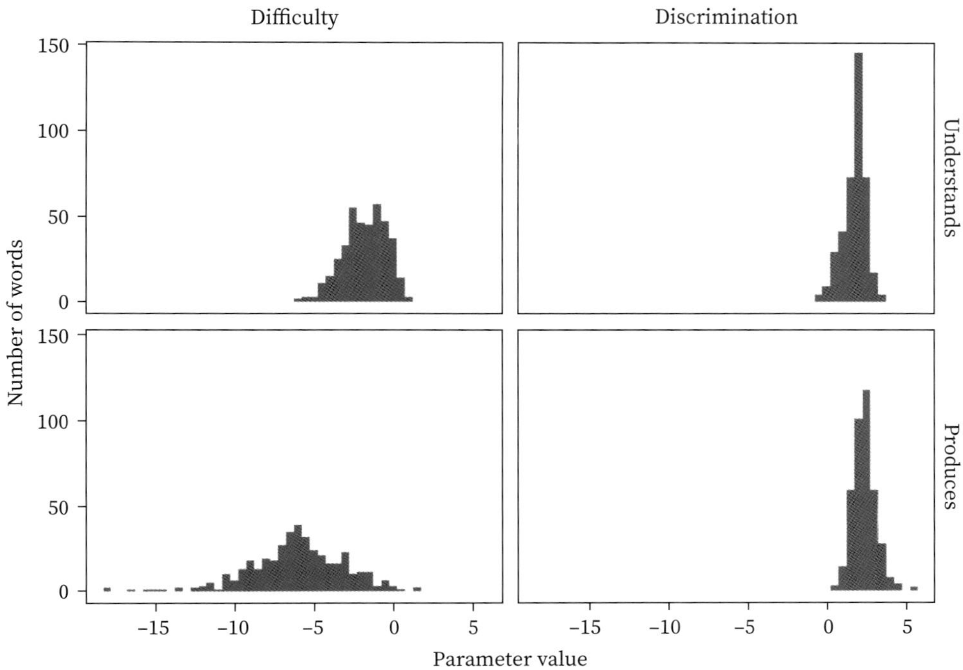

Figure 4.8
Histograms of words' difficulty and discrimination parameters for comprehension and production.

are also used for other purposes—such as the analyses we present in subsequent chapters—a larger set of items can be useful. We return to this general set of issues in chapter 18.

4.3.2 Production and Comprehension

We next use IRT to estimate whether there are differences between production and comprehension using WG data. To do so, we fit 2PL models to the WG data and examine the distribution of item parameters. In general, a good item distribution will have a range of difficulties so as to be sensitive to differences between children at a variety of levels. These items also should have relatively high discrimination, so that answers to individual items tend to provide relatively more information.

Figure 4.8 shows discrimination and difficulty parameter value distributions for WG production and comprehension. Difficulty is much higher (negative values) for production relative to comprehension, reflecting the expected asymmetry of production coming "after" (being more difficult than) comprehension. With respect to comprehension, several trends are visible. First, comprehension questions largely have positive discrimination parameters. Thus, these questions on the whole carry signal about children's latent linguistic ability. There do appear to be more items with low or even negative discrimination parameters, however, indicating more items that are not measuring ability appropriately (perhaps because they are difficult for all children or because they are too hard to assess). Mean discrimination is substantially lower for comprehension relative to production (1.8 versus 2.4).

Overall, this pattern is consistent with the hypothesis that production behavior is a clearer signal of children's underlying knowledge than assumed comprehension. Why? Perhaps parents are better reporters of production than comprehension, and hence these items are more discriminative of true behavior. The source of error in this case would be parents' mistaken belief that their child understands a word. Or perhaps comprehension is a fundamentally more variable construct, and hence, individual word knowledge consistent with understanding could be due to partial knowledge. Here the source of error is variance in how well children know the meanings of words. We cannot distinguish between these two models, but they have different underlying implications for the CDI.

4.3.3 Lexical Category Effects on Item Performance

One hypothesis that we have often speculated about is the question of whether there are special psychometric issues with particular word classes. For example, do parents struggle especially to identify whether children produce or understand function words?

Using 2PL parameter fits, figure 4.9 shows WS item difficulty and discrimination (as above) and the histogram of discrimination but broken down by lexical class (color). Many of the easy, non-discriminating items are found in the "other" section. In contrast, the hardest items tend to be function words. These items tend to have similar discrimination on average (2.4) compared with nouns (2.4), and modestly lower discrimination than adjectives (2.6) and especially verbs (3.0). The situation is not dire: all have a discrimination parameter above one. Thus, although function words are not the most discriminative items on the CDI WS, these items still appear to encode valid signal about children's abilities.

In our last analysis, we turn to the WG data. Figure 4.10 shows the mean (error bars show standard deviation) for discrimination parameter values. The only major trend is that there is a moderate level of discrimination for all classes except "other" (which includes items like *mommy* and *daddy* and a variety of animal sounds and social routines). One hypothesis about this finding is that, especially early on, parents are very generous in their interpretation of whether their child understands these specific words.

In sum, we do not find evidence that function words are particularly low-performing items from a psychometric perspective—even in comprehension

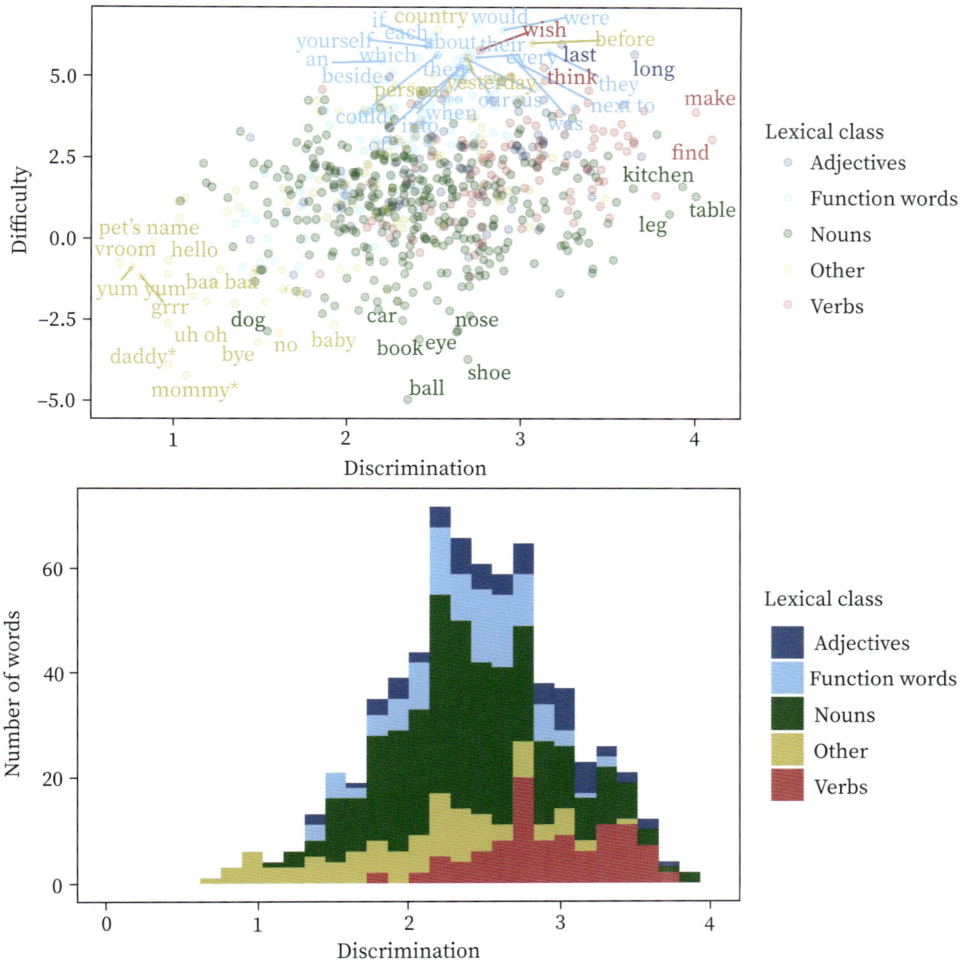

Figure 4.9
Lexical class effects on difficulty and discrimination for Words and Sentences. The top plot shows individual words plotted by their parameter values, with color representing the lexical class of the words. The bottom plot shows discrimination information in the form of a histogram.

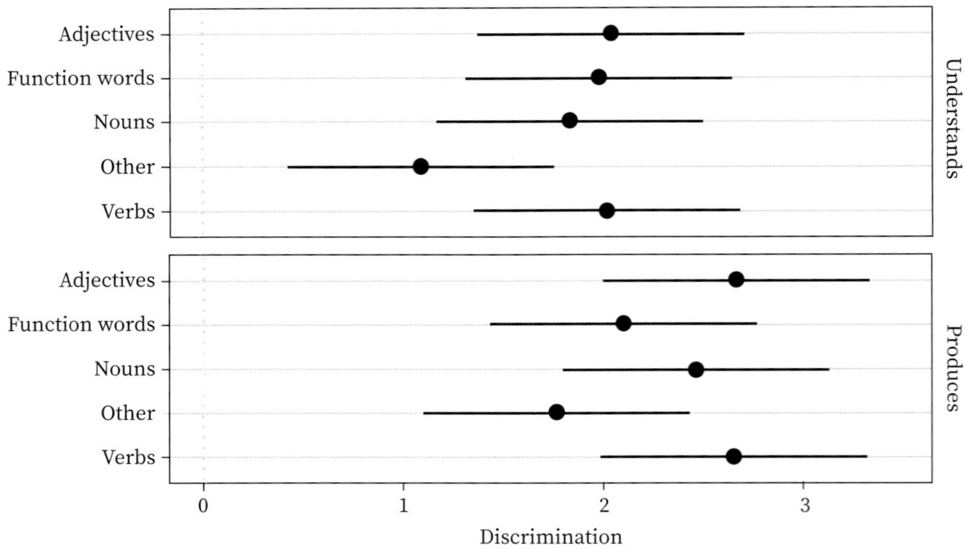

Figure 4.10
Mean discrimination values for individual words' in production and comprehension measures from the Words and Gestures form (error bars show standard deviation).

assessments! Rather, there are some low-performing items spread across all categories of the CDI form, and many of these likely perform poorly for the reasons described above—especially regarding difficulty in interpretation of very early behavior and variability in home experience.

4.3.4 IRT Models: Conclusions

One question regarding IRT-model derived parameters for individual children is whether they should be used in place of percentiles or raw scores for some of the measurement problems we encounter throughout the rest of the book. Although these latent ability scores might be overall better reflections of children's vocabulary than other measures, we do not find strong evidence to support that conclusion. For example, in the analysis above, we compared longitudinal correlations derived from raw scores, percentiles, and IRT ability parameters. While IRT parameters yielded higher correlations than raw scores, empirical percentiles performed better still (at least for Norwegian and English, two languages for which we have large amounts of data).

Further, there are other negatives associated with swapping an imperfect but straightforward measure (raw and percentile scores) for a model-derived measure (latent ability). Interpretation clearly suffers if we use the model-derived measure since readers will not be able to map scores back to actual behavior in terms of the checklist. In addition, model estimation issues across instruments introduce further difficulties in interpretation. Most obviously, model estimates with smaller datasets may vary in unpredictable ways; similarly, a greater presence of poorly performing items in certain datasets may lead to systematic issues in the latent estimates for those datasets. In the absence of clear solutions to these model-fitting problems, we choose the route of using the "sumscore" (Borsboom, 2006) while acknowledging its limitations.

4.4 Conclusions

In this chapter, we examined the measurement properties of the CDI from three perspectives. From a theoretical perspective, we reviewed why the design features of the CDI make it a reasonable tool for measuring child language, even if there are opportunities for error and bias throughout. (Of course, one of these design features is the style of administration for a particular study, so of course a poorly administered form will yield a dataset with lower reliability and greater bias). Then, we took advantage of the deep longitudinal data available for two languages and showed quite strong longitudinal correlations among CDI administrations. This pattern indicates that early language is a stable construct across development (Bornstein and Putnick, 2012). It also signals that measurement error between CDI administrations appears to be limited, at least when the span of time between administrations is not too great. Finally, we used item response theory to examine the measurement properties of individual items. While the CDI includes some items with limited measurement value (if all that the user cares about is a single ability score), most items show good psychometric properties. This analysis also revealed that comprehension questions and questions about function words do not appear to be particularly worse than other items, contrary to previous speculations. In sum, the CDI appears to be a reliable instrument for measuring children's early language, with measurement properties that support a range of further analyses.

5 Vocabulary Size

This chapter begins our substantive analysis of properties of language learning and their variation across languages and children. We start with one canonical view of Communicative Development Inventory (CDI) data, in which each child is represented by a single vocabulary score: the proportion of words that child knows out of the total in the form. We first quantify the median pattern of vocabulary growth observed in our data, and we then turn to characterizing variability across individuals in these data. In these analyses as well as in subsequent chapters, our inspiration comes from what we think of as the "Batesian" approach to variation.

> Far from simply reflecting noise in our measuring instruments or variability in low-level aspects of physiological maturation, the variations that we will document (in vocabulary development) are *substantial, stable,* and have their own *developmental course.* Because this variation is substantial, it is critical for defining the boundary between normal and abnormal development; because it is stable, it provides a window onto the correlates and (by inference) the causes of developmental change; and because it has its own developmental course, it can be used to pinpoint critical developmental transitions that form the basis for theories of learning and change. (Bates et al., 1995, 96)

We are interested in these theoretical uses of variability. But variability is only meaningful in the case that it is stable; that is, that it reflects signal about individuals (or cultures) rather than measurement error. With respect to the CDI, the strong evidence for the reliability and validity of the forms—reviewed in chapter 4 and in Fenson et al. (2007)—provides support for the contention that observed variability is meaningful. To foreground one of the strongest reliability studies, we note that Bornstein and Putnick (2012) collected multiple language measures at 20 and 48 months in a sample of nearly 200 children and used a structural equation model to estimate the stability of a single latent construct, language ability. Essentially all measures related strongly to this latent variable and the coefficient on its stability over time was $r = 0.84$, suggesting that early language is quite stable, at least when measured appropriately. Notably, the Early Language Inventory (ELI), a precursor to the CDI, was included in the measures at 20 months and was found to correlate with the 20-month latent construct at $r = .87$.

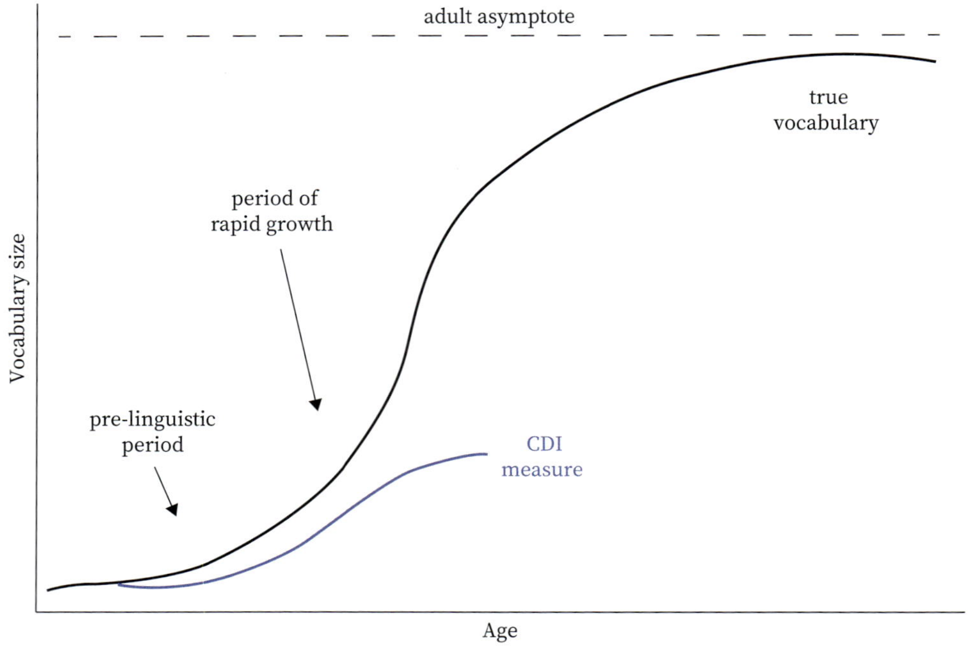

Figure 5.1
Schematic true vocabulary growth and vocbaulary growth as measured by the CDI.

This finding—along with the other evidence, mentioned above—justifies the implicit conceptual model of the following analyses. That model is that there is a single quantity, early language ability, that is stably measured by parent report and that can be approximated as the raw proportion of words a child “understands” or “understands and says” on CDI forms. Such evidence has primarily been collected for the English CDI, however. In this chapter we examine variability across the full set of languages, and it is worth remembering that we project the reliability and validity of the English instrument to its adaptations.

5.1 Central Tendencies

The first question we can ask about CDI data is about its central tendency—the median pattern of vocabulary growth. Our general expectation is shown in figure 5.1.

This schematic reveals a number of patterns that are explored in this and subsequent chapters. The CDI necessarily captures a small fraction of any individual’s true vocabulary, but even within the measured range there are a number of specific questions that can be addressed by different analyses. The question of the exact slope of children's growth, especially in the period immediately after the emergence of language, is treated in chapter 15—this question is sometimes posed as whether children undergo a vocabulary "spurt" (Ganger and Brent, 2004, but cf. McMurray, 2007). On the other side of the CDI curve, the question of the divergence between CDI-measured vocabulary and true vocabulary (and whether true vocabulary can be recovered via a statistical correction) is treated in work by Mayor and Plunkett (2011)—because of the English-specific nature of this work, we do not take up this issue here. In this chapter, we focus on the middle section of the CDI curve, in which children's vocabulary is neither at the floor or the ceiling of the instrument.

5.1.1 Commonalities across Languages

Figure 5.2 shows the median patterns of growth for early production. Rather than showing proportions, as we will do more standardly throughout the book, here individual item totals are plotted. In general, the median child before the first birthday is reported to produce a small number of words. (These data raise a number of questions about the specific reliability of very early parent reports, which we take up below.) Overall, however, these curves accord relatively well with our intuitive sense of early vocabulary development: they reveal that most children tend to speak at most only a few words before their first birthday, but that production accelerates across the second year.

This analysis also motivates the decision to omit production data from Words and Gestures (WG) forms from most of the analyses in subsequent chapters. Many WG forms end at 16–18 months, meaning that the median production is only around 50 words. For analyses of vocabulary composition or predictive modeling, these numbers are often too small to yield meaningful cross-item conclusions, although these data can be combined with Words and Sentences (WS) data (see appendix C).

The acceleration in early vocabulary is even clearer when looking at production reports from older children using WS. Figure 5.3 shows this pattern. In every language, the median child is reported to produce 50 words between 16 and 20 months (dotted line), though—as we will see below—this analysis masks tremendous between-child variability during this period. In addition, languages vary considerably in the absolute number of words reported. (As it is a major outlier, we have discussed the Beijing Mandarin WS data in chapter 3 in the section on difficult data). Nevertheless, there are still substantial consistencies in the shape and general numerical range across languages.

During the period of 24–30 months, we see curves leveling out. Presumably, this leveling does not reflect a slowing in the rate of acquisition, which most researchers assume continues unabated for many years (e.g., Bloom et al., 2001).

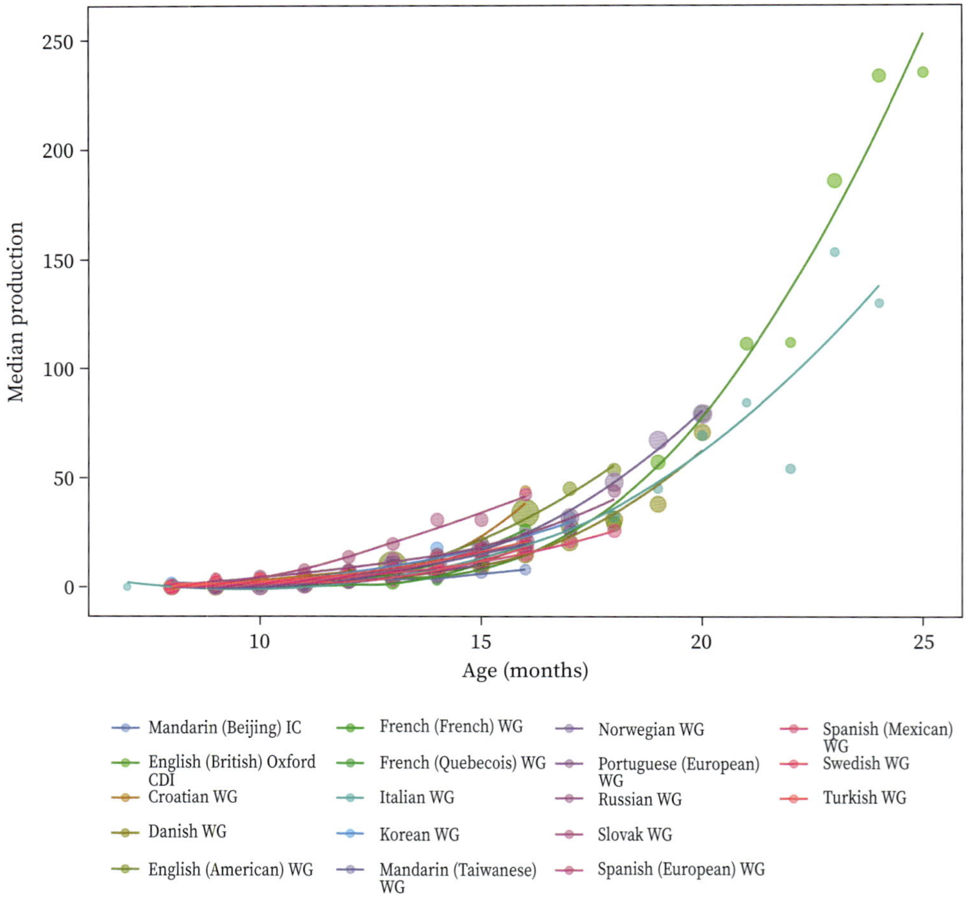

Figure 5.2
Median production using WG-type forms. Included are only languages where there are more than 200 administrations total.

Instead, it reflects the limitations of the CDI instrument, in that there are many possible "more advanced" words that children are likely learning, of which only a small subset are represented on any form.

We next turn to comprehension medians, shown in figure 5.4. Comprehension is queried only on the WG form. Reported comprehension increases much faster than production, so much so that most parents are reporting that their children understand most words on the form by 18 months (chapter 15 discusses differences in the balance between comprehension and production among children). As with the production data, we see substantial differences across languages in

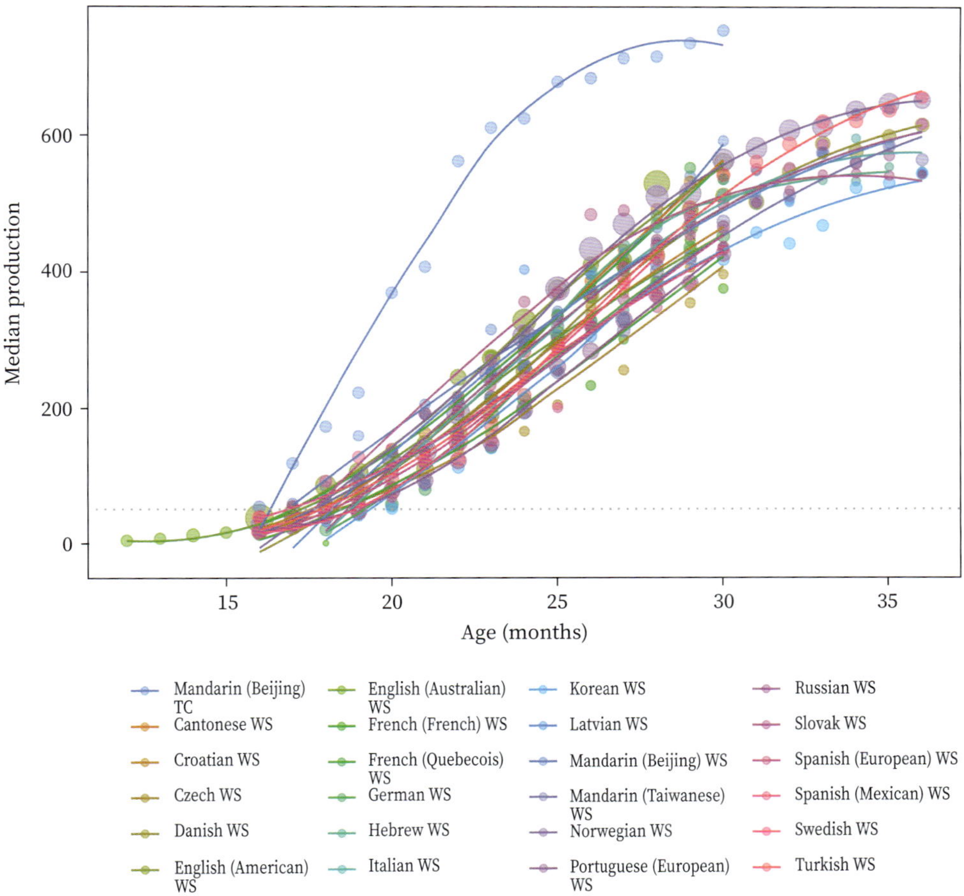

Figure 5.3
Median production using WS-type forms. Included are only languages where there are more than 200 administrations total.

reported vocabulary, discussed below (see chapter 3, section on difficult data, for more discussion of Taiwanese Mandarin comprehension data).

One striking aspect of the comprehension data is how early comprehension is reported. For example, from 8 months, we see parents reporting medians of 4.5 (Swedish) and 123.5 (Mandarin [Taiwanese]) words. To many researchers (and some parents) these high numbers feel unlikely. We are largely agnostic on this issue, but the literature does provide some support for early comprehension reports. A spate of recent infancy experiments suggests that, in fact, children in the second half of the first year do have some fragmentary representations of

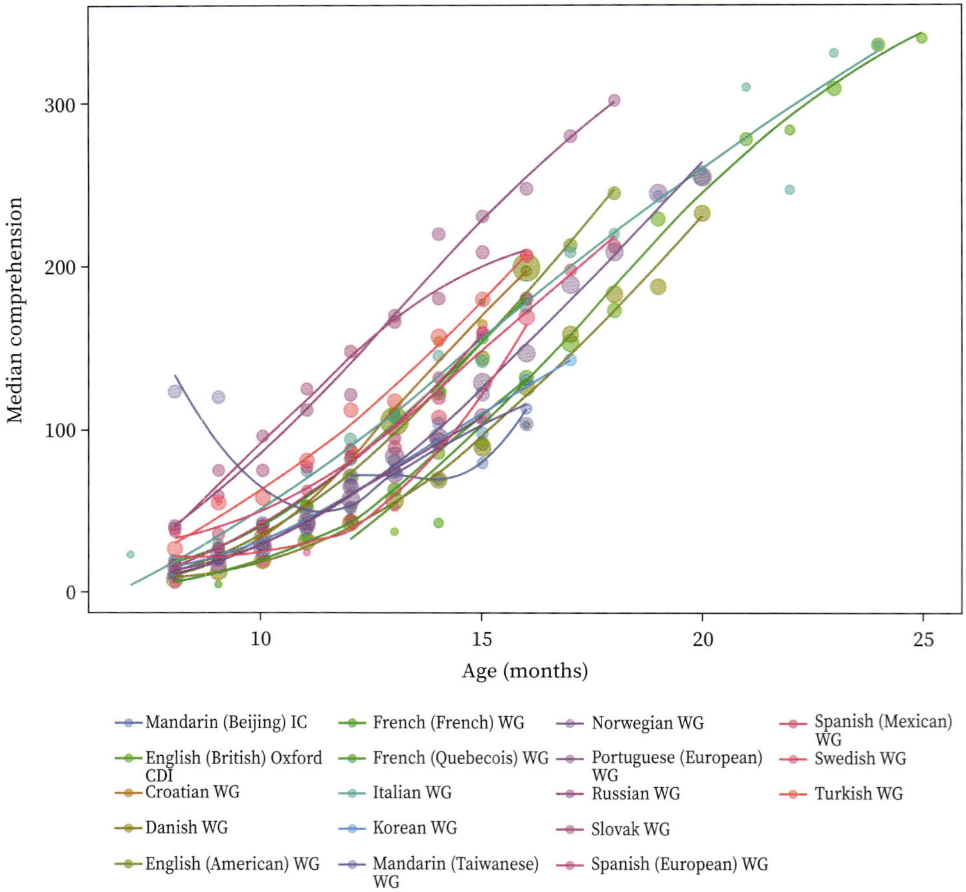

Figure 5.4
Median comprehension using WG-type forms. Included are only languages where there are more than 200 administrations total.

many common words available (e.g., Tincoff and Jusczyk, 1999, 2012; Bergelson and Swingley, 2012; Bergelson and Aslin, 2017). The representations revealed in these tasks are quite weak—often amounting to a 2–5 percent difference in looking to a target on hearing a word uttered. Despite its weakness, depending on the criterion used by parents, this knowledge may be what is detected in these early reports. Thus, these estimates may not be as far off as we initially supposed.[1]

1. An alternative possibility is that *both* accounts are true but unconnected: 8-month-olds could in fact know some common words, but parents could be overestimating their vocabulary based on observed behaviors—in essence, parents could be right but for the wrong reasons (Bergelson and Swingley, 2015).

5.1.2 Cross-Language Differences

Setting aside outlier datasets, there is other variation among languages apparent in the preceding analyses. To what extent is this variation meaningful? We explore this question next, focusing on production data from the WS form as these data are the densest and most reliable. We consider a range of explanations in turn.

1. Differences could be due to differences in form length. Medians for proportions and raw scores are highly correlated ($r(22) = 0.906$, $p < 0.001$), however, suggesting that the ordering of languages is not only a function of form length. Raw scores have a positive but nonsignificant correlation with form length ($r(22) = 0.336$, $p = 0.11$); this correlation changes direction and is not reliable for proportions ($r(22) = -0.079$, $p = 0.71$). In sum, it appears that there are form-length differences (motivating the use of proportions in general) but that there is still variation among languages even correcting for this issue. Figure 5.5 shows the relevant proportion trajectories, highlighting remaining differences between English and Danish (two languages for which we have substantial datasets with full demographic information). As Eriksson et al. (2012, p. 331) write, "Using raw data assumes that each form is exhaustive, while using percentages assumes that each form is equally exhaustive. Neither is correct and the truth lies somewhere in between."
2. Differences could also be due to form construction. For example, the Czech form could select relatively harder words for inclusion, leading to fewer words being checked. We cannot directly address questions about the difficulty distribution of items without moving to psychometric models (see chapter 4). These models in turn would need to be equated across forms to compare latent ability scores across instruments. While we have experimented with these procedures, there is a circularity to test equating procedures that makes us leery of proceeding. In particular, to equate across tests in standard IRT models, it is critical to have test items that are shared across instruments. But although we have *concepts* that are shared across instruments (see chapter 10), we do not believe the words that represent these concepts are necessarily equally difficult across languages—in fact, the premise of some of our later analyses is that they are not. Thus, assessing form difficulty across languages is a complex proposition that we do not address directly here.
3. Differences could also be due to demographic differences across samples. We have examined sample composition in chapter 3 and see that—to the extent we have access to demographic data—sample composition does vary in features that affect vocabulary (e.g., maternal education, birth order; see chapter 6 for fuller analysis). We are not yet in a position to conduct a full analysis of these differences, controlling for demographics, as data are sparse and the nature and extent of demographic differences *also* vary across cultures. Nevertheless, we can note, for

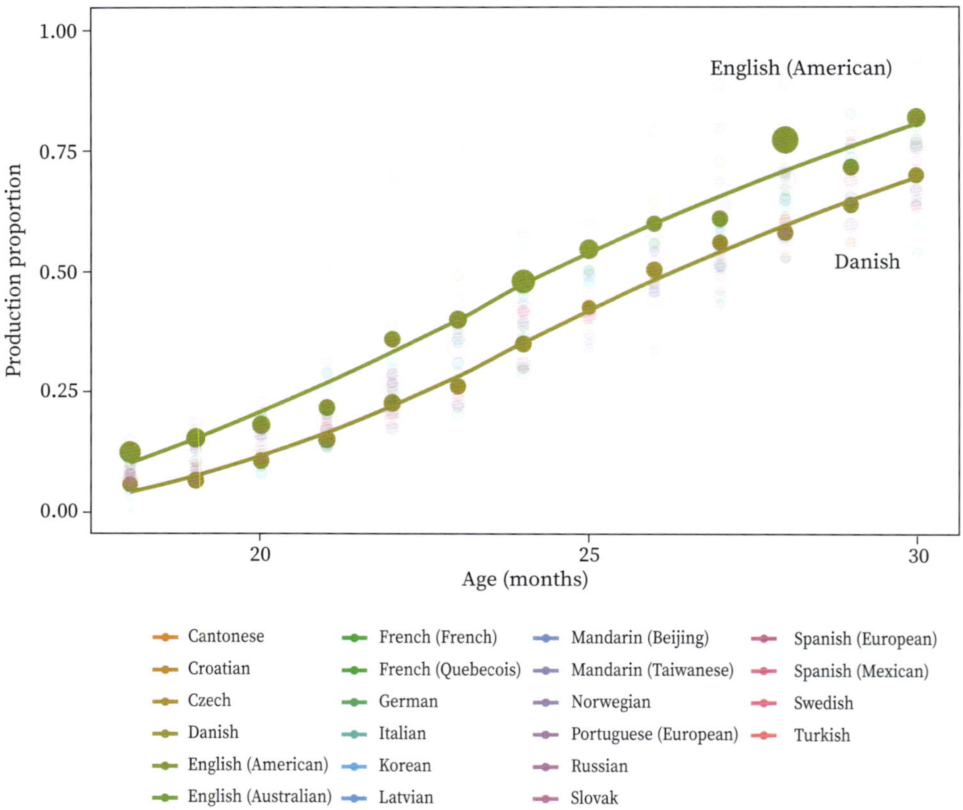

Figure 5.5
Cross-linguistic production data, proportions plotted by age. English (American) and Danish are highlighted.

example, that differences between Danish and English (American) children look quite similar (though noisier) in female, first-born children of college-educated mothers (figure 5.6). This would suggest that demographic differences cannot fully explain the cross-linguistic differences that are observed.

4. Differences could be due to cohort effects, in which older sets of data show differences from newer datasets. Most of our data date from the period 2005–2015, but some of our English and Spanish data are older as they date to the period in which CDI forms were first being designed (the early 1990s) (Fenson et al., 1994). Unfortunately, we do not have reliable information about the collection date for many datasets, so we cannot use this variable in regression analyses. Naively, we would expect later cohorts to show higher vocabularies, consistent with the Flynn effect (Flynn, 1987). Yet, the Danish data, for example, are relatively recent and

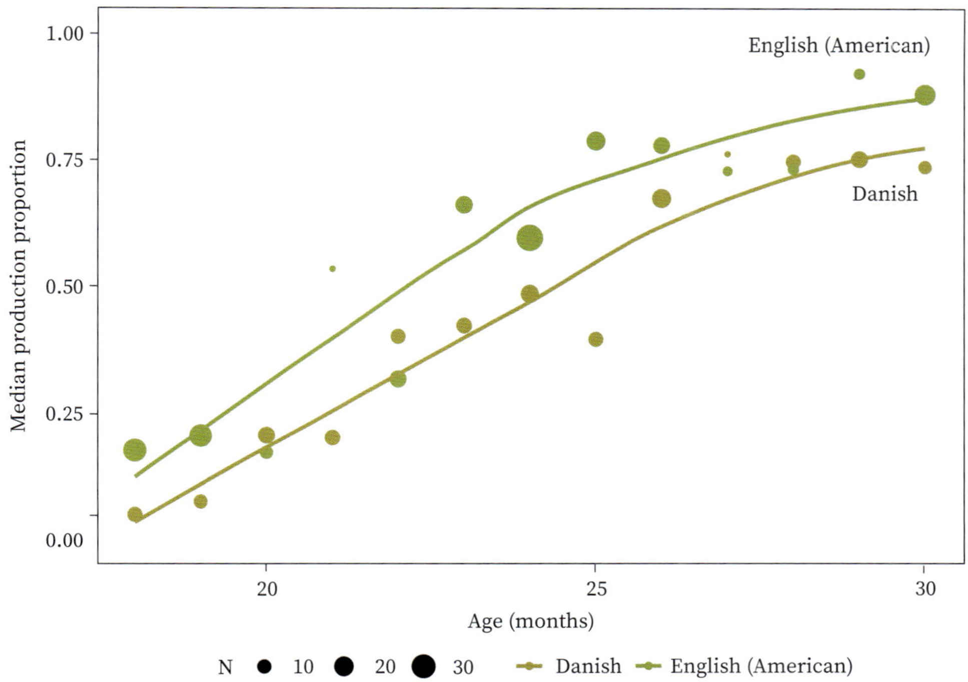

Figure 5.6
Proportion production plotted by age for Danish and English samples, now subsetting to the first-born female children of college-aged mothers.

were collected online using standardized instructions. Danish is subject to its own issues (see below), but the Norwegian data are also relatively recent and are quite comparable with the English data.

5. Differences may relate not to demographics of the sample but to the procedure at administration. These differences are not transparent to us in all cases, and so, similar to cohort effects, we cannot control for them statistically. For example, instructions at administration—whether written on the form or given by experimenters—might have been more liberal in the case of Slovak or English (American) samples. Such instructions could have emphasized completeness in reporting vocabulary. Or the circumstances of administration could have been different—for example, Danish data were collected online while most English (American) data were collected using paper and pencil forms. English (American) data are contributed by many different labs, so there are likely many different administration styles represented.

6. Differences could be due to cultural or experimental differences in reporting bias. Slovak parents might have a lower criterion for reporting knowledge of a word. Following our discussion in chapter 2, we could speculate that Slovak parents' models of children's overall competence might be higher. (Such an explanation could be true in principle for the case of the Mandarin data discussed above as well, though this would be a case of extreme differences!) This explanation is an extension of the discussion above of administration and instructions—perhaps the relevant differences are in cultural expectations for what it means to be producing a word or for how verbal children are expected to be.

7. Finally, differences could be due to true differences in language acquisition. While this explanation is a possibility, we hope we have emphasized that it is only one among the many enumerated here. Nevertheless, one reason we have emphasized the Danish/English comparison above is that many researchers working on Danish believe that, due to its phonological properties, it truly is a difficult language to learn (Bleses et al., 2008, 2011). In particular, Danish is characterized by some highly distinctive phonological reduction processes which greatly reduce the frequency of obstruents and more generally lead to "an indistinct syllable structure which in turn results in blurred vowel-consonant, syllable and word boundaries [where] … word endings are often indistinctly pronounced" (Bleses et al., 2008, p. 623). The authors of this study are also able to provide evidence against the alternative view that Danish parents are simply more reluctant to respond "yes"—there were no differences on either gestures or word production. They conclude that the phonological structure of Danish produces an initial obstacle to breaking into the stream of speech and is reflected in overall patterns of vocabulary development. While this generalization may in fact be true, research is still needed to explore what other factors might converge to make language acquisition relatively easy versus hard in some languages than others.

In summary, differences among languages in the sheer number of words reported are unlikely to be accounted for purely by differences in form size or demographic differences among samples. In our (very speculative) synthesis of the preceding discussion, they likely result from a combination of cultural attitudes toward children's language, differences in administration instructions, and real differences in learning across languages. Partialing out these differences would likely require better-controlled data collection that included constant administration and sampling methods. For this reason, in the remainder of our analyses, we attempt to avoid interpreting overall differences in vocabulary size as much as possible and limit ourselves to quantities that can be effectively normalized.

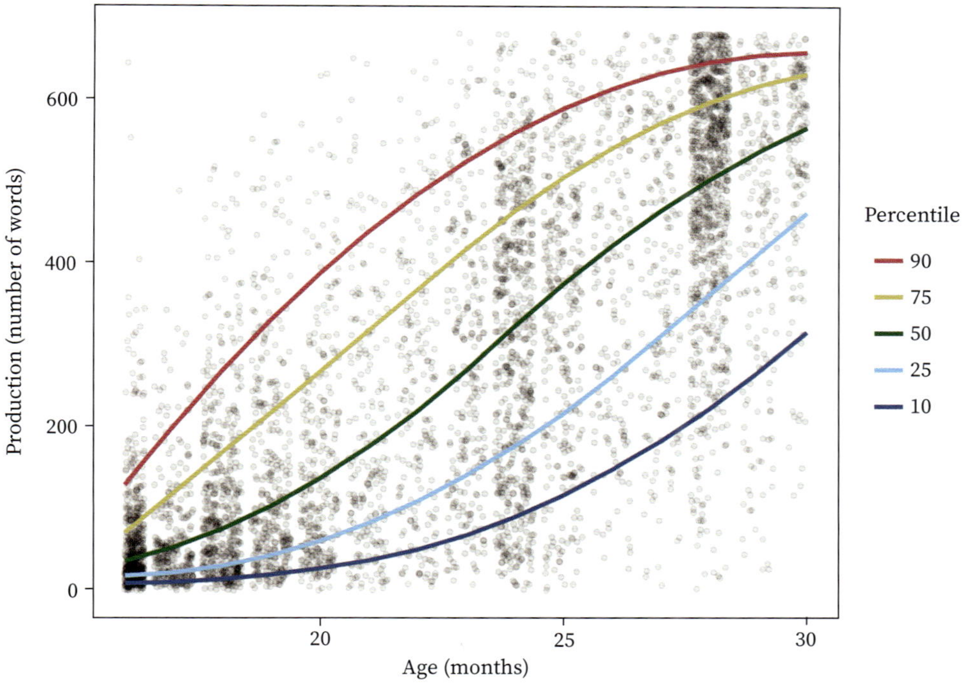

Figure 5.7
Raw production scores for English (American) production data. Dots show individual participants, while lines give standardized percentiles, computed via spline-based quantile-regression.

5.2 Variability among Individuals

We next turn from central tendencies in early vocabulary to the issue of variability, one of the most important features of early vocabulary development (Fenson et al., 1994). Across every language in the database there is huge variability in the vocabulary sizes reported for individual children, even within an age group. How can we characterize this variability? That is the question addressed by our next set of analyses.

5.2.1 Quantifying Variability

As an example, we zoom in on the English (American) production data from the WS form. The canonical view of these data is given in figure 5.7. It is very clear that variability is the norm! There are children close to the floor and ceiling of the instrument at almost all age groups.

We zoom in even further to consider just a single age group, 24-month-olds. A histogram of production vocabulary for this group is shown in figure 5.8. The

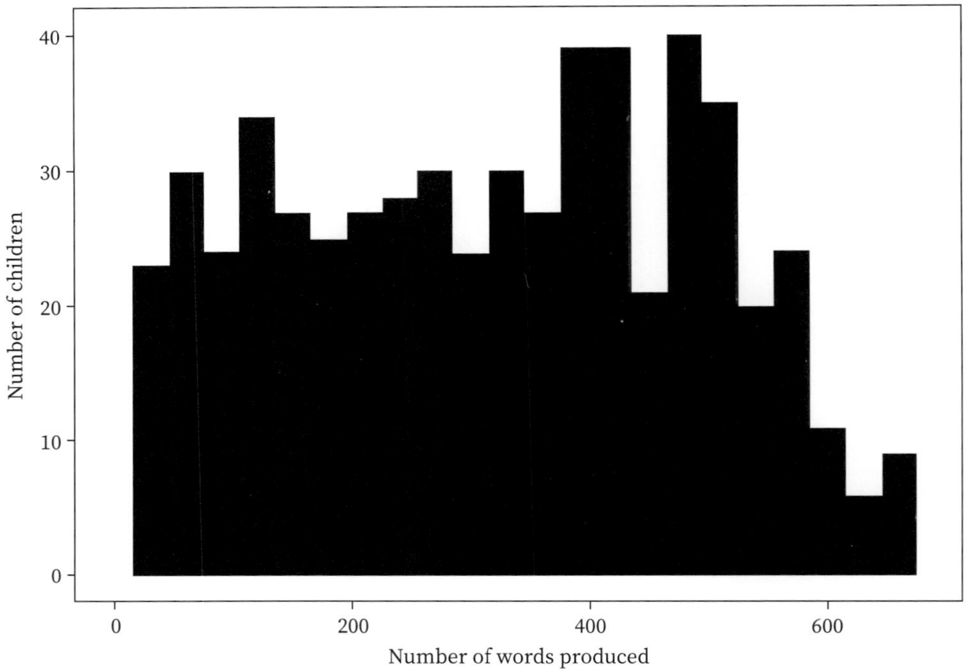

Figure 5.8
Histogram of English (American) production values for 24-month-olds.

distribution of vocabularies across children is far from normal, with many children at the very bottom of the scale and almost as many at the top. In other words, quite a few two-year-olds on their second birthday are reported to produce only a handful of words, and others are producing nearly all of the 680 listed on the form.

One way to describe these data is to consider the relationship of the variance to the central tendency. The "coefficient of variation" (CV) is a common measure used for this purpose (with σ denoting the sample standard deviation and μ denoting the sample mean):

$$CV = \frac{\sigma}{\mu}.$$

This statistic allows us to compare variability across measurements with different scales, an important concern when we want to compare forms with very different numbers of vocabulary items. For example, for 2-year-olds, the mean productive vocabulary is 319 words and the standard deviation is 175 words, leading to a CV of 0.55.

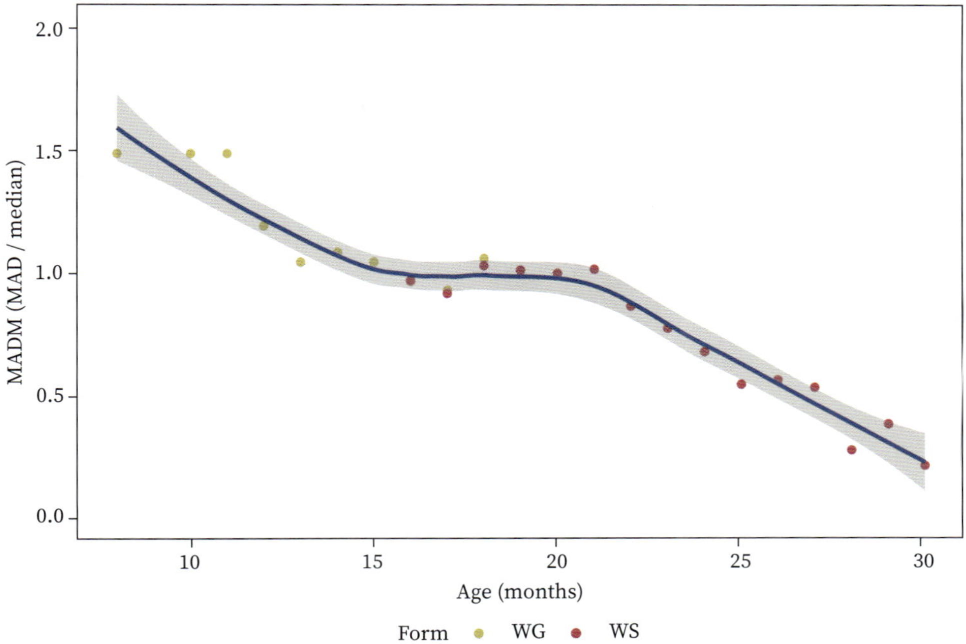

Figure 5.9
MADM values plotted by age for English (American) production data, across forms. The smoothing line is produced by a locally weighted (loess) smoothing function.

But again, as seen in figure 5.8, the distribution of productive vocabulary scores is far from normal. And distributions deviate even more from the standard normal distribution at younger and older ages. Thus, a nonparametric approach is more appropriate. Accordingly, we compute the mean absolute deviation from the median (MADM) statistic, the nonparametric equivalent of the CV (Pham-Gia and Hung, 2001). In MADM, the mean μ is replaced by the median $m(x)$, and the standard deviation σ is replaced by the mean absolute deviation (which captures how far away values are from the median):

$$MADM(x) = \frac{\frac{1}{n}\sum_{i=1..n}|x_i - m(x)|}{m(x)}.$$

Appendix B demonstrates that, although MADM is more appropriate for our data, CV and MADM are very highly correlated with one another across CDI datasets.

Figure 5.9 shows the MADM value for American English production data, plotted by age. In these data, the MADM is actually close to 1.0 from age one until almost age two, suggesting that the average *difference* from the median is actually as big as the median itself!

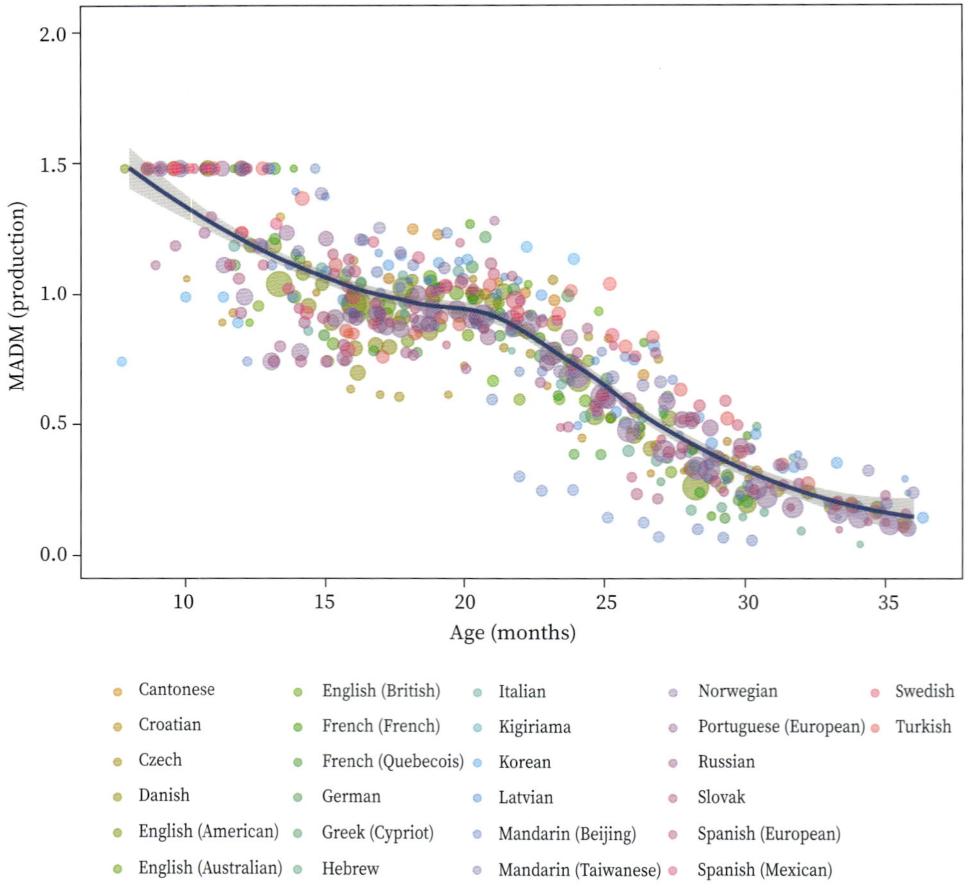

Figure 5.10
MADM for production, plotted by age group, for the full sample of languages in our dataset.

To get a sense of this variability, it can help to have a smaller dataset to consider. Imagine groups of three children. A group where one produced 30 words, one produced 100, and another produced 170 would have a MADM of 0.99. In contrast, one where they were more closely grouped—say, 70, 100, 130—would have a MADM of 0.44.

Does the general level of variability observed in American English hold for other languages? Figure 5.10 shows the MADM for production across languages and instruments. This similarity in variability is quite striking. Between the first and second birthdays, children's early language is remarkably variable, but this variability itself is quite consistent across languages.

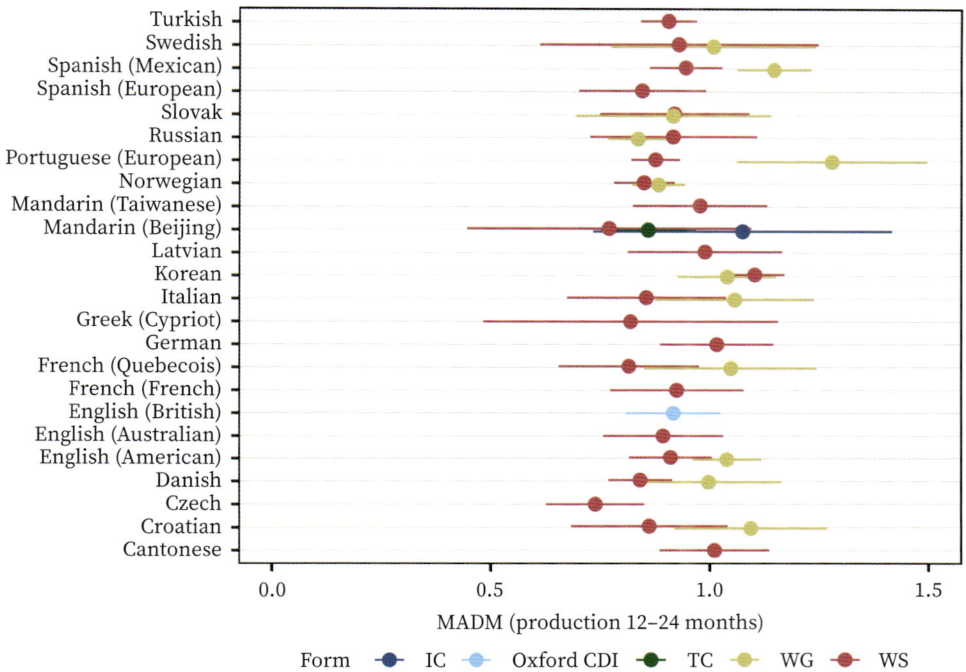

Figure 5.11
MADM values from 12–24 months for all languages and forms.

We summarize the MADM in the second year of life by taking its mean across that age range. This summary is shown in Figure 5.11. The MADM mean is close to 1.0 for almost every language and form for which we have data. In sum, confirming the analysis above, we see strong cross-linguistic consistency in the variability of children.

One question that could be raised regarding the analysis above is the extent to which variability is caused by *variability across children* versus *variability in reporting*. The extreme values seen in the English data, for example, could conceiveably be the result of a mixture of lazy parents who stopped answering the form with overly diligent parents who misunderstood and checked every box for a word they thought the child had been exposed to. But, to the extent these biases are the source of variability, they are extremely *consistent* across languages—which, recall, is the exact opposite argument from the one we considered above (where parent diligence was supposed to be variable enough across samples to lead to differences between languages).

The same analysis for comprehension data is shown in figure 5.12. For comprehension, we see a gradual decrease in variability throughout development.

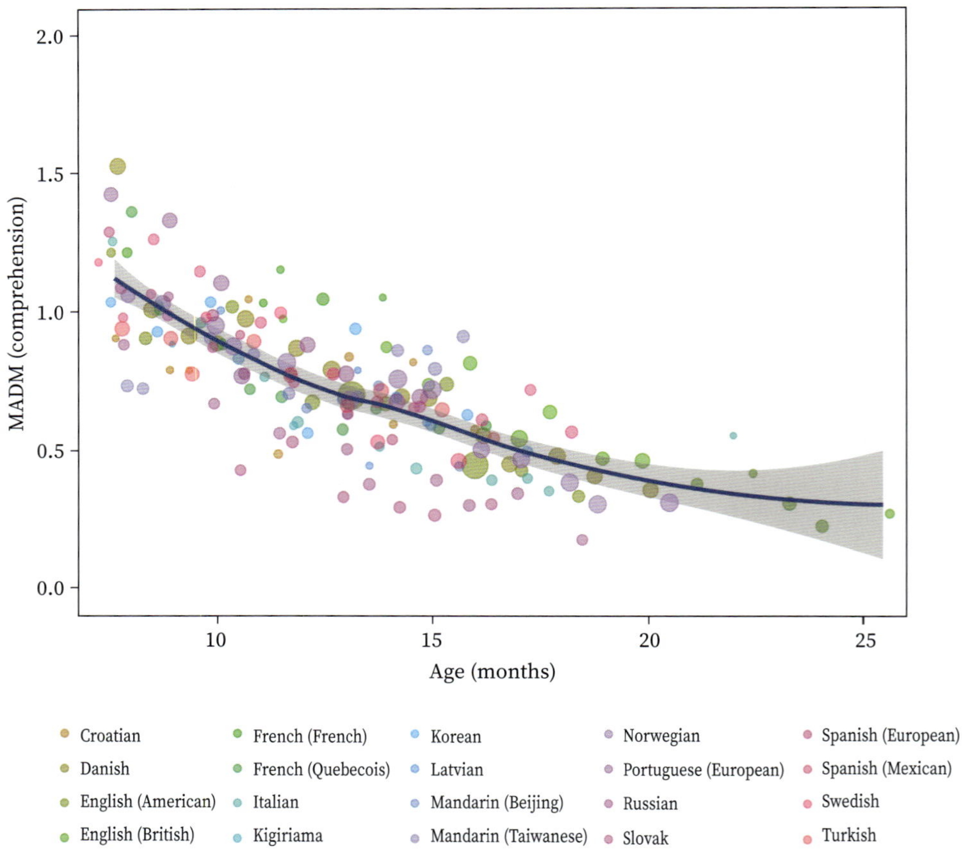

Figure 5.12
MADM for comprehension, plotted by age group, for the full sample of languages in our dataset.

The intercept for the 12–18 month period appears to be lower than that observed in production, despite—or perhaps due to—higher comprehension scores. This observation matches one made by Mayor and Plunkett (2014), namely that, across children, production vocabulary appears more idiosyncratic than comprehension vocabulary. One speculative explanation for this difference would be the tremendous differences in speech-motor development (as well as general differences in loquacity) between toddlers (for an example, see Clark, 1993). This variability would then carry over into production vocabulary size. Another possibility, however, is that true variability is masked by the overall lower reliability of comprehension items (see chapter 4). Our data do not allow us to distinguish between these two explanations.

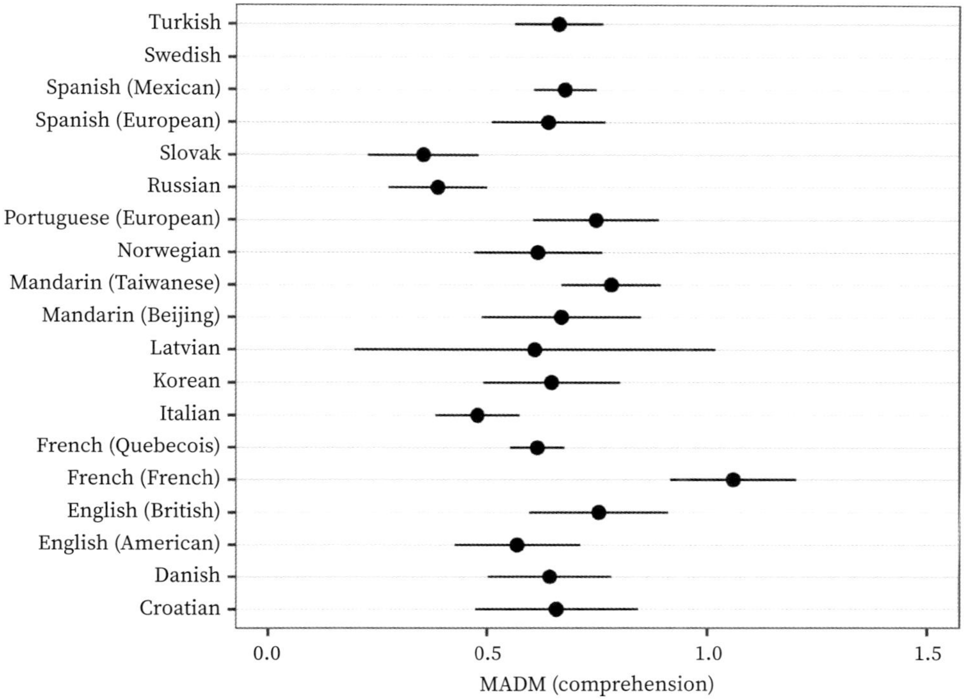

Figure 5.13
MADM values from 12–18 months for all languages and forms.

The final plot in this sequence is shown in figure 5.13, which shows 12 to 18-month average comprehension MADM values. These are slightly lower and slightly more variable than the production values shown above but still display a quite consistent level of variability.

5.2.2 Is There a Ceiling to Variability?

The analysis above suggests that variability between individuals decreases. But this conclusion is compromised by ceiling effects: once children begin to reach the ceiling of the CDI form, variability is necessarily truncated. No analysis can completely eliminate these effects, but the use of item response theory (IRT)–based analyses can partially address the issue by estimating variation in latent ability rather than variation in raw scores themselves.

Chapter 4 provided a summary of our approach to using IRT with CDI data. In brief, IRT provides a framework in which the full test (the CDI) is analyzed as a series of items, with each having its own logistic model predicting the response for a particular child on the basis of their latent ability. In chapter 4, we examined the parameters of individual items with respect to their properties, but fitting an IRT model also implies estimating a set of latent ability parameters for individual participants. These latent ability parameters are logistic regression coefficients and hence are not bounded in the same way that individual responses (and hence raw scores) are. Thus, we can examine their variability as a way of dealing with ceiling effects.

We fit a four-parameter logistic (4PL) IRT model to each dataset and estimated the resulting latent ability scores for each child. We then conducted the analysis above, substituting latent ability scores for raw vocabulary scores. Figure 5.14 shows the normalized standard deviation of latent ability scores, plotted by age. The absolute size of the standard deviations is not easy to interpret—the range of latent ability scores on a form is a function of how consistently difficult or easy the items are (and as we noted above, we do not think it is trivial to equate these scores across tests). Thus, the standard deviation of these scores is not comparable across models.

On the other hand, age-related trends in the standard deviation are interpretable and can be used as an index of whether variability in ability stays constant, increases, or decreases. Across datasets, age slopes for the variability of latent ability tend to be flat or even increasing (with Mandarin, which has extreme ceiling effects, being the exception). This finding suggests that the developmental decreases in variability observed above are very likely due to ceiling effects. When we remove these ceiling effects, we find that variability is constant—or perhaps even increasing—throughout the full measured range of early language.

5.2.3 Discussion

We have observed a striking consistency in the individual variability of children's vocabulary during their second year and perhaps beyond. Across languages and forms, it appears to be the norm that toddlers vary.

What does it mean to have such a high level of variability? For one comparison, we compare age of walking onset (as measured by a Norwegian national survey with parent 47,515 respondents) and age of achieving production and comprehension milestones (also in Norwegian). Walking data are from Størvold et al. (2013).

Table 5.1 shows the 25th and 75th percentiles for walking and several language milestone behaviors. The spread of achieving walking (defined as taking a step independently) is quite tight with a mean of 12.9 months and a spread of only a month between 25th and 75th percentile. Very early language comprehension and production are relatively similar with 2- and 3-month spreads. In contrast, construed as milestones in this way, production and comprehension of larger

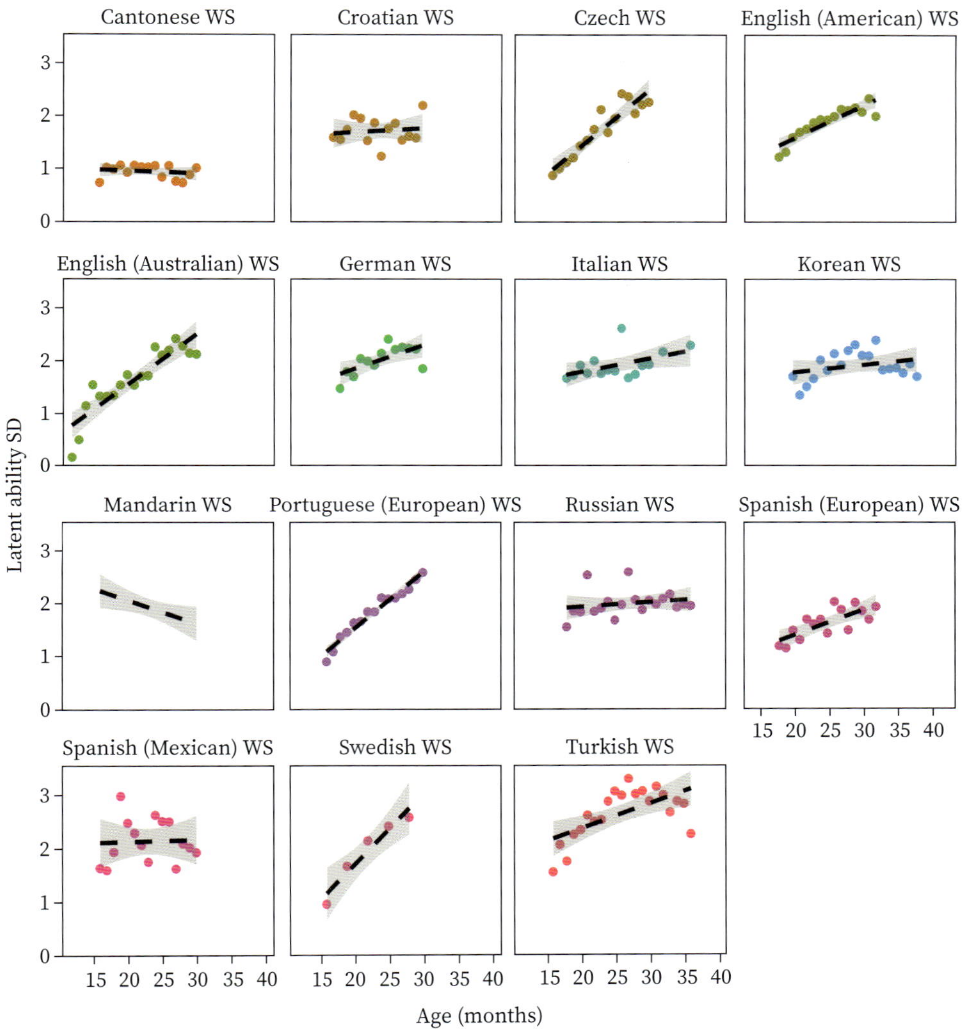

Figure 5.14
Standard deviation of latent ability scores from 4PL IRT models fit to each WS-type dataset. Panels show individual languages. Smoothing lines are linear model fits.

Table 5.1
Comparison of language milestones with walking. The columns 25th and 75th give the ages at which that percentile ranking is achieved; ranges give the interval between 25th and 75th percentile in months and as a proportion of total age.

Response	25th	75th	Range (months)	Range (prop)
Walking	12	14	2	0.15
Produces 10 words	13	16	3	0.21
Produces 50 words	17	20	3	0.17
Produces 100 words	18	23	5	0.25
Understands 50 words	10	15	5	0.42
Produces 200 words	20	26	6	0.26

numbers of words each have quite a large spread in comparison to walking (even as a percentage of age).

In sum, our results echo the conclusions of Bornstein and Cote (2005), who based on their comparative study of children learning Spanish, English, and Italian. They noted that "individual variability is probably a universal feature of early language acquisition" (p. 311).

6 Demographic Effects on Vocabulary Size

Chapter 5 examined cross-linguistic consistency and variability in the size of children's reported vocabulary. In this chapter, we follow up these analyses by beginning the process—which continues throughout the book—of attempting to understand the nature and sources of these differences. In particular, we take advantage of the sample diversity described in chapter 3 to explore differences in the median trajectory of vocabulary growth across demographic characteristics, focusing on three variables that are available in much of our data: sex, maternal education, and birth order.

While some demographic differences in vocabulary are quite consistent—substantially so, as it turns out—the overall proportion of variance in vocabulary that they capture is still relatively limited, as noted by previous analysts of variability (e.g., Fenson et al., 1994; Feldman et al., 2000; Eriksson et al., 2012). We will examine a number of different perspectives on how to quantify this relationship, moving back and forth between emphasizing consistency in differences in the central tendency and emphasizing the limited size of these effects relative to the variability documented in chapter 5.

Our analyses in this chapter are limited to a subset of languages, as demographic data for many contributed datasets were not available. We begin with sex, the variable for which the most analysis has already been done and for which we have the most data, then move on to birth order, and finally, turn to maternal education. Throughout, we focus on earlier comprehension differences from Words and Gestures (WG)-type forms and later production differences from Words and Sentences (WS)-type forms. For the sake of length, we omit analysis of production data from WG-type forms. As discussed in chapter 5, these data tend to show limited variability due to floor effects.

6.1 Sex

Our first analysis examines how vocabulary development differs by children's sex.[1] The literature on cognitive differences due to sex is vast, controversial in many places, and difficult to summarize (see Miller and Halpern, 2014, for a useful recent review). Focusing on development, Maccoby and Jacklin (1974) began the enterprise of systematizing and summarizing gender differences. Their conclusions were largely deflationary but did imply some differences in aggression and verbal ability (which were suggested to emerge in the period of early adolescence). The claim about verbal ability is most relevant for our analysis, but has been controversial as well.

Using meta-analytic tools, Hyde and Linn (1988) found that differences in verbal ability were minimal, but more recent studies have suggested consistent verbal ability differences. For example, Stoet and Geary (2013) found differences in reading ability across nations in a massive elementary education dataset (the Program for International Student Assessment [PISA]), with variance in the magnitude of difference but with girls very consistently showing an advantage. Similarly, Robinson and Lubienski (2011) reported consistent differences in reading ability (favoring girls) at the onset of kindergarten in a nationally representative US sample. A potential prediction from this literature is thus that we might observe a modest but consistent female advantage in early vocabulary.

Of course, a complication of our analysis is the potential presence of caregiver reporting bias added to any true sex differences. In contrast to these findings suggesting modest and consistent female advantages, there is substantial cross-linguistic variation in gender stereotypes (Nosek et al., 2009). Thus, a second plausible speculation is that, if stereotype-based reporting bias plays a major role in gender effects, the cross-national variance should be high and correlated with gender stereotypes.

Despite these predictions, it almost goes without saying that any finding from our analyses here is subject to the full range of possible explanations articulated in the literature. These range from caregiver and academic socialization (e.g., self-fulfilling expectations that girls are more verbal) to "self socialization" in which affiliative differences produce differences in behavior all the way to biological explanations.[2] While descriptive data of the type cited above (and reported in

1. Throughout, we will assume that parents report on children's assigned sex at birth rather than their gender identity. In the absence of any data that would tease apart the biological, psychological, and social aspects of sex and gender, we use interchangeably the terms sex/gender, female/girl, and male/boy.

2. As an illustrative example, some literature has implicated fetal testosterone in sex differences in verbal ability. Lutchmaya et al. (2001) used CDI measures and recovered an effect somewhat similar to

our analyses below) can be more or less consistent with some of these theories, conclusive evidence will not be forthcoming.

Our analyses below replicate and extend the results of Eriksson et al. (2012), who used an overlapping sample of CDI data from twelve languages to explore sex effects on vocabulary size.[3]

6.1.1 Comprehension (WG)

We begin by examining data from WG-type forms using comprehension measures. Figure 6.1 shows our approach. Each subplot shows median reported comprehension for each age and sex group. Smoothing lines show the predictions of a robust generalized linear model (GLM) (we selected a robust GLM to avoid some pathological effects from outliers in a small subset of situations).

Visual inspection of the data suggest limited sex differences, but a female advantage is present in some languages (most pronounced in Korean, Latvian, and Hebrew). Note that many authors do not find gender differences in early comprehension. For example, using an overlapping 12 language dataset, Eriksson et al. (2012) concluded that there were no major comprehension differences. And in an earlier study, Feldman et al. (2000) also did not find gender differences in comprehension using a large, relatively representative American dataset, though this study included data only from younger children (10–13 months).

We can examine statistical models to get a clearer picture. For each language, we fit a robust GLM predicting vocabulary size (number of words understood out of the total) based on age and the interaction of age and sex. We specified this simple model so that the coefficient estimate for the age by sex interaction (as shown in figure 6.2) provides a convenient summary of the difference in vocabulary growth across groups. Despite the small magnitudes of the coefficients, 16 of 22 languages had a female advantage. Conversely only two showed a male advantage, and the remainder did not show a significant sex by age interaction.

As a contrast to the model-based method above, we can look for a measure of effect size (similar to that used in chapter 5). Effect size quantifies the size of the difference between groups in terms of the variability, producing a scale-free measure of difference that is appropriate for comparison across languages. Normally we would use a measure like Cohen's d here, where mu_1 and mu_2 index the means for each group and σ_{pooled} indicates the pooled standard deviation across groups:

---

ours with a small sample ($d = 0.64$ at 18 months with $N = 87$), and $d = 0.60$ at 24 months for a subsample). They found some relationship with fetal testosterone across sexes, but it did not hold up within sex (perhaps due to small samples). The mechanism by which testosterone translates into vocabulary growth is unclear, however.

3. An earlier version of this analysis was reported in Frank et al. (2016a).

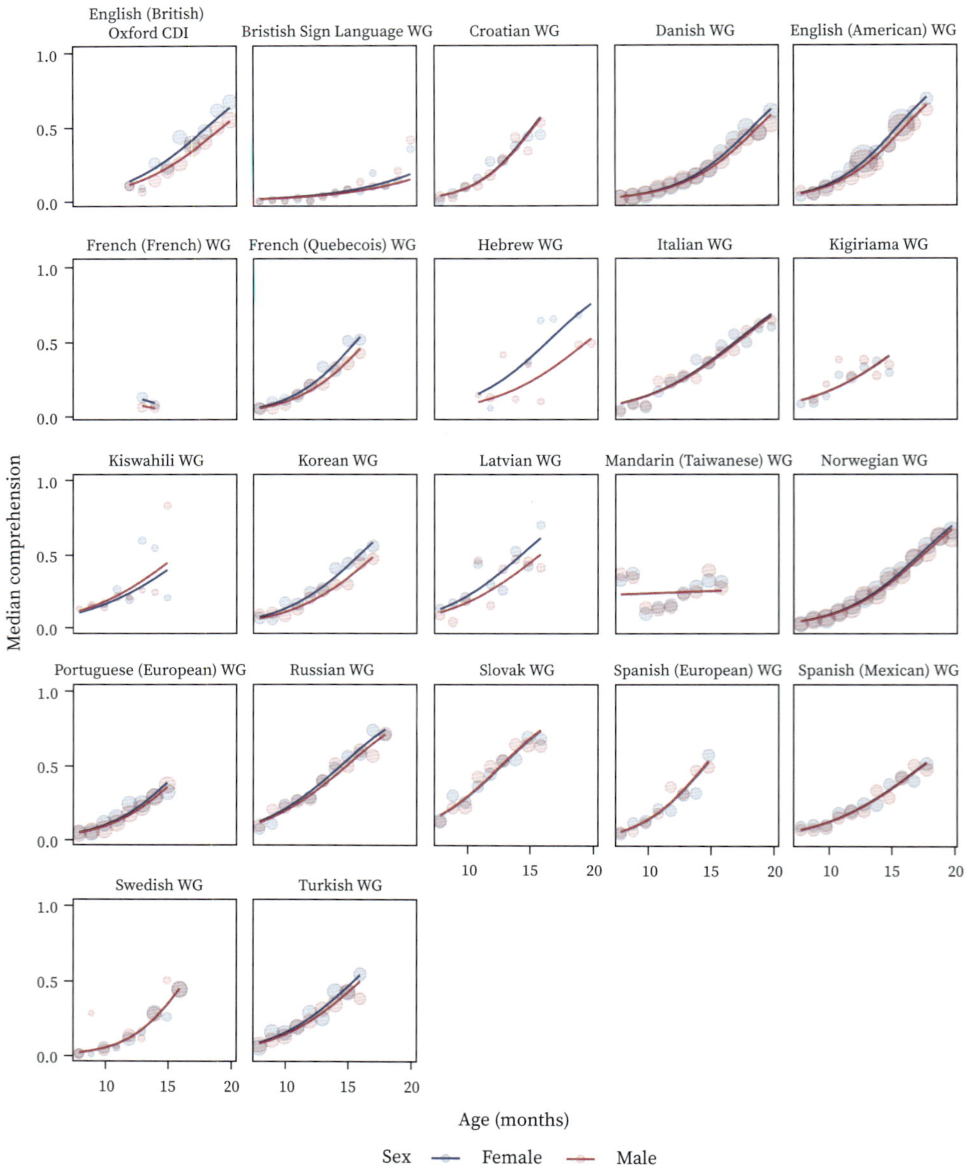

Figure 6.1
Differences in WG comprehension scores by sex, plotted across age by language.

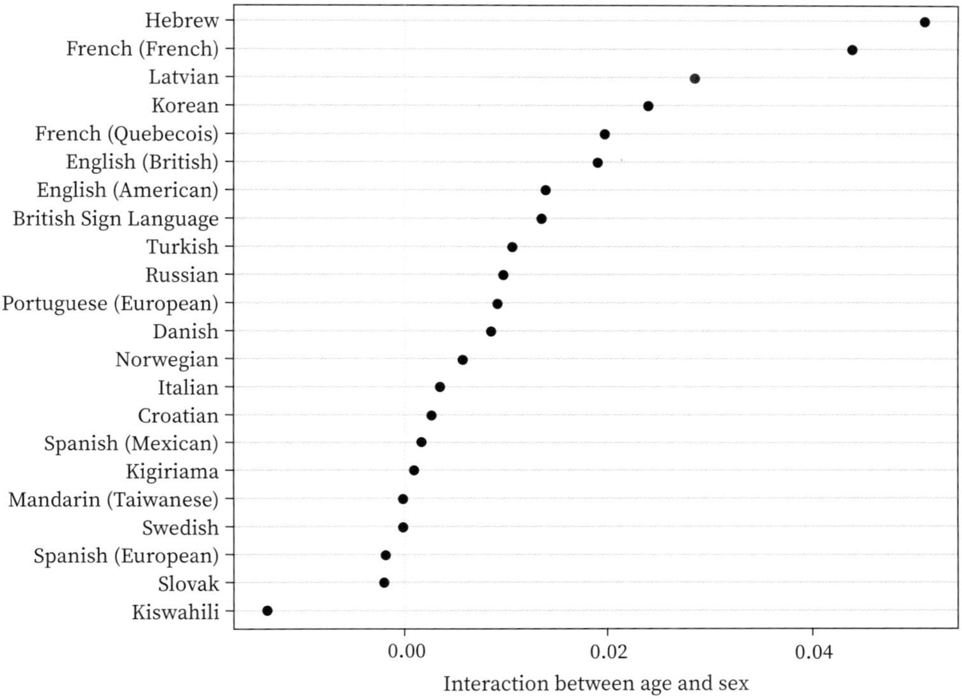

Figure 6.2
Interaction term between age and sex for WG comprehension data in each language.

$$d = \frac{\mu_2 - \mu_1}{\sigma_{pooled}}.$$

But, as in chapter 5, we have the problem of non-normal distributions. To circumvent this issue, we use a non parametric measure derived from the same components: the difference between medians, divided by the median absolute deviation (MAD). (We call this the median of the maximum absolute deviation [MMAD]; appendix B shows that it is comparable to Cohen's *d* as a measure of effect size.)

Applying this measure to the data on comprehension, we see a quite small average female advantage that appears relatively constant across age (figure 6.3). For those languages with dense enough data, we can take a weighted average of this pattern across ages, which reveals substantial variability (figure 6.4). The overall median for these 18 languages is quite small as well—0.08. In summary, there is some evidence for a modest female advantage in comprehension.

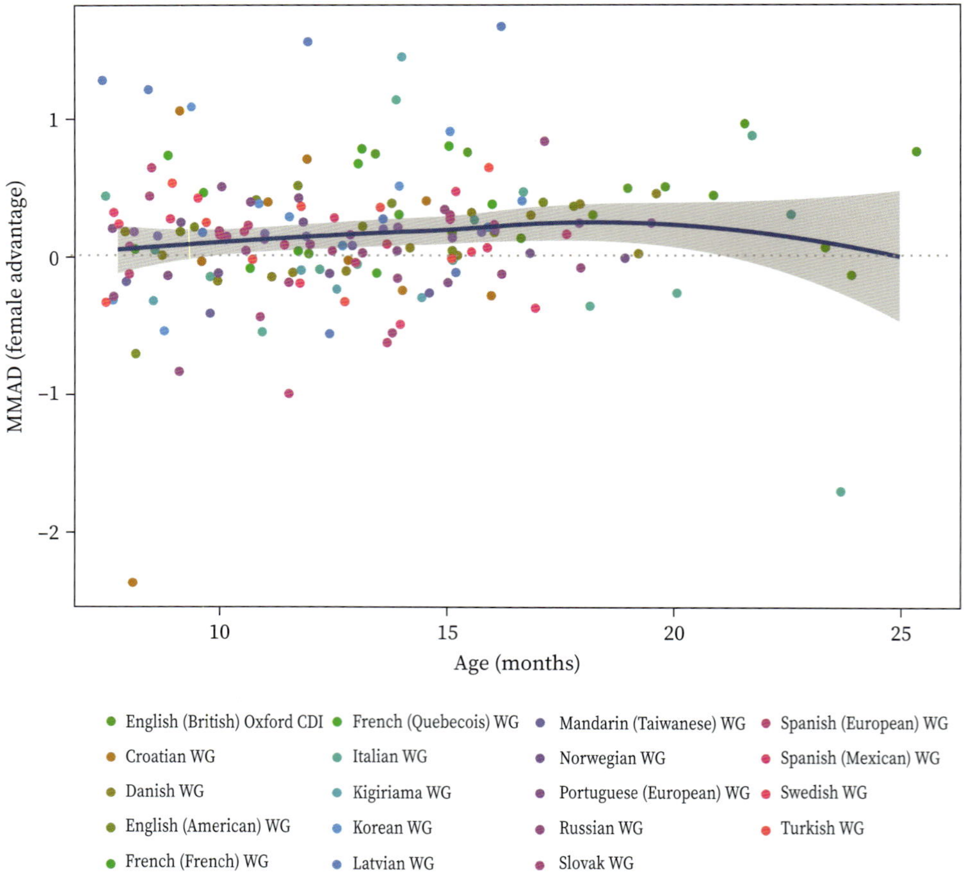

Figure 6.3
MMAD female advantage for WG comprehension data in each language across age.

6.1.2 Production (WS)

Now we turn to production data on the WS instrument. Figure 6.5 shows curves for each individual language. Visual inspection confirms a female advantage in almost every case, and an analysis of the fitted models (see figure 6.6) shows that 25 of 26 languages show a statistically significant female advantage!

We next move to the MMAD effect size measure (see figures 6.7 and 6.8). Here we see a relatively consistent difference across ages with perhaps a slight downward trend in effect size with age. This downward trend might be a function of ceiling effects on the form, however, as seen in the model-fit curves, for example, for Danish. When variability is limited by the form ceiling, effect size estimates

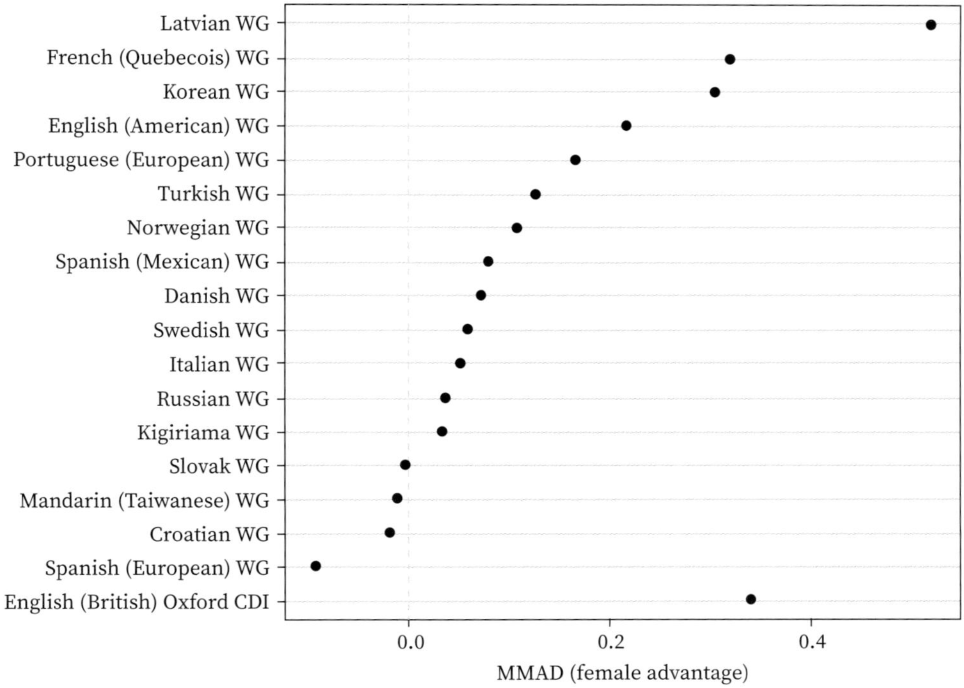

Figure 6.4
MMAD female advantage for WG comprehension data in each language averaged over age.

will necessarily be depressed. The median female advantage is 0.40, substantially larger than that seen in early comprehension.

6.1.3 Reporting Bias?

Do these differences reflect differences in measurement that are unique to the CDI? One way of addressing this question is to examine other studies of gender differences. Unfortunately, many of the studies reporting differences themselves rely on the CDI or similar measures, likely for the reasons reviewed in chapter 2 (e.g., Bauer et al., 2002; Fenson et al., 1994; Feldman et al., 2000). For example, Feldman et al. (2000) collected CDIs with a large dataset of low-income American English speakers at 12 and 24 months. In those data, early comprehension showed no significant gender differences, but production at 24 months showed a difference comparable to what we observed here ($N = 2156$, $d = 0.35$, as recomputed from provided summary statistics). These data, while providing a replication in an independent dataset, do not speak to whether reporting biases contributed to or created the observed sex effects.

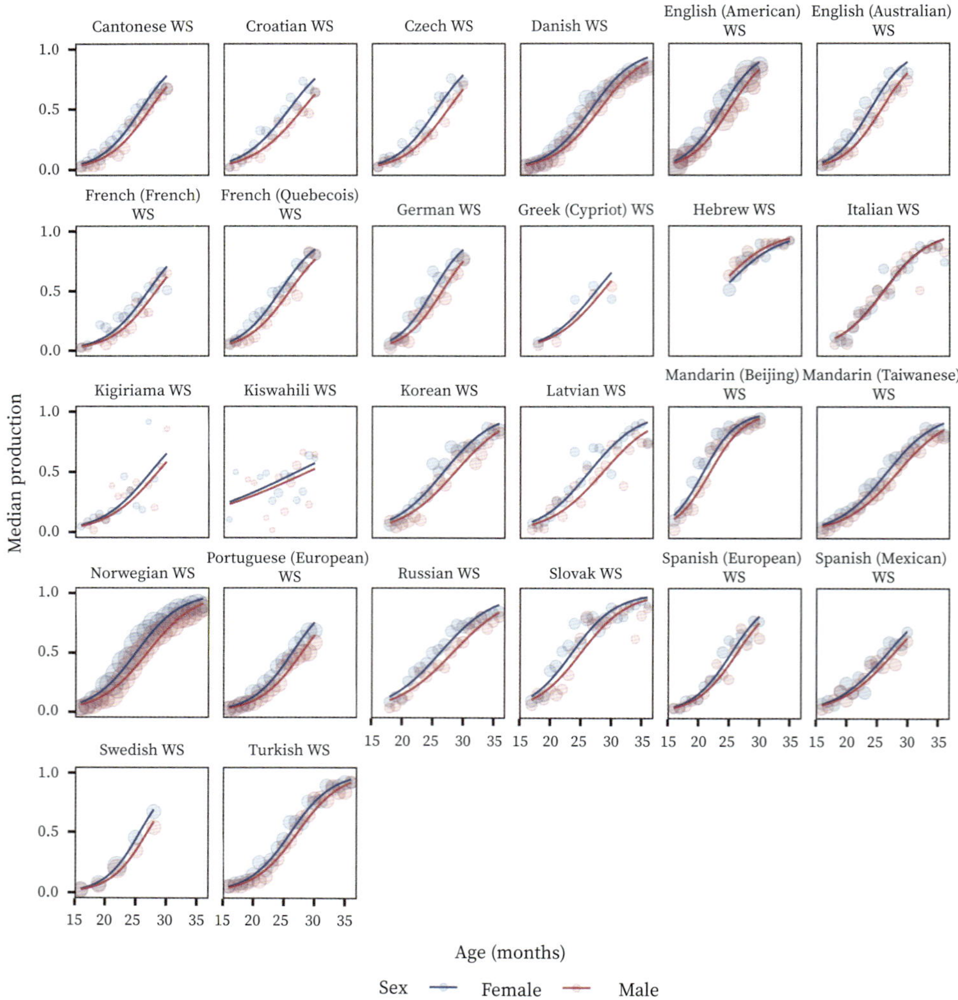

Figure 6.5
Differences in WS production scores by sex, plotted across age by language.

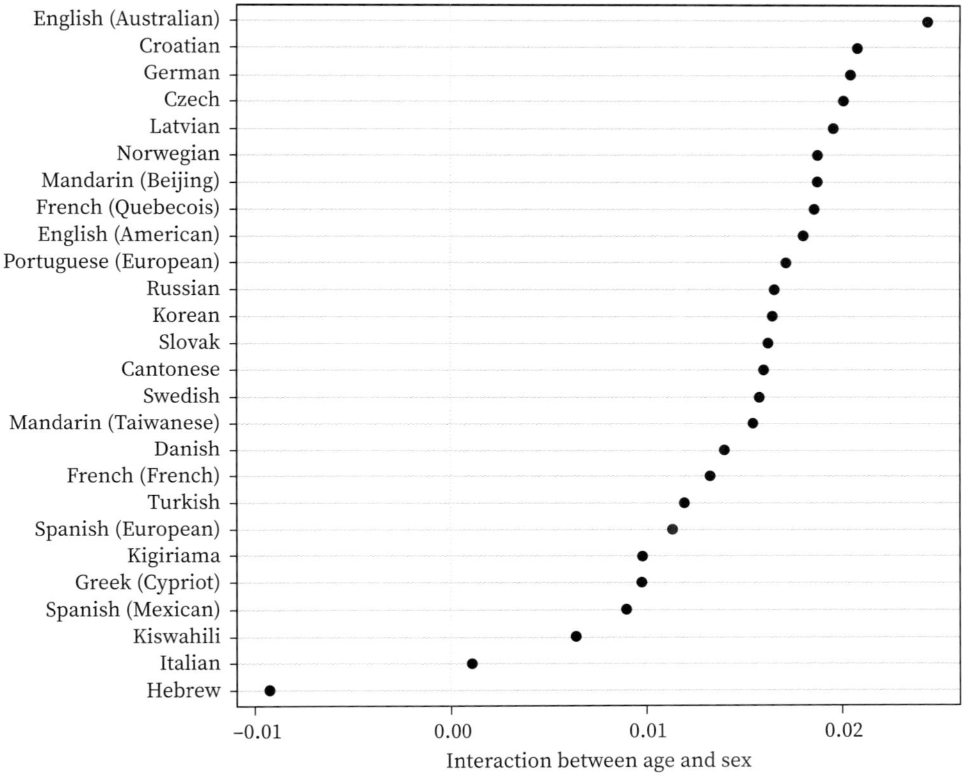

Figure 6.6
Interaction term between age and sex for WS production data in each language.

For external validation, we turn to two other studies that provide more objective (non-parent-report) measurements of early language. First, a seminal study by Huttenlocher et al. (1991) measured gender effects in vocabulary production as estimated from a naturalistic language sample, finding substantial differences in vocabulary growth favoring girls. Although the measures from this study are not comparable with those used in the current analysis, the effects were quite large (and were relatively unaffected by controlling for maternal language exposure).

Second, Bornstein and Putnick (2012) used a particularly powerful study design to examine stability in early language estimates across different measures. They gathered longitudinal data at 20 and 48 months using a wide range of standardized and parent-report measures and then used structural equation modeling to model shared variance due to parent report as well as to standardized latent language ability at each age. We digitized data from their figure 2 to examine the size of

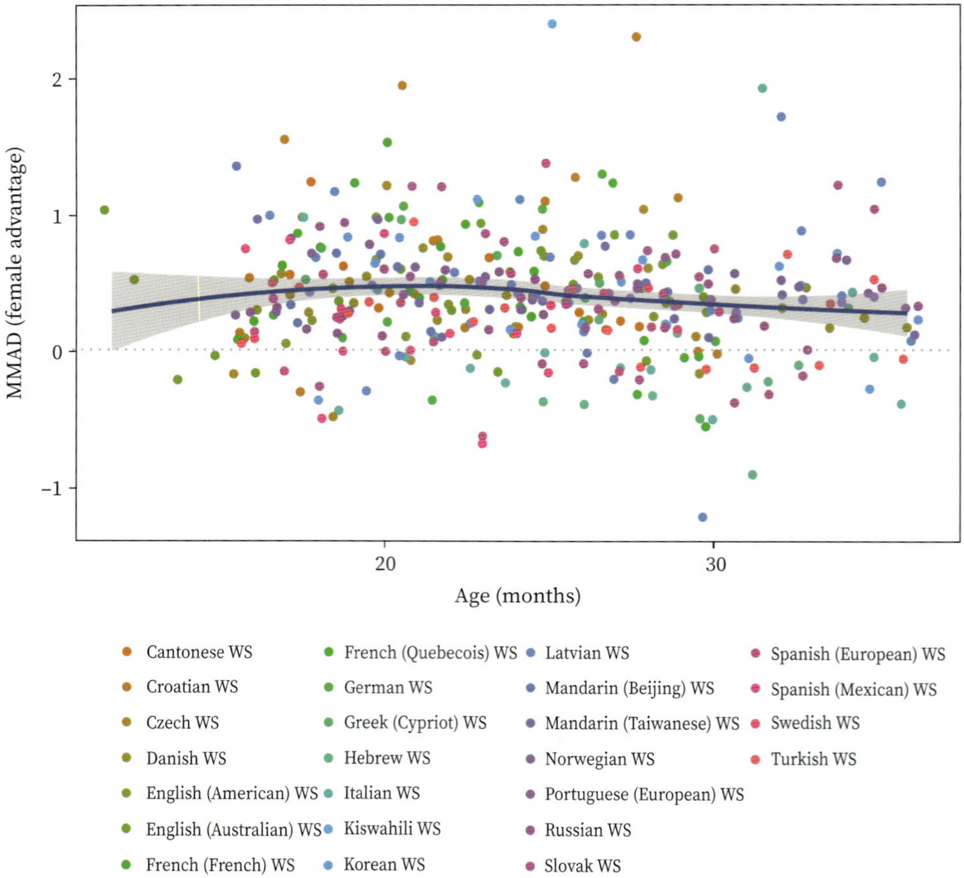

Figure 6.7
MMAD female advantage for WS production data in each language across age.

the gender differences in the latent vocabulary construct that they recovered (see figure 6.9).[4] As these scores are standardized, we can examine the difference in means and recover an estimate of the standardized effect size for gender in the data, which is 0.62 standard deviations. Since this measurement is *greater* than that found using the CDIs (the comparable American English measurement was 0.35), this observation gives us reason to doubt that our observed effect is due solely to reporting bias.

4. These data are not error free; we have double-digitized one point.

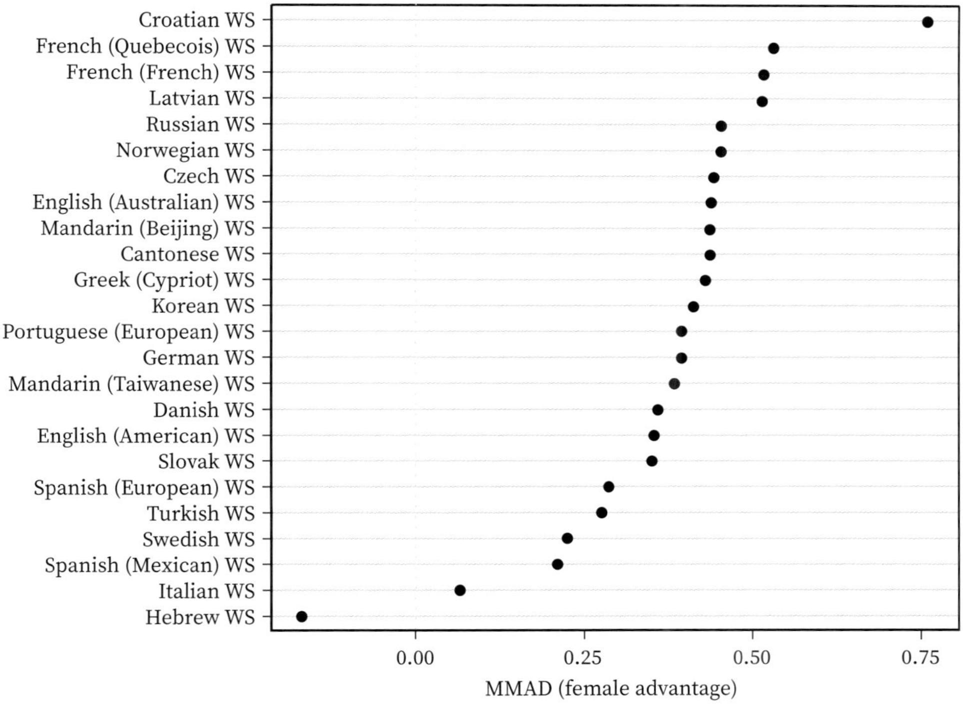

Figure 6.8
MMAD female advantage for WS production data in each language averaged over age.

6.1.4 Discussion

In summary, we found a considerable and strikingly consistent cross-linguistic female advantage in early language production (replicating and extending Eriksson et al., 2012). A much smaller but still relatively consistent female advantage was reported in comprehension. We suspect that neither of these effects are due to parent reporting bias. First, our review of the literature suggests that studies using direct assessments yield similar effects to the extent that we were able to compare. Second, comprehension is very likely to be the measure *more* affected by reporting biases as it is likely to be more subjective (Feldman et al., 2000; Fenson et al., 2000a), yet we find a much smaller gender effect in comprehension. As noted above, we remain agnostic about the causes of these differences. In chapter 16, we speculate about inferences from consistency in sex differences across languages.

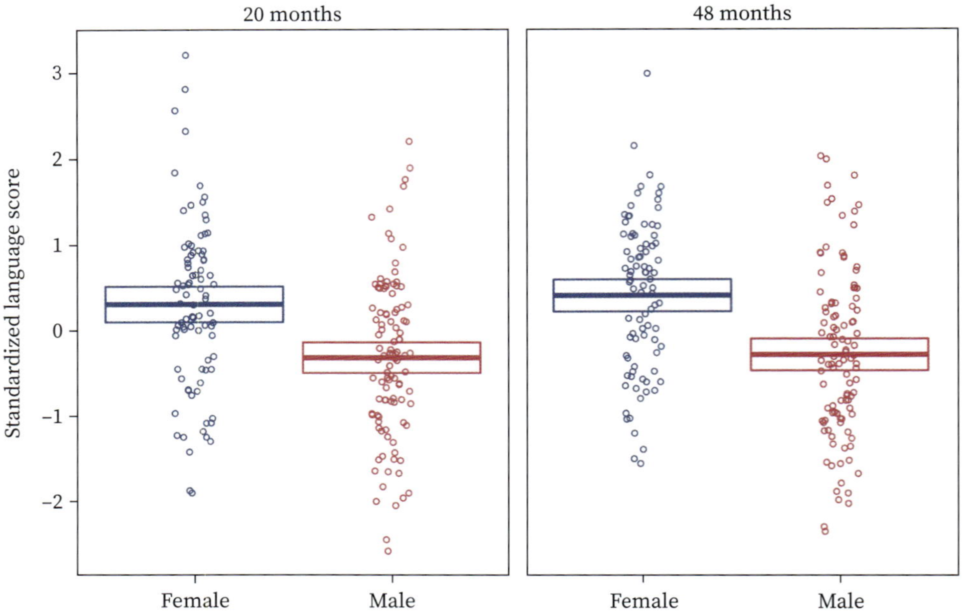

Figure 6.9
Latent vocabulary scores from Bornstein and Putnick (2012) by age and sex (crossbars show means and 95 percent confidence intervals).

6.2 Birth Order

Another factor that likely contributes to individual differences in children's vocabulary development is birth order. The literature suggests some evidence for a firstborn advantage in early vocabulary development, but these differences are small and tend to be most evident early in development. For example, Bornstein et al. (2004a) found that mothers report larger receptive and expressive vocabularies in their firstborns. Using naturalistic language samples, Berglund et al. (2005) found that firstborn children reached the 50-word milestone earlier than later-born children, but that birth order differences diminished later in development. Finally, Hoff-Ginsberg (1998) found that firstborn children were more advanced in vocabulary development than later-born children, but that later-born children were more advanced in their conversational skills.

One dissenting report comes from Oshima-Takane et al. (1996), who found that second-born children were relatively equivalent in general language proficiency. They also found that second-born children were more proficient with personal pronouns, a suggestion that we investigate in chapter 9. They used this equivalence to argue for second-born children's ability to learn from overhearing. Although we agree with this general suggestion, their failure to observe a firstborn advantage for overall language proficiency is likely due to the small sample size in their study ($N = 20$).

Here, we can examine birth order effects in early vocabulary comprehension and production in a subset of languages in our sample: languages have birth order data, with data available for 11 languages for WS, and 8 languages for WG.

6.2.1 Comprehension (WG)

We perform the same set of analyses as for sex, shown in figures 6.10, 6.11, 6.12, and 6.13.

6.2.2 Production (WS)

The parallel set of analyses for WS Production data are shown in figures 6.14, 6.15, 6.16, and 6.17.

6.2.3 Discussion

Analyzing birth order effects, we see a relatively consistent cross-linguistic pattern: earlier-born children show larger vocabularies in production (though not in comprehension for the most part). This general finding is consistent with previous literature reporting a firstborn advantage for individual languages. Our results suggest that the same pattern appears in most languages, with only a few showing different magnitudes. While the current dataset cannot rule out reporting-related reasons for these demographic differences, this explanation does seem unlikely for two reasons (mirroring discussion of sex differences above). First, our results largely mirror non-parent-report findings in the literature (Berglund et al., 2005). Second, reporting bias would be relatively *more* likely to influence comprehension relative to production vocabulary.

We can only speculate as to the cause of birth-order-related differences in early language given our current data. That said, it seems very reasonable to assume that parents speak more to firstborn children as the addressee, due to the pure statistical fact of having a second possible addressee for other utterances (cf. Zajonc and Markus, 1975). And although it is certainly possible to learn from overheard speech under optimal conditions (e.g., Akhtar et al., 2001; Akhtar, 2005), a variety of studies suggest that speech directed to a particular child is the best predictor of that child's learning outcomes (Weisleder and Fernald, 2013; Shneidman and Goldin-Meadow, 2012).

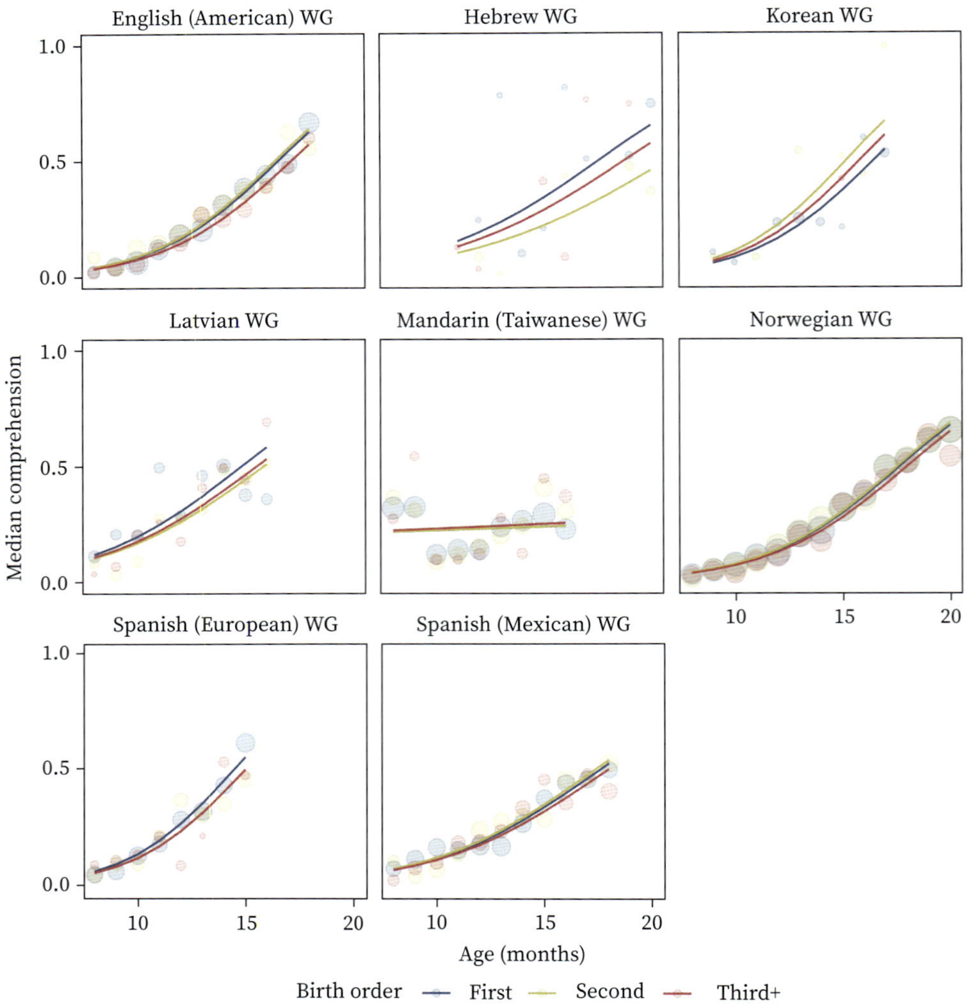

Figure 6.10
Differences in WG comprehension scores by birth order, plotted across age by language.

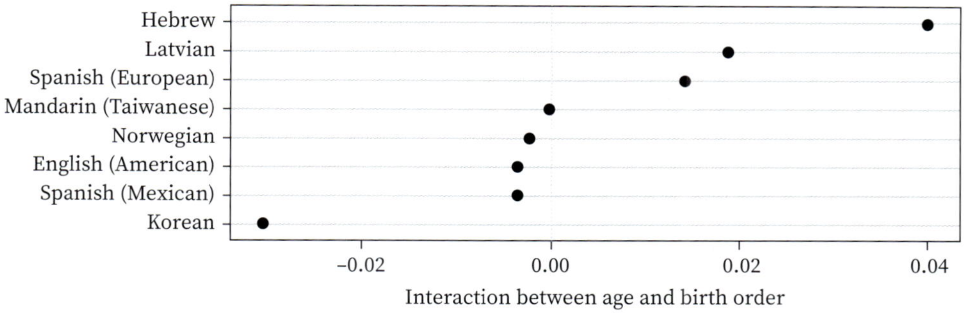

Figure 6.11
Interaction term between age and birth order for WG production data in each language.

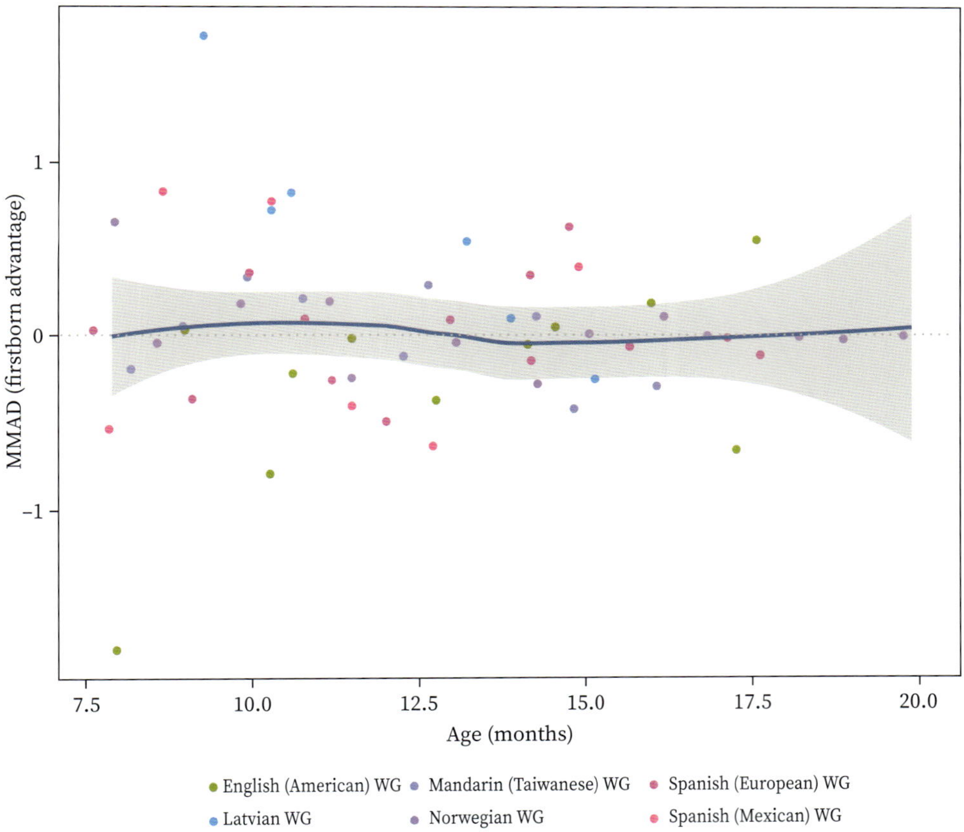

Figure 6.12
MMAD firstborn advantage for WG comprehension data in each language across age.

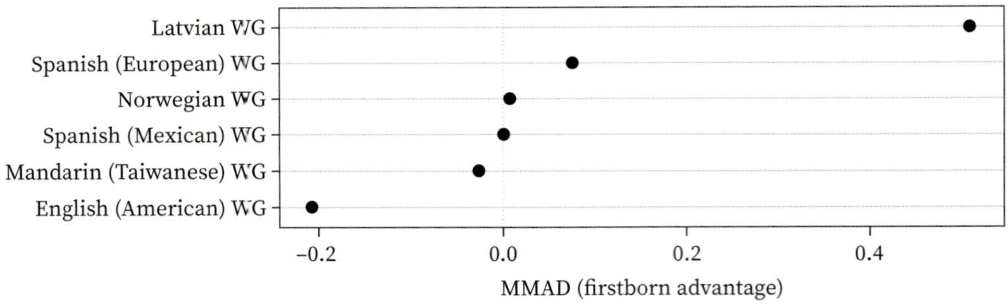

Figure 6.13
MMAD firstborn advantage for WG comprehension data in each language averaged over age.

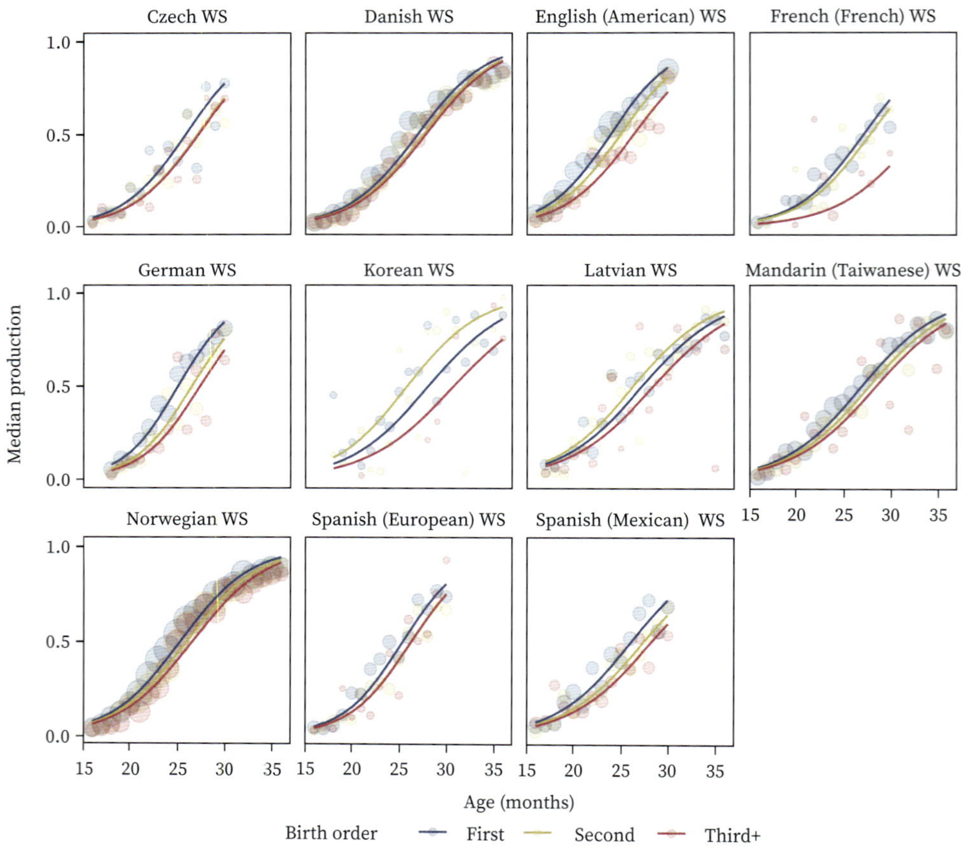

Figure 6.14
Differences in WS production scores by birth order, plotted across age by language.

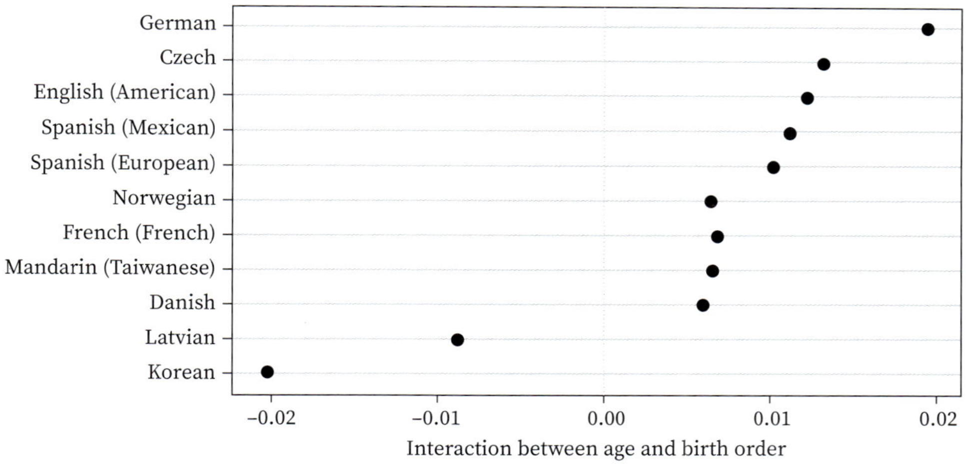

Figure 6.15
Interaction term between age and birth order for WS production data in each language.

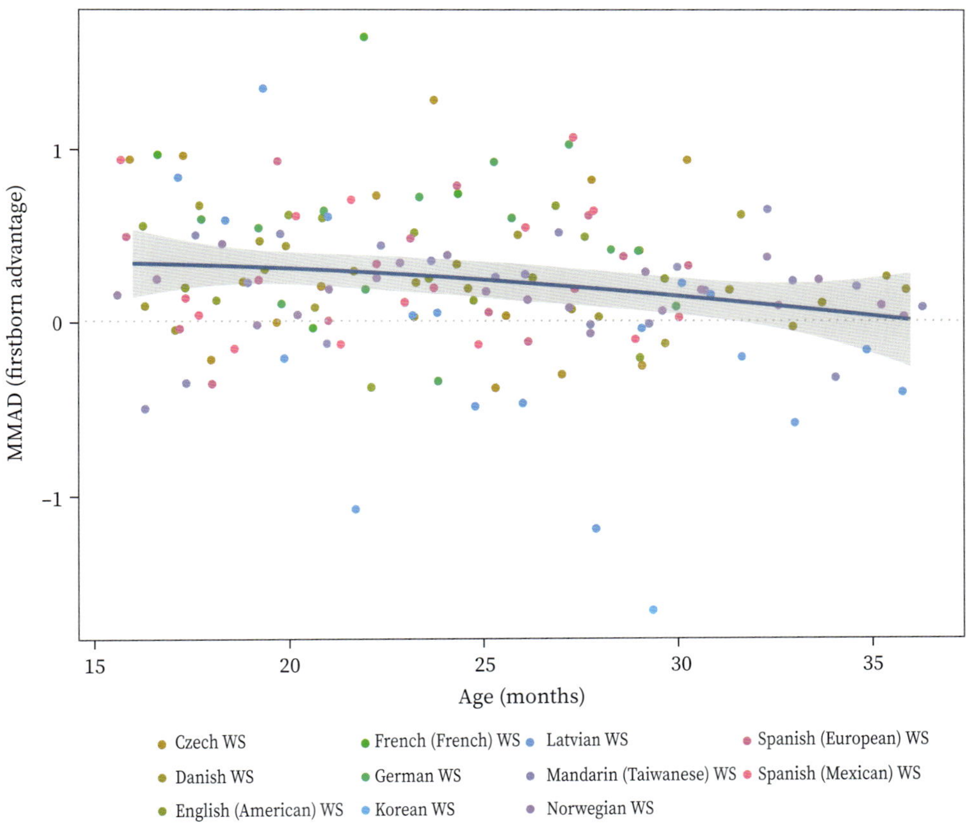

Figure 6.16
MMAD firstborn advantage for WS production data in each language across age.

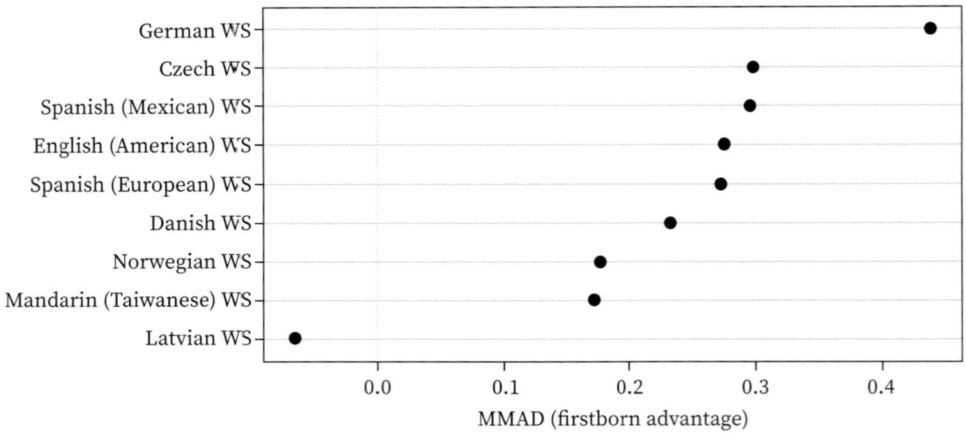

Figure 6.17
MMAD firstborn advantage for WS production data in each language averaged over age.

6.3 Socioeconomic Status

From health to education, children from lower socioeconomic status (SES) backgrounds tend to be at higher risk for a variety of negative developmental outcomes, compared to their higher-SES peers (Bradley and Corwyn, 2002). A large literature documents specific relations between SES and children's early language abilities, especially oral vocabulary, which is in turn related to children's later reading outcomes (e.g., Hart and Risley, 1995; Hoff, 2003; Fernald et al., 2013). The parent-report method allows the assessment of the influence of SES on vocabulary outcomes earlier in development than is possible with direct assessments. Using the CDI WS form, Arriaga et al. (1998) compared the language skills in 103 very low-income toddlers with a sample of middle-income toddlers from the Fenson et al. (2007) norming sample, that were matched on age and sex. They found that the vocabulary production scores for the low-income group were consistently about 30 percent lower than those for the middle-income group. The size of these effects suggest that differences in SES are evident from the very earliest phases of language development.

Environmental explanations of these SES effects are often given in terms of indirect factors that affect life opportunities or experiences, such as nutrition and access to health care, as well as more direct factors that impact daily life, such as smoking during pregnancy or access to quality childcare. One well-studied factor is the quality of the interactions that caregivers spend in interactions with their young children, with studies showing that talk from caregivers mediates the effects of SES on child outcomes (e.g., Weisleder and Fernald, 2013; Hoff, 2003). Alternatively, even early language shows a significant genetic component, raising the possibility that SES-vocabulary links may instead be genetically mediated (Hayiou-Thomas et al., 2012). That is, parents of higher-SES backgrounds may use language more extensively and at a higher level due to their genetic endowment, which is then passed on to their children. Some initial DNA-based evidence for a genetic component to the SES-language link is provided by Trzaskowski et al. (2014).

In the current dataset, we use maternal education as a proxy for SES, following previous work suggesting that maternal education is strongly related to SES variation, particularly early in development (Bornstein et al., 2003; Hoff, 2003). The presence of this variable gives us the opportunity to explore the extent of SES effects on vocabulary across several language communities, albeit using an imperfect proxy measure. While different language communities may differ in their distributions along the SES gradient, cross-language comparisons may nevertheless shed light on the factors that lead to relations between SES and children's vocabulary outcomes. On the one hand, relatively constant relations across language communities that vary widely in indirect and direct factors that shape learning would provide prima facie support for genetic explanations. On the other hand a greater degree of cross-language variability would point to the origins of SES effects in aspects of children's early environments that vary with SES to differing degrees across countries (e.g., Fernald et al., 2012).

6.3.1 Comprehension (WG)

We again perform the same set of analyses, shown in figures 6.18, 6.19, 6.20, and 6.21.

These results suggest a small—and perhaps negative—relationship between maternal education and early comprehension vocabulary. We discuss this result below.

6.3.2 Production (WS)

The next analyses are shown in figures 6.22, 6.23, 6.24, and 6.25.

In contrast to the comprehension results, there were robust positive maternal education differences in nearly every dataset we examined, though the magnitude varied across datasets.

6.3.3 Discussion

The relationship between maternal education and children's vocabulary is variable across countries in our data, but we did observe some kind of college advantage for production in nearly every dataset that we examined. The observational nature of

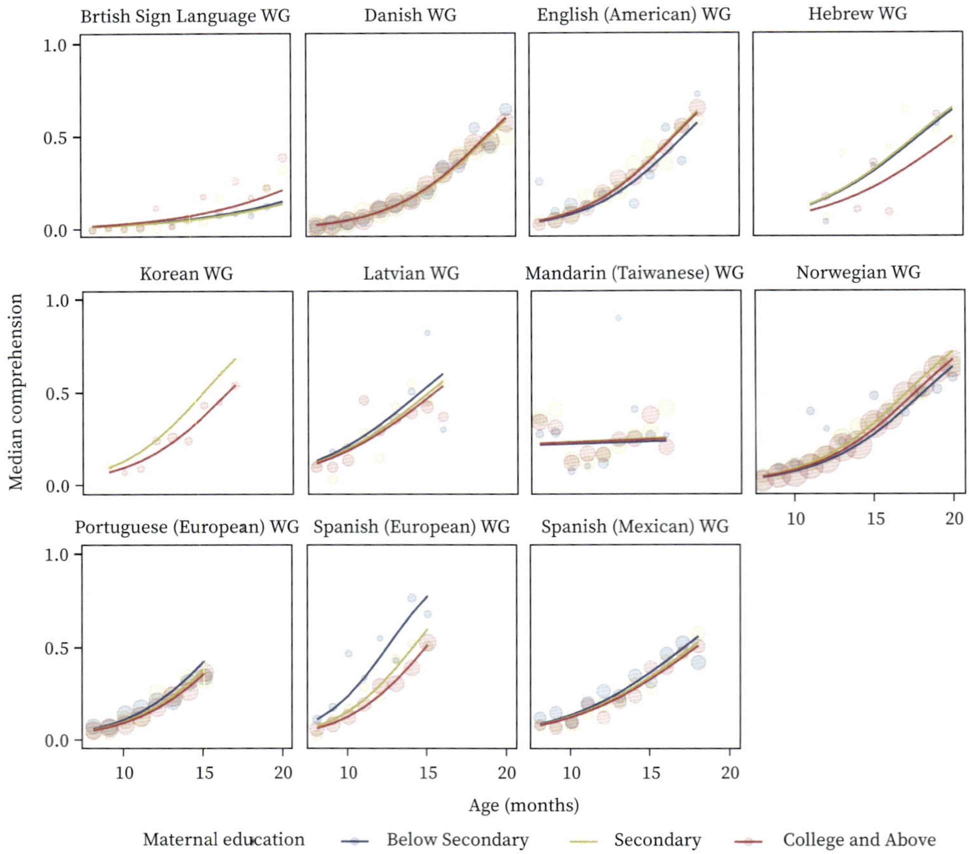

Figure 6.18
Differences in WG comprehension scores by maternal education, plotted across age by language.

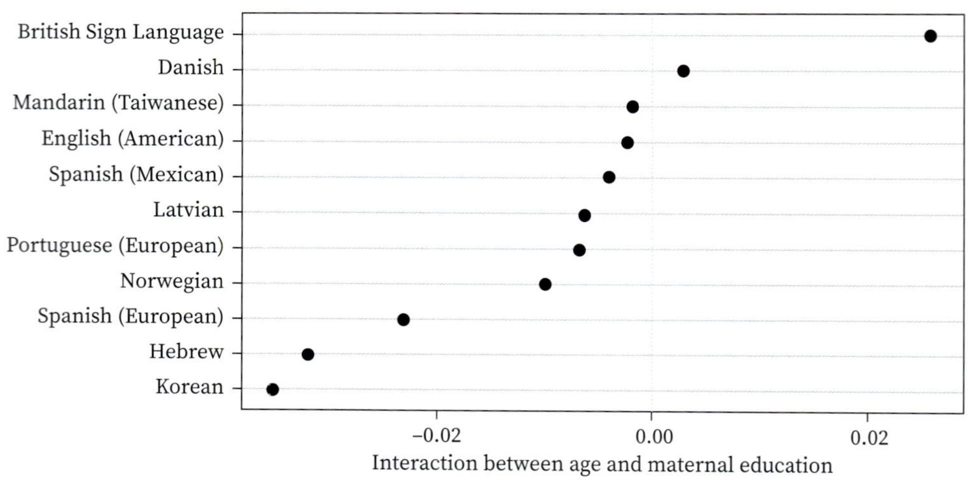

Figure 6.19
Interaction term between age and maternal education for WG comprehension data in each language.

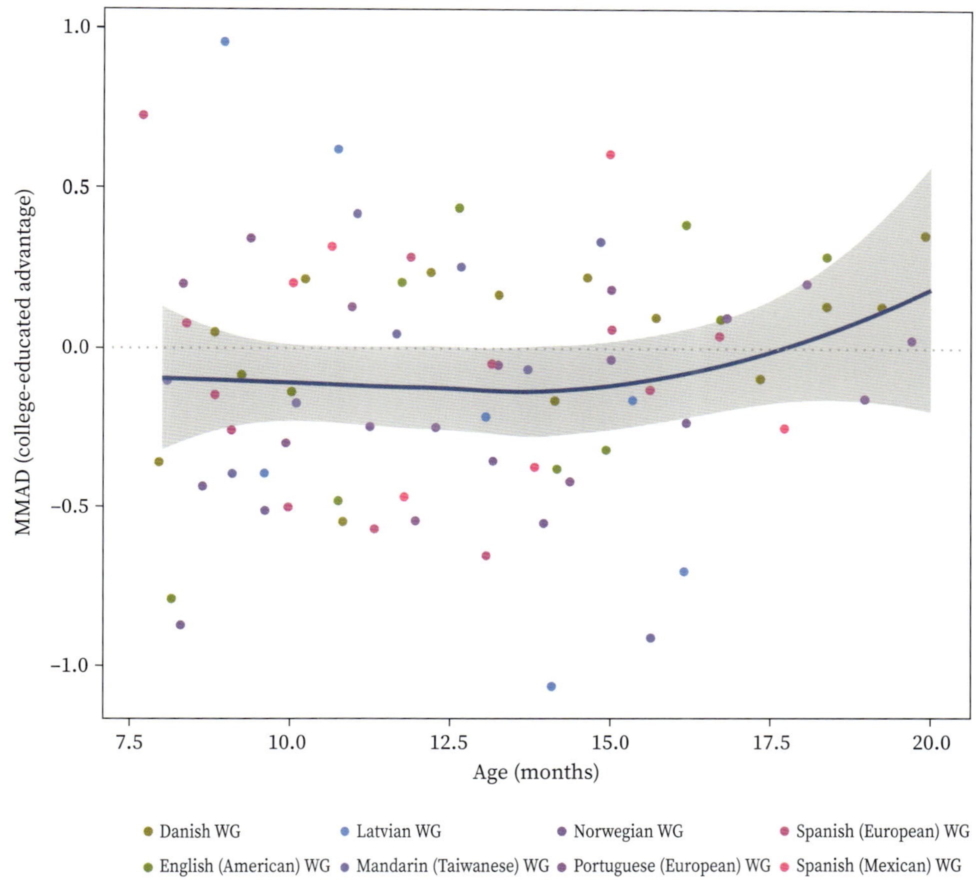

Figure 6.20
MMAD college-educated advantage for WG comprehension data in each language across age.

our data precludes strong inferences about the precise factors that lead to these differences. Speculatively, the magnitude of the differences across countries suggests a role for environmental factors in shaping variation in child outcomes even before the age of 3 years, although of course genetic and other biological factors may also play a role.

While lower scores on the vocabulary checklists can reflect authentic differences across children from different SES groups, it is also possible that some SES effects are the result of differential reporting biases (for discussion, see Fenson et al., 2007). The direction of such effects is unknown, however; it is not a given that low SES parents would under report their children's language. For example, in Feldman et al. (2000) study of more than 2,000 children, vocabulary comprehension scores

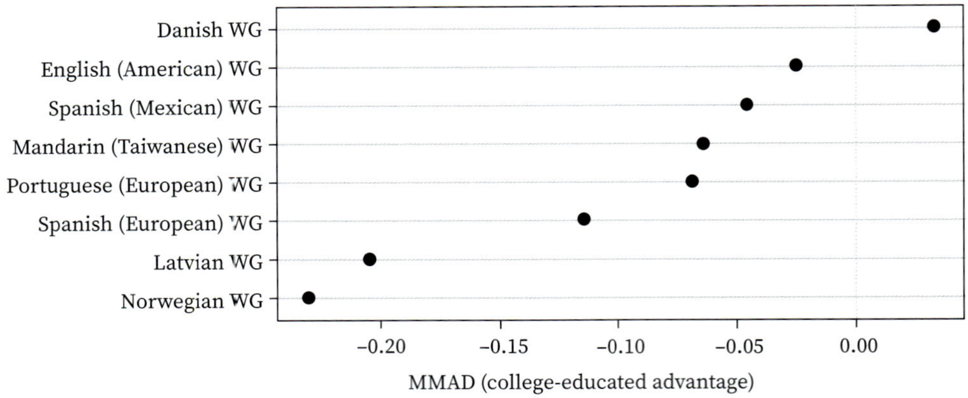

Figure 6.21
MMAD college-educated advantage for WG comprehension data in each language averaged over age.

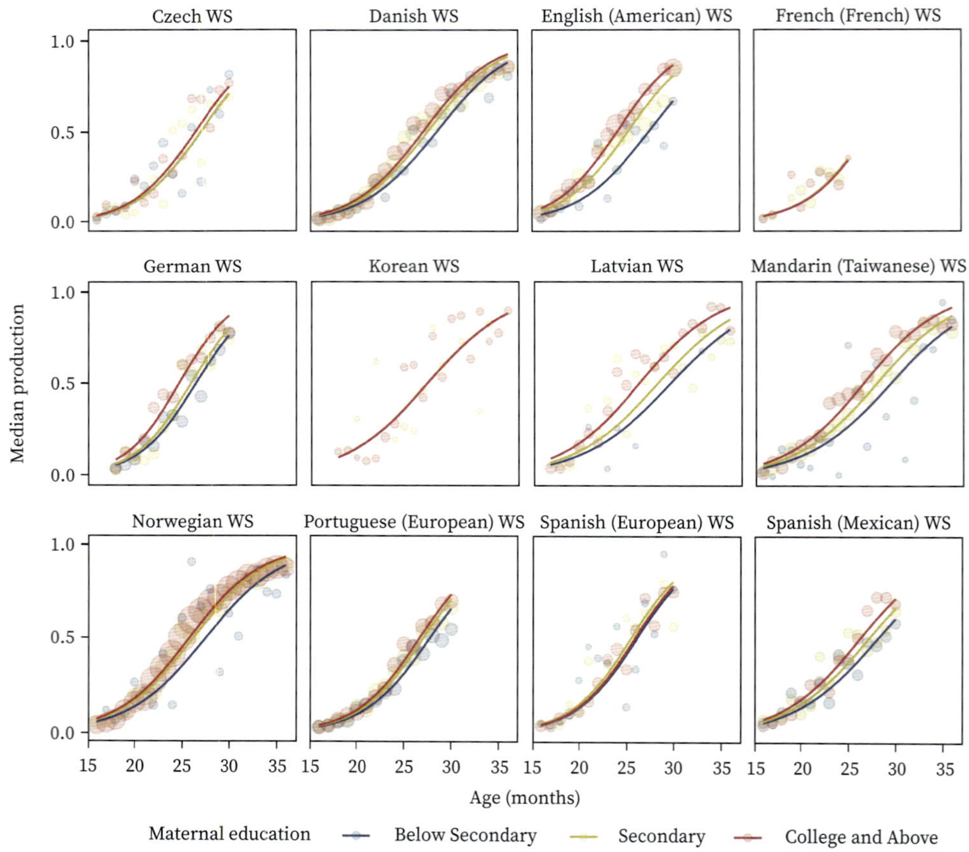

Figure 6.22
Differences in WS production scores by maternal education, plotted across age by language.

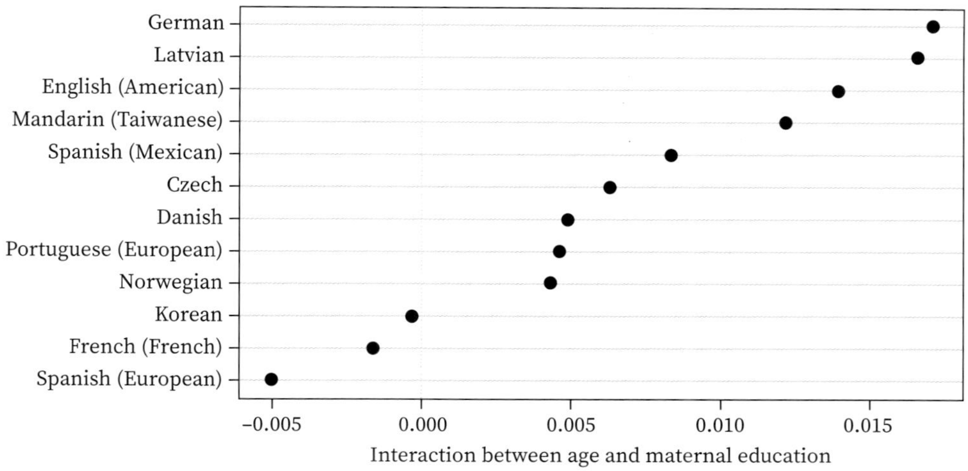

Figure 6.23
Interaction term between age and maternal education for WS production data in each language.

on the WG form were *higher* for caregivers with lower education than for caregivers with higher education, whereas the opposite relation was found for vocabulary production and later grammar skills from the WS form. In that study, parents with low educational and income levels seemed to overestimate their child's comprehension abilities. Overreporting of comprehension by low-SES parents could account for the different effects of maternal education that we observed in comprehension compared with production—on this view, we should put most stock in the production results.

Based on these findings and apparent failures of administration for comprehension (e.g., the Taiwanese Mandarin WG form; see chapter 3, the section on difficult data), we suspect that comprehension measures—while reliable when used appropriately—are nonetheless more vulnerable to bias. In order to report on comprehension, parents have to separate evidence of comprehension of a word in isolation from evidence for comprehension in highly constrained contexts where the child can "do the right thing" without truly understanding the linguistic content. In addition, some parents may recall times when the words were used in the child's presence and may confuse exposure with understanding. These difficulties may explain why parental reports of verbal comprehension are sometimes higher than results of direct testing (Tomasello and Mervis, 1994; Bergelson and Swingley, 2015) and why correlations between parental reports and comprehension scores from other methods are low for very young children (Goldfield and Reznick, 1990; Bergelson and Swingley, 2015).

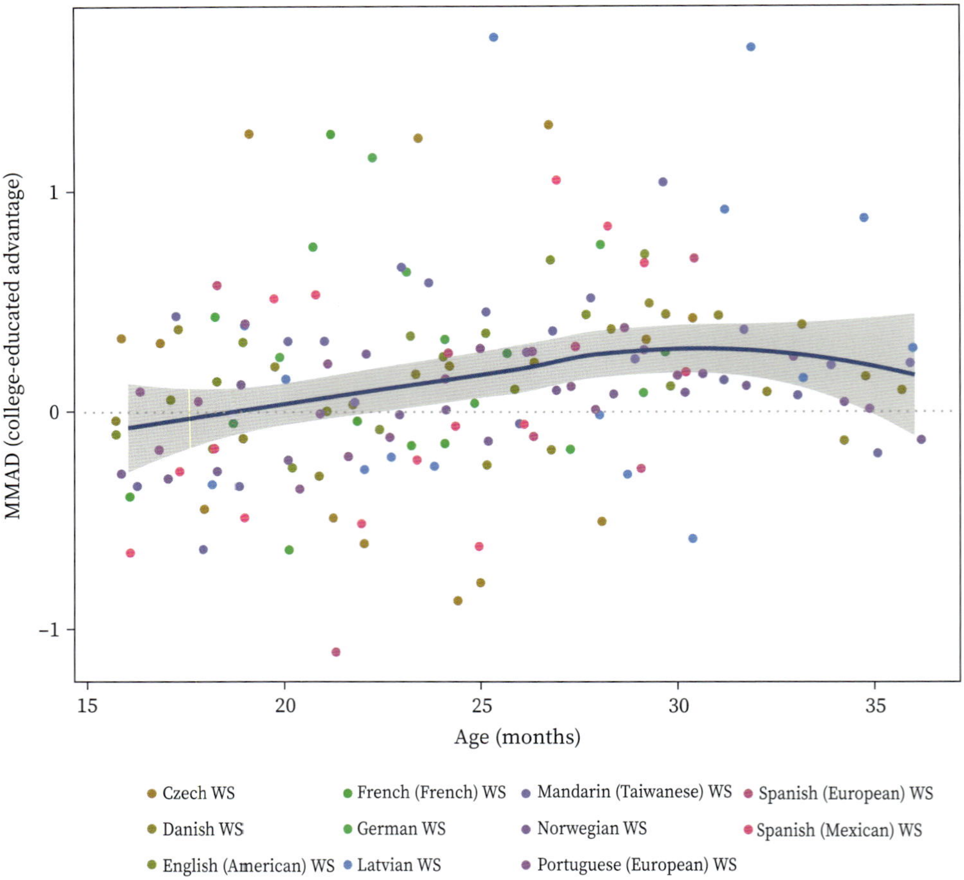

Figure 6.24
MMAD college-educated advantage for WS production data in each language across age.

6.4 Discussion

In this chapter, we quantified the relationships between vocabulary and sex, birth order, and maternal education. Although all of these have some potential to be influenced by reporting bias, in all cases we see some reasons from prior literature to suspect that similar effects are present (at least in English) for non-parent-report measures. The one likely exception is in the case of maternal education, where comprehension reports may be a less reliable signal of SES-related differences due to differences in the interpretation of the comprehension measure. With respect to production measures, in contrast, we see no reason to discount CDI-based measurements.

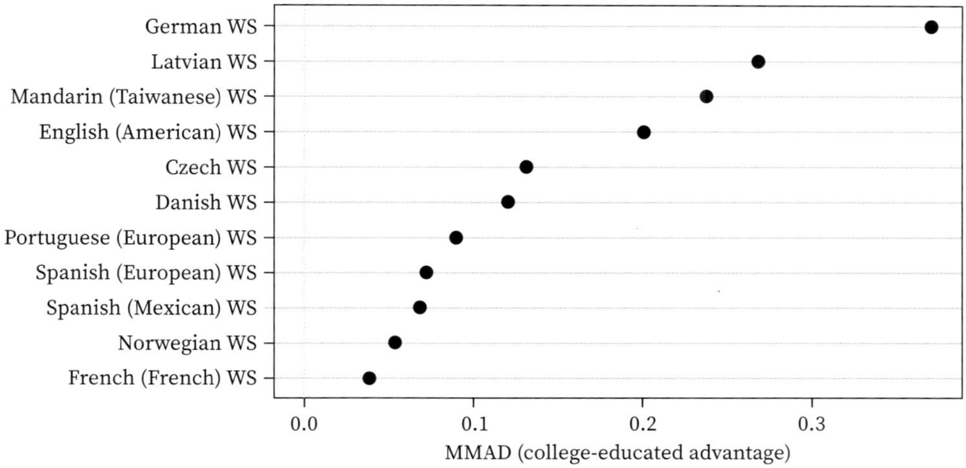

Figure 6.25
MMAD college-educated advantage for WS production data in each language averaged over age.

The relationships we observed were more prominent in every case for production than comprehension. This prominence could be a function of the relatively greater psychometric reliability of production compared with comprehension (see chapter 4)—the reporting bias issue (at least in the case of SES). It could also reflect demographic differences increasing over developmental time, since the production data we examined are largely from older children than the comprehension data. (We did not include early production in our analyses as we found that the data were noisy and hard to interpret due to children's small production vocabularies for much of the measured range).

We could only speculate about the origins of the relations we found, but the directionality of these relations was similar across languages, and they showed reasonable consistency across the full language sample (especially for sex differences). For sex differences, this consistency leads us to speculate about the cognitive origins of verbal ability differences, which are found quite consistently outside of the CDI as well (e.g., Maccoby and Jacklin, 1974). In contrast, birth order and maternal-education-related differences were slightly more variable than sex differences and have been argued in previous work to relate to children's input. We return to the interpretation of demographic differences in vocabulary in chapter 17.

7 Gesture and Communication

In addition to containing early words, Communicative Development Inventory (CDI) forms designed for infants (the Words and Gestures [WG] family) also contain a set of items designed to probe early gestural communication. This chapter explores these items. Our goals are to examine (1) the robustness of the measurement properties of these nonverbal parent-report measures, (2) the degree of cross-linguistic consistency and variability of reporting milestones like first pointing as well as social routines like waving hi and playing peekaboo, and (3) the relationships between gestural development and linguistic development.

7.1 Introduction

Children's most recognizable early linguistic accomplishments are surely their first words—a topic we turn to in chapter 8. However, even before infants approach this important milestone, they are already communicating through another modality: gesture. For example, a child who extends their hands and opens and closes their fist likely wants something to be given to them. A child who points to a bird up in a tree likely wants to get their caregiver's attention so that they can share in the delight together. Children's early vocalizations are also sometimes accompanied by gestures; for example, a child might raise both of their hands in the air and say "up!" The social and communicative routines that these gestures enable may themselves form part of the supportive context in which early language learning happens (Bruner, 1985). In sum, gestures are an important aspect of children's early communicative development.

Early gestures have long been thought to have a common mental status with later-developing linguistic accomplishments because both may reflect the child's understanding of symbols—that is, that a name or gesture can "stand in" for a thing in the world. The classic theories of Piaget (1962) and Werner and Kaplan (1963) proposed that all symbols have their origins in actions carried out on objects, and moreover, such symbols can be manifested in either the vocal or the gestural domain. These proposals suggest a common underlying mental function that is critical to the development of all symbolic skills, both in language and in certain types of gestures.

While the strong representational claims in these theories may be too extreme by modern standards, they do correctly predict developmental continuity between early gesture use and children's later lexical and syntactic development (e.g., Bates et al., 1975; Thal and Bates, 1988). For example, children's ability to point to distant objects is linked to the onset of the production of first words (Fenson et al., 2007; Brooks and Meltzoff, 2008), and children with delayed onset of pointing are likely to also be delayed in first word production (Clark, 1977; Butterworth, 2003). In addition, children's early gesture use is correlated with their later comprehension abilities (Bates et al., 1991), and children's use of gestures in combinations with words is linked to the later production of multi-word combinations (e.g., Goldin-Meadow, 1998; Iverson et al., 1994).

These early correlational findings could simply reflect a common cause: children who use gestures might also be better at learning words. More recent studies have demonstrated more specific links between early gesture use and later lexical and syntactic development, however (e.g., Rowe and Goldin-Meadow, 2009). For example, the particular lexical items that enter a child's vocabulary are more likely to be names for those objects that are earlier labeled using gestures (Iverson and Goldin-Meadow, 2005). Moreover, early gesture vocabulary is specifically linked to later word vocabulary, whereas early gesture plus word combinations are linked specifically to children's later word combination skills (Rowe and Goldin-Meadow, 2009). Taken together, the pattern of predictive correlations suggests that children's early gestures provide an important social, communicative, and linguistic foundation for later language development.

Early gestures serve many different functions. Children typically first begin by using "deictic gestures"—for example, giving, pointing, and showing (e.g., Volterra and Caselli, 1985). Such deictic gestures are clear precursors to important linguistic and communicative functions, including establishing reference and promoting shared attention (Carpenter et al., 1998). However, these deictic gestures do not necessarily have symbolic content per se (i.e., they do not stand for objects in the world; Bates et al., 1980). Early on, pointing gestures may serve a general imperative function—for example, to request something from an adult. In contrast, later pointing is more likely to be referential, directing a caregiver's attention to another object or person (Bates et al., 1975; Masur, 1990; Vygotsky, 1980).

Children also use gestures as part of social activities, for example, by waving "bye bye" or signaling "all done." At first, social gestures might occur simply as imitations, but then later, children are able to produce these social gestures spontaneously in appropriate communicative contexts. Children's social gestures can also reflect their ability to engage in activities during pretend play—for example, talking on a pretend phone or pretending to stir a soup.

Finally, older children's gestures also take on a "true" symbolic meaning as a child might use a conventional gesture to recognize or classify objects as an instance of a category (e.g., pretend to drink from a cup or sniff a flower). Some theorists have hypothesized that children's ability to use gestures in this symbolic way may reflect a common underlying "vocabulary" in both the verbal and gestural domain (e.g., Acredolo and Goodwyn, 1985; Bates et al., 1980). These gestures may thus have a similar relationship to language as children's pretend play does (Bergen, 2002; Smith and Jones, 2011).

In sum, both early deictic gestures and later symbolic gestures have played an important role in shaping theories of communication and language learning. Thus, their cross-linguistic variability and their relationship to other aspects of language learning are important topics for investigation.

7.2 Measurement Properties of CDI Gestures

Following our work in chapter 4, we begin by analyzing the degree of measurement signal in the gesture items. The available dataset is unfortunately significantly smaller than is available for vocabulary items—a number of users of the CDI forms did not measure (or at least did not provide us with data) on gesture items. Consequently, we focus first on the American English WG form and use face validity as our primary criterion rather than a more sophisticated metric.

7.2.1 Measuring the Development of Gesture

Unlike the word items on the CDI, which typically ask parents to make a binary decision about whether a word is in their child's vocabulary (although comprehension and production are separate decisions), the First Gestures on CDI forms ask parents to make a three-way decision, determining if their child produces a given gesture "often," "sometimes," or "not yet." We begin by asking whether parents' responses are sensitive to this distinction as the choice of whether to treat all three levels as meaningful impacts downstream analytic decisions. We perform this sensitivity analysis on the American English CDI as it is the inventory for which we have the best a priori intuitions.

Figure 7.1 shows the proportion of American English learning children who give each of these responses. If each of the three responses is meaningfully different, the developmental trajectory of each should be distinct and predictable. The proportion of children whose parents indicate that they do "not yet" produce

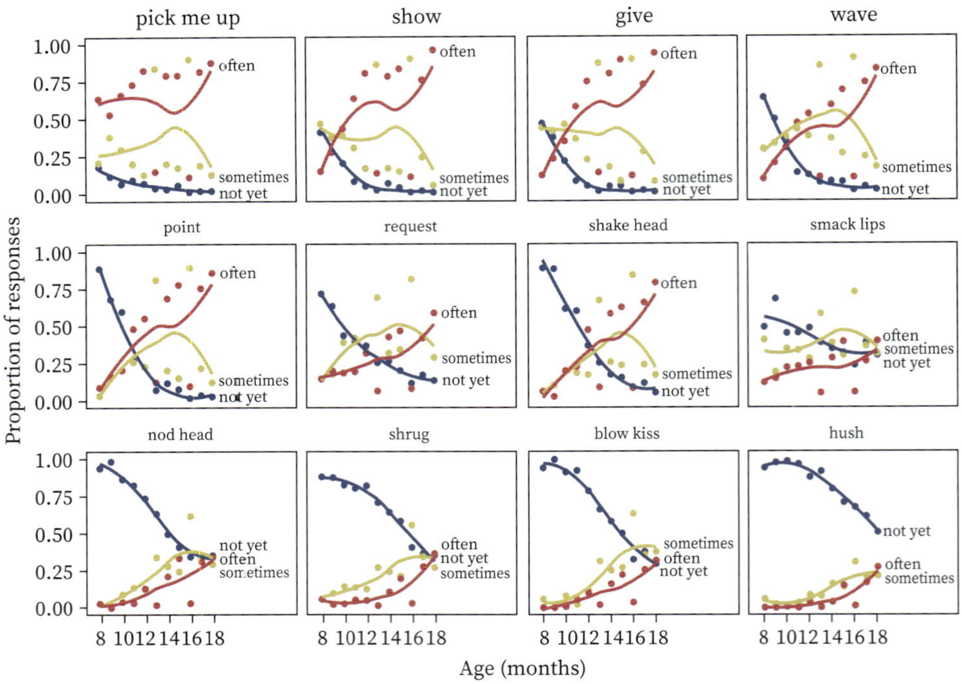

Figure 7.1
Proportion of each response type over age on each First Gestures item for American English.

each gesture declines predictably over development. However, the other two responses—“sometimes” and “often”—do not appear to have reliably different trajectories. Perhaps they are used differently by different parents or in different samples.

For comparison, we collapse the “sometimes” and “often” into a single value and plot the proportion of children at each age whose parents report that they produce each gesture (figure 7.2). The trajectories look generally smooth and prima facie reasonable, with the potential exception of the *smack lips* gesture for which there is very little developmental change (this gesture, which corresponds to the vocalization *yum yum*, may be unusual or less stereotyped).

While these gestures are categorized on the CDI as “first gestures,” the form also asks parents about a variety of other kinds of gestures that children produce, including those involved in games and pretend play. Do these gestures have similar trajectories? Figure 7.3 plots developmental trajectories for these other categories of gesture.

While some are clearly learned later than the early gestures, a number of these—for example, peekaboo and pretend play with cups and spoons—appear to be

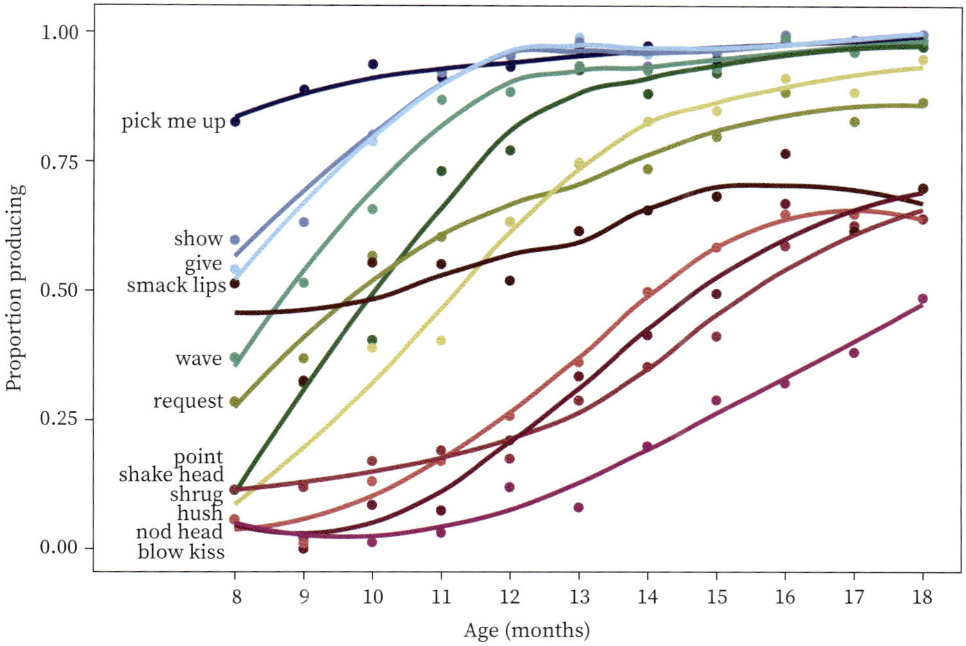

Figure 7.2
Trajectory over age of each First Gestures item for American English.

learned quite early as well. They all also appear to have generally smooth and increasing trajectories with the exception of *so big* (from Games and Routines) that, like *smack lips* from the First Gestures section appears to be either less stereotyped, more difficult to identify, or more variable across children.

Taken as a whole, it is clear that almost all of the gesture items have developmental trajectories, not unlike word items, and that they thus have the potential for informative analyses. Further, trajectories look qualitatively similar across categories. Consequently, for general cross-linguistic analyses, we will consider all of the gestures together and compress "often" and "sometimes" into a single affirmative choice.

To estimate the coherence of these categories, we compute age of acquisition estimates for each of the American English gestures by gesture type: First Gestures (e.g., *pick me up, point*), Games and Routines (e.g., *play peekaboo, chase*), Actions with Objects (e.g., *brush teeth, push car*), Imitating Other Adult Actions (e.g., *type, use pen*), and Pretending to Be a Parent (e.g., *sweep, feed from a spoon*). These estimates were produced by fitting a robust linear regression to the proportion of children who produce each gesture and estimating the age at which 50 percent

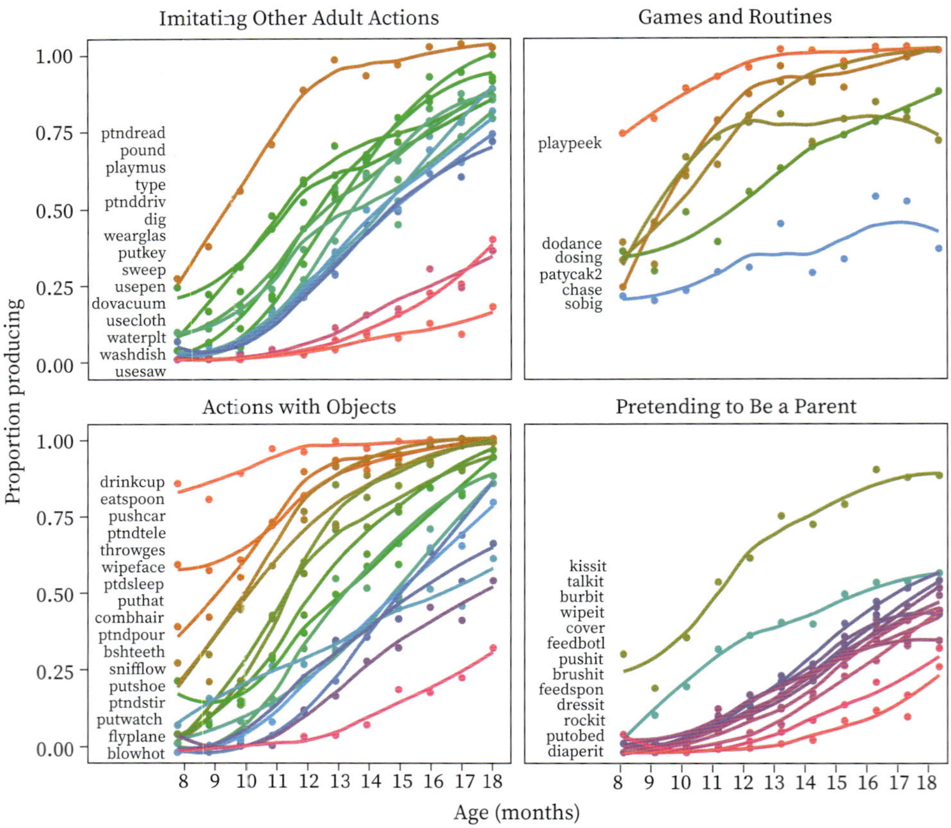

Figure 7.3
Trajectory over age of each gestures item by type for American English.

of children produce the given gesture (see appendix D). The resulting ages of acquisition for each gesture type are shown in figure 7.4. These categories vary in coherence, but overall First Gestures and Games and Routines tend to be produced early and Imitating Other Adult Actions and Pretending to Be a Parent—more representative of pretend play—are produced relatively later (many not reaching 50 percent within the administration range of the form). The object gestures vary substantially in their ages of acquisition.

7.2.2 Consistency of the First Gestures

While the First Gestures are not universally learned before the other gestures measured, they are among the earliest learned. Because of the particular theoretical importance of these early communicative gestures (e.g., deictics like pointing and

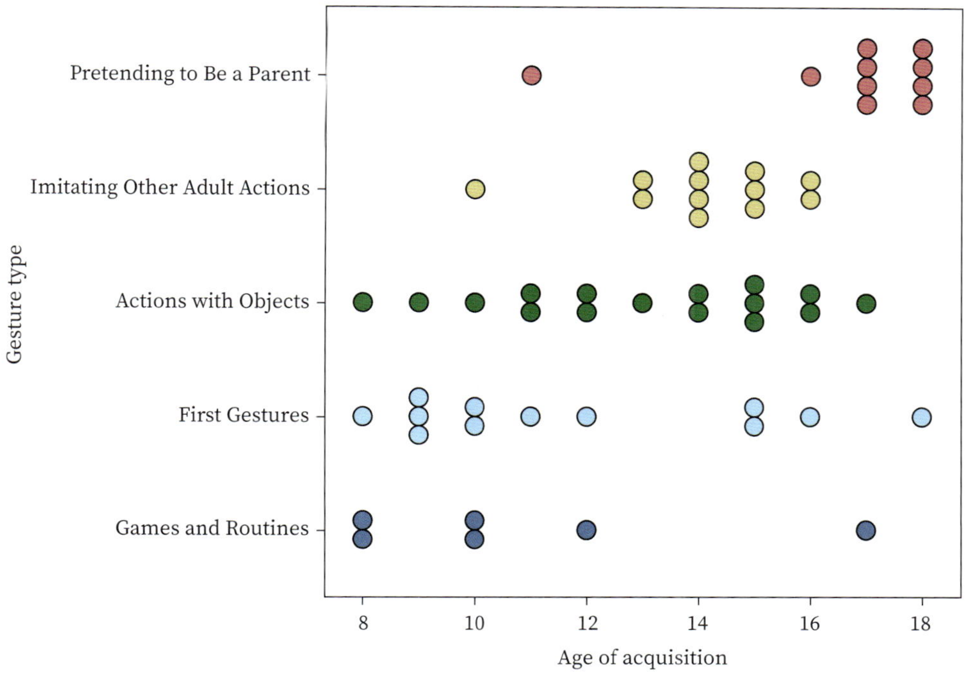

Figure 7.4
Age of acquisition for each gesture item by type for American English.

showing, routines like *pick me up*), we analyze the cross-linguistic consistency of these at the item level.

Table 7.1 and figure 7.5 show both consistency and variability across items. As in the learning of words, the means and variances of these ages of acquisition were correlated ($r = 0.81$; Mollica and Piantadosi, 2017). The primary outliers were *request*, which appears to be produced surprisingly late in American English, and *shrug*, which was produced surprisingly late by French-learning infants. In general, however, most of the cross-linguistic differences appear to be consistent across the gestures (i.e., French-learning infants were reported to perform all gestures later). It is difficult to tell from this small sample of mostly European languages whether these differences are driven by linguistic factors or rather by properties of our samples or variability in parents' interpretation of the form. Nonetheless, they provide some evidence for consistency in the process of gestural development cross-linguistically. To get additional leverage on this process, we next consider the full set of gestures.

Table 7.1
Summary statistics for ages of acquistion for each First Gestures item across languages.

Item	Mean	SD	N
Show	9.12	0.99	9
Give	9.60	0.89	7
Shake head	10.29	1.60	8
Wave	10.29	1.60	8
Point	10.40	1.67	8
Pick me up	10.50	0.71	7
Smack lips	13.75	2.06	7
Request	14.33	3.21	8
Nod head	14.50	3.78	8
Hush	15.17	2.99	7
Shrug	15.71	3.82	7
Blow kiss	16.00	1.90	6

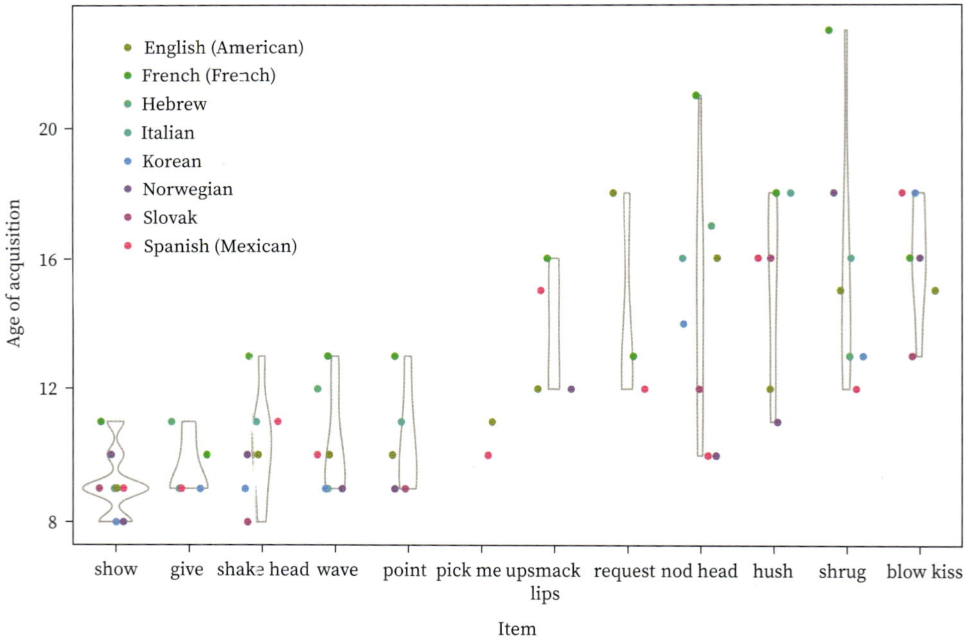

Figure 7.5
Distribution of ages of acquistion for each First Gestures item for each language.

Table 7.2
Summary statistics for intercorrelations between gesture categories for each language.

Language	Mean	SD
English (American)	0.600	0.113
French (French)	0.560	0.178
Hebrew	0.605	0.141
Italian	0.682	0.104
Korean	0.839	0.129
Norwegian	0.572	0.110
Slovak	0.687	0.080
Spanish (Mexican)	0.668	0.087

7.2.3 Intercorrelation among Gestures

Given both the similarity and the variability in the developmental trajectories of different gestures as well as the cross-linguistic variability in First Gestures, a natural next step is to quantify the relationship of gestures to each other. We begin by computing the average intercorrelation among each of these gesture categories. In this analysis, we take gesture categories in pairs (e.g., Imitating Other Adult Actions and First Gestures) and ask how the proportion of items that children know in one predict the proportion of items they know in the other. For American-English-learning children, the proportion of gestures they know across categories is correlated at $r = 0.60$—nearly identical to the value of approximately 0.6 reported by Fenson et al. (1994). For comparison, the same intercorrelation computed across categories of words (e.g., "animals" and "places") yield 0.64 for production and 0.57 for comprehension.

This cross-category intercorrelation is quite similar cross-linguistically, ranging from 0.56 in French (French) to 0.84 in Korean. The full set of intercorrelations is shown in table 7.2.

7.3 The Relationship between Language and Gesture

A critical theoretical question in early communicative development concerns the relationship between language and gesture. As alluded to above, a number of early influential theories (e.g., Piaget, 1962; Werner and Kaplan, 1963) held that gesture and language should be intimately related because of their reliance on a shared system of symbolic reasoning. To the extent that they are underpinned by the same system, words and gestures should have related developmental trajectories—children who gesture early should also speak early and vice versa (Bates et al., 1991). Following in the footsteps of Fenson et al. (2007), we ask this question at

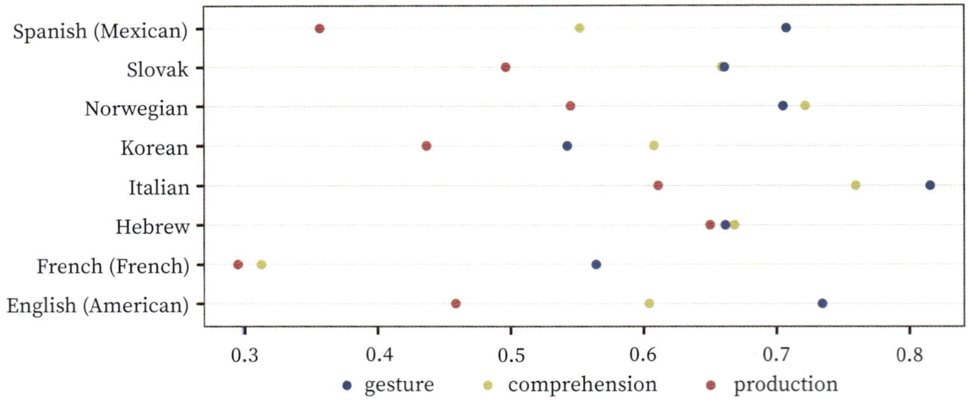

Figure 7.6
Correlations between age and subscales (gesture, comprehension, production) across languages.

larger scale and cross-linguistically. To assess this relationship, we will look at the correlations between children's gestural and linguistic vocabularies.

To first provide a baseline, however, we compute the correlation between children's language and gesture development and their age. As figure 7.6 shows, gesture shows as much or more developmental change than comprehension and production—at least over the age range measured by the CDI WG forms. The variability in the correlation with age in all three measures hangs together within languages; languages where there is more developmental change in linguistic development also tend to have more gestural development. Production typically shows the lowest correlation, presumably because many children are still at floor on this measure for most of the age range being measured.

But what is the relationship between language and gesture? As both children's lingusitic and gestural vocabularies increase over development, we should expect to see a positive correlation between language and gesture regardless of any interesting assocations between them. One way of getting leverage on their relationship is to exploit the difference between comprehension and production. As noted previously, comprehension and production do not proceed in lockstep, and comprehension generally outpaces production. This is in part because production requires additional control over the developing motor systems necessary for speech. But it may also be driven by individual differences in children's personalities or communicative goals: some children may want to talk more than others (see chapter 15).

To the extent that gesture and language are related by their shared reliance on symbolic understanding, their correlation should be highest when only this

Table 7.3
Correlation with gesture subscale for each vocabulary subscale in each language.

Language	Comprehension	Production
English (American)	0.75	0.53
French (French)	0.67	0.37
Hebrew	0.78	0.65
Italian	0.81	0.50
Korean	0.51	0.45
Norwegian	0.69	0.50
Slovak	0.63	0.54
Spanish (Mexican)	0.76	0.55

shared system is tapped. In this case, we should predict that gesture *production* and language *comprehension* are more tightly correlated than gesture *production* and language *production*. In contrast, if the correlation is due primarily to a shared desire to communicate and engage socially with caregivers, we should predict a stronger correlation between gesture *production* and language *production*. Across these 8 languages, children's production of gestures is consistently more highly correlated with their comprehension of language than their production of language (table 7.3). This analysis provides at least some initial evidence for the role of symbolic understanding in driving the correlation between language and gesture. An alternative explanation is also plausible, however—comprehension may simply show more variability than production in this age range, allowing for a higher level of correlation.

As a second, more specific test of this hypothesized relationship, we consider explicitly the deictic gestures. Deictic gestures are most closely associated directly with communicative goals rather than symbolic content: pointing, showing, and giving (Bates et al., 1991). Because all of these gestures are produced by children relatively early and thus contain less signal for older children, we compare these with the remaining First Gestures plus Games and Routines that appear in at least 6 languages: wave, pick me up, shake head, nod head, hush, request, blow kiss, shrug. We excluded "smack lips" because of its lack of developmental change (see above). We then applied the same analytic method as earlier—computing the correlation between children's gestural and language development—seperately for these deictic and non-deictic early gestures. Because there are many more non-deictics, we take all 56 possible combinations of 3 of these gestures at a time.

Figure 7.7 shows deictic gesture-vocabulary correlations. As before, correlations between gesture and comprehension are higher than gesture and production, but in all cases language is more correlated with the children's non-deictic gestural development rather than their deictic gestural development. This finding

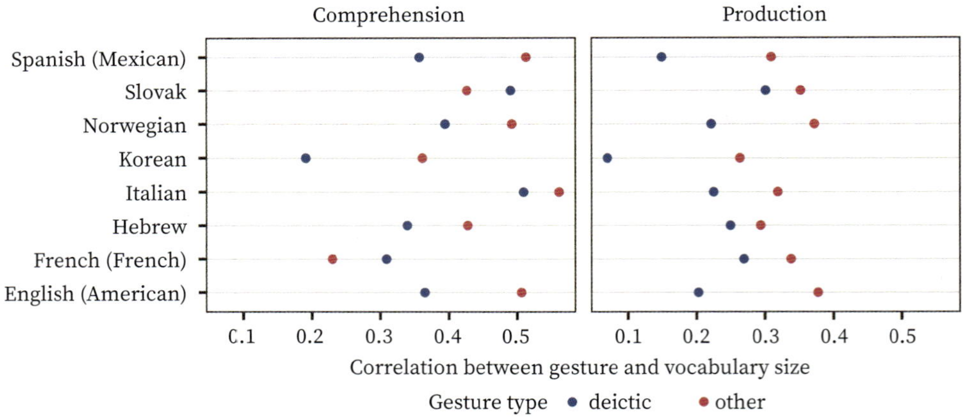

Figure 7.7
Correlations between gesture and vocabulary size for both deictic and non-deictic gestures, split by production and comprehension vocabulary.

provides a second piece of evidence that the relationship between linguistic and gestural development is driven more strongly by reliance on a shared symbolic format rather than reliance on individual children's motoric development or desire to communicate. Although this finding adds additional specificity to our analysis, we still mark this conclusion as tentative due to possible confounding with measurement issues.

7.4 Conclusions

Gestures appear early in young children's communicative repertoires, with an understanding of the deictic purpose of pointing being found as early as 12 months of age (Liszkowski et al., 2007). The frequency of gestures in children's input—as well as the precociousness of their own gesture productions—is also reliably correlated with their linguistic development (Rowe and Goldin-Meadow, 2009; Brooks and Meltzoff, 2008). Here we extended the analyses developed in Fenson et al. (2007) to look at these relationships cross-linguistically using large-scale CDI data.

In the first part of the chapter, we evaluated the measurement properties of the gesture items as well as the coherence of the groupings of items on the CDI forms. Our analyses confirm that parents are likely able to report reliably on their children's gesture development. We found a high degree of consistency in development of different gestures—children who learn some gestures early are also likely to learn other gestures early. Further, we observed a high degree of cross-linguistic consistency in the age of acquisition of different gestures.

Finally, we examine the relationship between gesture and language directly. We found that individual differences in gesture development were highly correlated with language development and that gesture was more related to comprehension than production. Further, development of deictic gestures is less predictive of language development than development of non-deictic gestures. One speculative conclusion from these analyses is that precociousness in the gestural domain may be related to children's developing understanding of symbolic communication systems rather than shared motoric development or a strong desire to communicate with their caregivers.

8 Consistency in Early Vocabulary

Which words do children learn first? In spite of tremendous individual variation in rate of development (see chapter 5; Fenson et al., 1994; Hart and Risley, 1995), the first words that children utter are reported to be quite consistent. We examine this claim both qualitatively and quantitatively, focusing on the first ten words initially and then zooming out to examine this claim in typological and developmental context.

8.1 Introduction and Methods

Based on the examination of diary studies, a number of early studies noted the similarities in children's first words across languages (e.g., Clark, 1973; Slobin, 1970; see also Schneider et al., 2015). This observation formed the basis for a number of theories of usage (including, e.g., Clark's influential semantic feature hypothesis). While we return briefly to the question of *why* we see similarity across languages in the contents of early vocabulary, in this chapter we seek to establish a firmer empirical understanding of both the similarities and differences in early learned words across languages.

Our approach primarily follows the lead of a systematic examination of the early vocabularies of children learning English, Mandarin, and Cantonese (Tardif et al., 2008). Tardif and colleagues found that children's first 10 words in all three languages tended to be about important people in their life (*mom, dad*), social routines (*hi, uh oh*), animals (*dog, duck*), and foods (*milk, banana*). Here we attempt to generalize this analysis, asking more broadly whether words tend to be learned in the same order across languages.

One challenge is that the precise words that children learn in different languages are (of course) language-specific. We would really like to ask whether the *concepts* that are being talked about are the same—or at least similar. As detailed in the chapter 2, the items on each language's form are adaptations and not translations:

They are intended to capture the spirit of the items on the English form rather to replicate them exactly. Thus, not all words appear on all forms. In addition, conceptual mappings across languages are also subject to cross-cultural variation (the "tortilla problem" we discuss earlier). In what follows, we acknowledge this caveat, but assume for simplicity that *dog, chien,* and *perro* name (roughly) the same concept. Our approach is thus to take advantage of when translation equivalents appear on multiple forms and examine variability in how quickly these words are acquired across languages. This analysis is in some sense a "rough draft" of our more systematic quantitative approach in chapter 10, focusing on first words specifically.

To estimate the similarity of each item's trajectory, we use a single measure of its difficulty: age of acquisition—the age at which 50 percent of children in each language are estimated to have acquired it (appendix D). We analyzed consistency in age of acquisition for both comprehension and production, using Words & Gestures forms to estimate age of acquisition in comprehension, and stitching across Words & Gestures and Words & Sentences forms to estimate age of acquisition in production. Because of this strategy of combining forms, we were restricted to the 15 languages for which data for both forms were available.

In total, we estimated ages of acquisition for 945 total words spread across the 15 languages. Unfortunately, not every word appeared on all forms. Figure 8.1 shows the cumulative proportion of forms on which every word appears. For our consistency analysis, we considered only the 335 words that appeared in at least 8 of the 15 languages.

8.2 The First Ten Words

Following Tardif et al. (2008), we begin by examining the first 10 words acquired by children across the 15 languages we measured (tables 8.1 and 8.2). Similar words appeared in the top 10 across languages, especially in children's earliest productions. In production, 12 of the 36 words appeared in the top ten earliest words of every language (0.33), and all but 2 appeared in at least ten languages (0.94). In comprehension, 13 of the 40 words appeared in the top ten earliest words of every language (0.32), and all but 4 appeared in at least ten languages (0.90). These words consist primarily of important family members (*mommy, daddy, grandma*), social routines (*hi, bye, peekaboo*), and sounds (*yum yum, vroom, woof woof*).

Strongly ratifying the conclusions of Tardif et al. (2009), similar words appeared in the top 10 across languages. Similarities were especially prominent in children's earliest productions. These words consist primarily of important family members (*mommy, daddy, grandma*), social routines (*hi, bye, peekaboo*), and sounds (*yum yum, vroom, woof woof*).

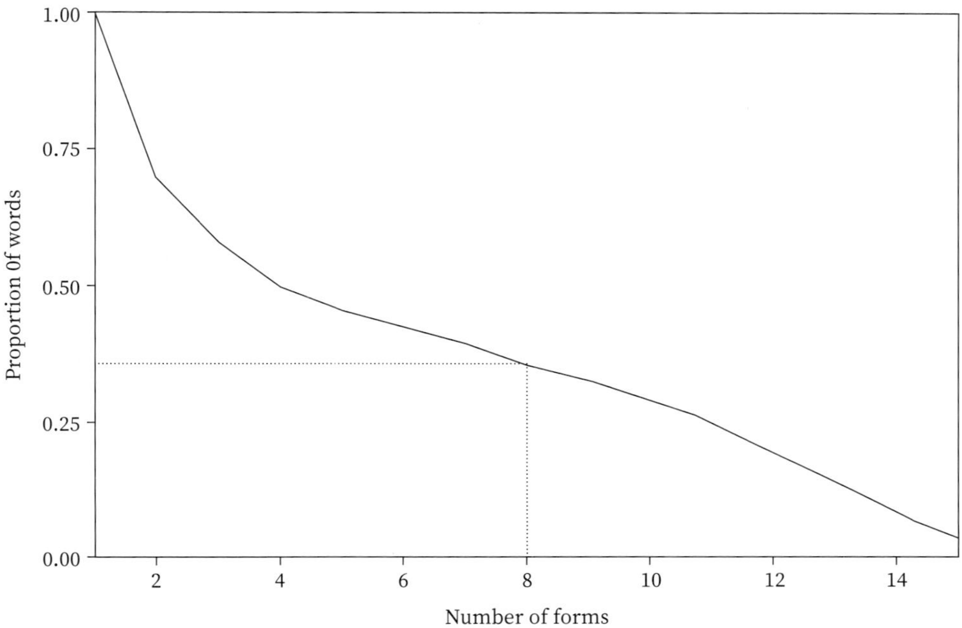

Figure 8.1
The proportion of words found on at least some number of languages' CDI forms (e.g., all words appear on at least one form, 111 words appear on at least 2 forms, and so on, with 34 words appearing on all 15 languages' forms). The dotted line shows the cutoff value we chose as (8).

Unfortunately, we cannot determine if the greater consistency found in early production is a real regularity about children's lexical development, or is instead a measurement artifact arising from the greater difficulty of reporting on a child's comprehension (see chapter 4).[1] It may be that early communicative needs drive the first words children produce to be even more similar than the first words they comprehend.

Despite these differences between comprehension and production, words that are reported to be acquired early on one measure are also generally reported to be acquired early on the other. Figure 8.2 shows the relationship between the mean age of acquisition in production and the mean age of acquisition in comprehension for each of these 335 words across the 15 languages. The correlation between the two measures was quite high: $r = 0.80 (p < 0.001)$. Thus, apparent inconsistencies in

1. This finding is *prima facie* inconsistent with another recent analysis comparing variability in comprehension and production vocabularies (Mayor and Plunkett, 2014). This analysis noted that comprehension vocabularies tend to be less idiosyncratic across children—rather than across languages—than production vocabularies.

Table 8.1
The 10 earliest words that children produce in each language. Bolded words appear in at least 6 languages.

Croatian	Danish	English (American)	French (French)	French (Quebecois)	Hebrew	Italian	Kiswahili	Korean	Norwegian	Russian	Slovak	Spanish (Mexican)	Swedish	Turkish
mommy	hi	**mommy**	**daddy**	**mommy**	**mommy**	**mommy**	**mommy**	**mommy**	**vroom**	meow	**mommy**	**mommy**	**mommy**	**mommy**
daddy	**woof woof**	**daddy**	**mommy**	**daddy**	yum yum	**daddy**	**daddy**	**daddy**	**mommy**	**daddy**	**daddy**	**daddy**	**daddy**	yum yum
grandma	thank you	**ball**	**baby**	**no**	grandma	**woof woof**	car	peekaboo	yum yum	**woof woof**	**woof woof**	water	thank you	brother
bye	**mommy**	**bye**	**bye**	**bye**	**vroom**	grandma	cat	**woof woof**	hi	grandpa	grandma	yum yum	**woof woof**	**woof woof**
woof woof	**no**	hi	thank you	**baby**	grandpa	water	meow	cracker	**daddy**	aunt	**vroom**	**woof woof**	hi	**baby**
baby	**bye**	**no**	bread	**ball**	**daddy**	hi	motorcycle	water	**bye**	**mommy**	food	bread	peekaboo	**vroom**
no	**daddy**	dog	peekaboo	**vroom**	banana	grandpa	**baby**	**baby**	thank you	grandma	yum yum	**no**	drawer	**bye**
yes	**vroom**	**baby**	**ball**	sock	this	meow	bug	yes	**woof woof**	**bye**	**bye**	**bye**	meow	water
grandpa	yes	**woof woof**	sock	peekaboo	**bye**	**no**	banana	**ball**	yes	cereal	dog	**baby**	moo	**ball**
aunt	food	banana	shoe	moo	car	shoe	baa baa	**no**	peekaboo	**ball**	car	yes	**no**	doll

Table 8.2
The 10 earliest words that children comprehend in each language. Bolded words appear in at least 6 languages.

Croatian	Danish	English (American)	French (French)	French (Quebecois)	Hebrew	Italian	Kiswahili	Korean	Norwegian	Russian	Slovak	Spanish (Mexican)	Swedish	Turkish
grandma	**daddy**	bottle	**no**	milk	**yum yum**	**mommy**	baa baa	food	**child's name**	**yum yum**	**woof woof**	**yum yum**	**child's name**	water
mommy	**child's name**	**daddy**	**mommy**	**mommy**	cat	**daddy**	meow	**daddy**	**mommy**	meow	**yum yum**	water	**mommy**	**ball**
bye	**mommy**	**mommy**	**daddy**	**daddy**	doll	**peekaboo**	car	**mommy**	**daddy**	cat	dog	milk	**daddy**	up
daddy	**peekaboo**	**child's name**	**peekaboo**	**child's name**	balloon	**child's name**	bug	**peekaboo**	**bye**	eye	car	**mommy**	**peekaboo**	**yum yum**
peekaboo	**yum yum**	**bye**	**bye**	**bye**	**ball**	**hi**	cat	**no**	**hi**	**daddy**	**ball**	**daddy**	**bye**	**peekaboo**
vroom	**no**	**no**	bath	**peekaboo**	head	**woof woof**	doll	dirty	**peekaboo**	**grandma**	food	**bye**	bath	**mommy**
grandpa	**bye**	**peekaboo**	**hi**	**no**	light	dog	**ball**	bath	**no**	grandpa	pacifier	**no**	**hi**	**daddy**
cat	**hi**	**hi**	good night	bathtub	**daddy**	water	milk	**ball**	**yum yum**	**hi**	**mommy**	**woof woof**	lamp	**child's name**
child's name	**woof woof**	dog	yes	bath	**mommy**	bottle	medicine	cracker	good night	**woof woof**	**daddy**	dog	**grandma**	**bye**
woof woof	**ball**	**ball**	meow	**ball**	grandpa	**grandma**	spoon	water	**grandma**	dog	**grandma**	cookie	**no**	bottle

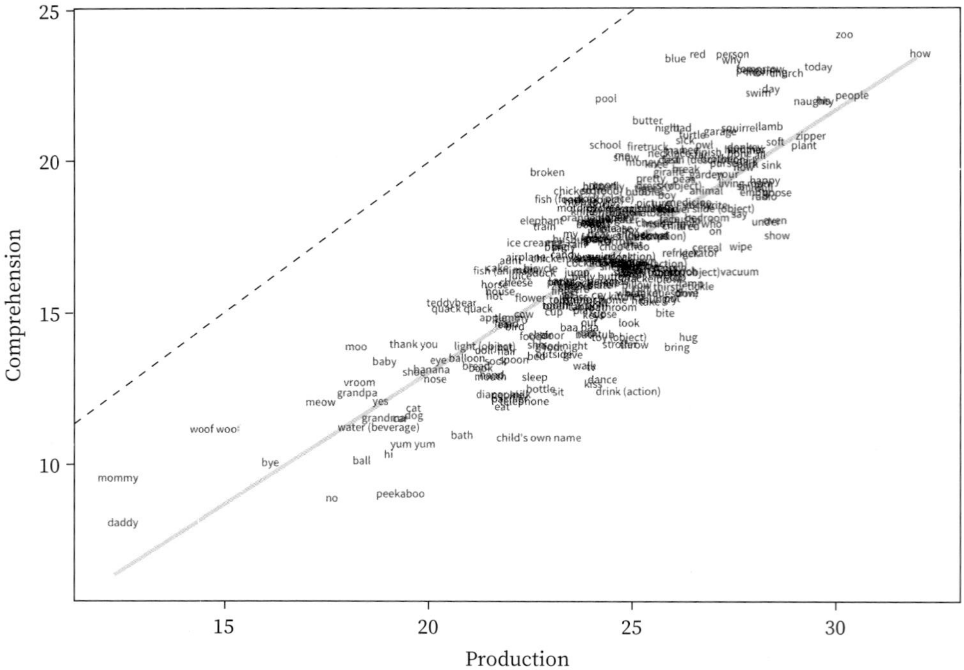

Figure 8.2
Average age of acquisition in comprehension and production for each measured word. Dashed line provides a reference with slope = 1 (identical age of acquisition).

first words for comprehension may be more a function of measurement errors in comprehension than any systematic difference.

Taken together, these analyses suggest that children's earliest words, and by inference the processes that underpin them, are highly similar across languages. The source of this similarity is hard to pin down, however. One possibility is that the difficulty of learning a word is determined predominantly by the complexity of the concept denoted by that word, and thus that variability in linguistic (e.g., phonological and syntactic complexity) and cultural (e.g., styles of parental interaction with children) features play a relatively small role in determining the difficulty of learning a word (Gentner and Boroditsky, 2001). Alternatively, the primary driver of difficulty could be linguistic, but the dimensions of linguistic variability could be orthogonal to the difficulty of learning. For instance, verbs may be more difficult than nouns because they are relational, and thus learning nouns makes learning verbs relatively easier than learning verbs makes learning nouns (Gleitman, 1990). In this case, the linguistically relevant dimensions would be relatively invariant across languages (Snedeker et al., 2007). Finally, it is worth noting that because the words on the Communicative Development Inventory (CDI) are not a random sample of words in each language, these correlations may overestimate the degree of cross-linguistic similarity, even though they are consistent with earlier diary studies.

In chapter 10 we begin to take up these questions using predictive models. Prior to taking this step, however, we consider cross-linguistic ordering more holistically. In the remainder of the chapter, we address this problem from two directions: (1) Is similarity in order of acquisition for two languages related to the degree of similarity between the two languages, and (2) Does similarity in order of acquisition change over development?

8.3 Acquisition Similarity and Linguistic Similarity

Unfortunately, the 15 languages in our analyses are both a small and non-representative sample of the world's languages, and thus do not have sufficient power to detect typological features of language that might be responsible for differences in the similarity of acquisition across languages (Piantadosi and Gibson, 2014). Nonetheless, the languages do come from different languages families, and do vary in their phylogenetic distance. We leverage this variability to ask whether similarity between two languages is related to similarity in how quickly words for the same concepts are learned in those two languages.

Instead of correlating the average similarity of age of acquisition across all languages, we consider the pairwise similarities in the age of acquisition of each of the 335 words in each language. Figure 8.3 shows these pairwise correlations for production as a matrix in which each cell shows a single pairwise correlation. This correlation matrix appears to contain a significant amount of structure, with languages that are from the same language family (e.g., Norwegian and Danish) showing higher correlations in their ages of acquisition for the same concepts. Perhaps unsurprisingly from the high average correlation between production and comprehension, pairwise correlations were nearly identical for production and comprehension ($r = 0.98$, $p < 0.001$); we omit the comprehension matrices for brevity. Figure 8.4 shows a dendrogram produced by hierarchically clustering these pairwise correlations.

These dendrograms show high similarity within the North Germanic, Slavic, and Romance language families. Some relationships resist straightforward linguistic explanations (e.g., the relationship of Quebec French to other languages). These may be due to non-uniform sparsity of data across these languages, or may instead reflect interesting cultural variation or other sources of variability. Despite these cases, the order in which words are acquired appears to a high degree to reflect the structure of the languages that children learning these words speak. To confirm

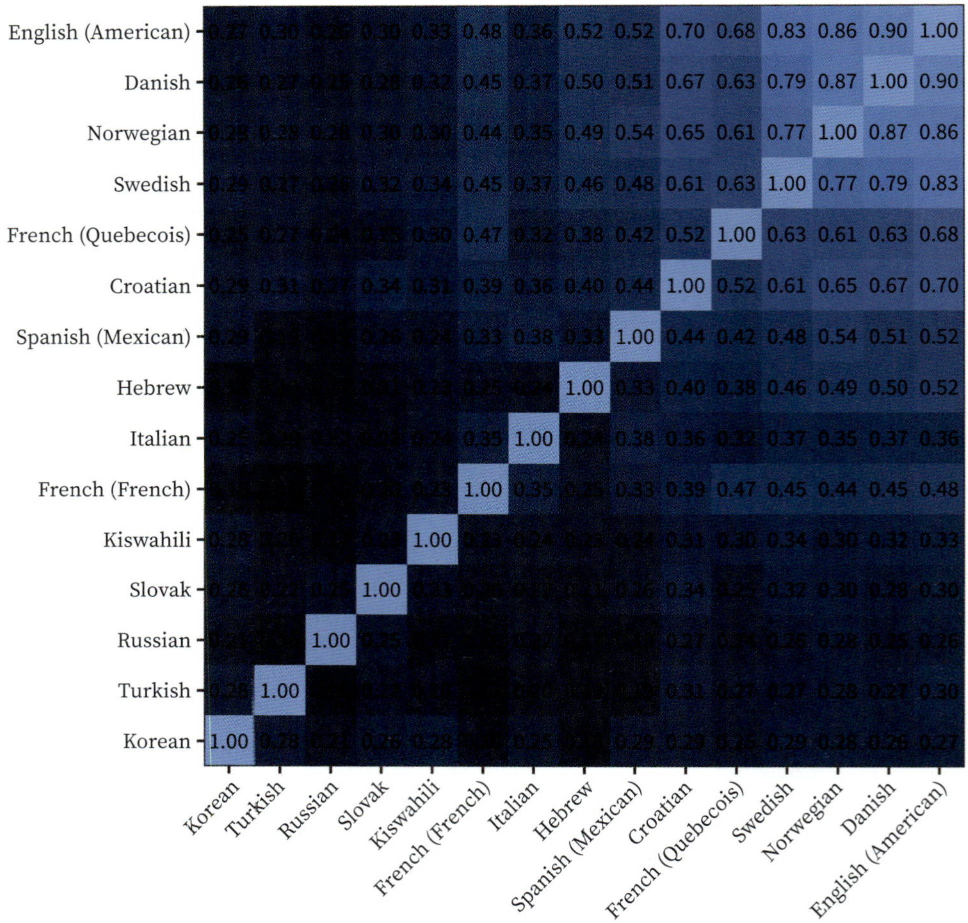

Figure 8.3
Correlation matrix showing pairwise correlations in words' age of acquisition. Languages that are more similar have more similar acquisition orders.

this observation quantitatively, we borrowed an established measure for measuring linguistic similarity: the similarity in lexical form of words expressing the same meaning (Wichmann et al., 2010).

Using a set of 40 words for meanings common to all of the words languages, Holman et al. (2008) were able to use a string-edit distance metric to recover linguistic similarity estimates that correlated highly with geographic distance and also several typological systems. This method is appealing for our purposes as it is relatively agnostic as to the processes of language contact and change that have produced modern-day languages and instead tracks the similarity of word forms themselves. The language distance measures produced by this method were

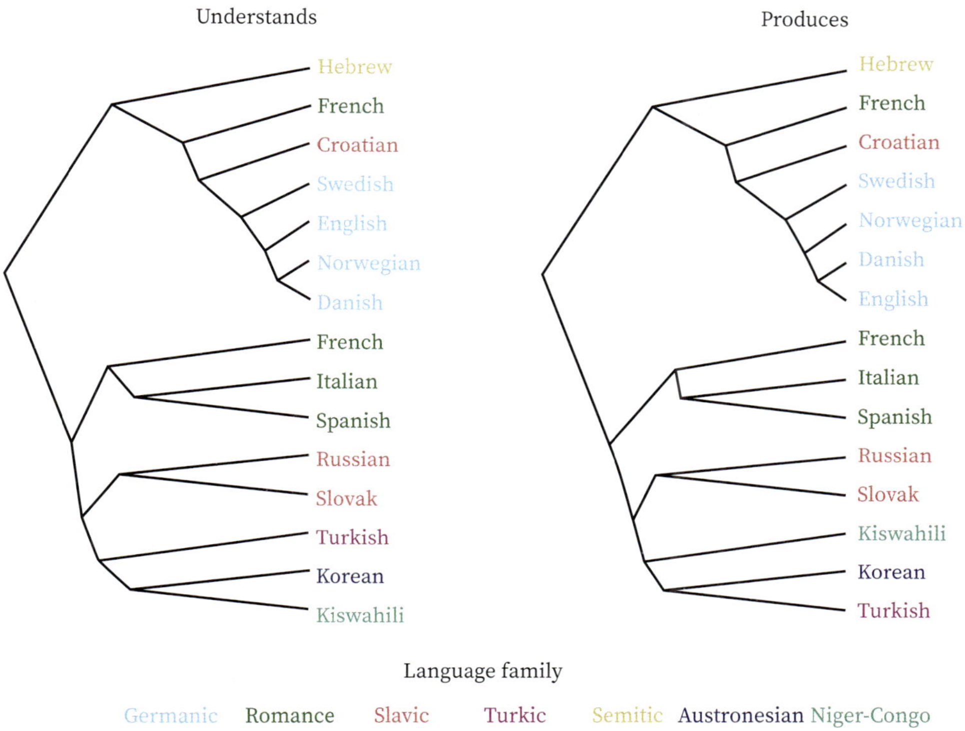

Figure 8.4
Dendrograms of the similarity in the ages of words' first production cross-linguistically.

highly correlated with pairwise correlations in acquisition trajectories for both production ($r = -0.44$, $p < 0.001$) and comprehension ($r = -0.41$, $p < 0.001$).

We also applied this same analysis to the words on the CDI themselves. For each language, we computed the average normalized Levenshtein (1966) distance between words for each of the 335 common words in our analyses.[2] This measure was even more highly correlated with pairwise acquisition trajectories than similarity computed using the 40 words identified by Holman et al. (2008), with relatively high correlations for both production ($r = -0.57$, $p < 0.001$) and comprehension ($r = -0.52$, $p < 0.001$).

Because this analysis likely overestimates the dissimilarity of languages written in different scripts—as every word receives a normalized Levenshtein distance of

2. Levenshtein distance is a measure of the minimum number of insertions, deletions, or substitutions required to transform one string into another. For instance, the distance between the Italian and Norwegian words for dog (*cane* and *hund*) is three. We computed this measure pairwise for all words, and then divided it by the number of characters in the longest word in order to get the edit distance per character (0.75 for *cane* and *hund*).

1 in this case—we replicated this analysis at the phonemic level. We used eSpeak to compute phonetic transcripts of each word and repeated the same analysis on distance between words' phonetic units in the International Phonetic Alphabet (IPA; Decker et al., 1999). These correlations between IPA distance and pairwise age of acquisition trajectories were again reliable, although slightly attenuated for both production ($r = -0.29$, $p = 0.01$) and comprehension ($r = -0.26$, $p = 0.02$). The robustness of these correlations across a variety of methods suggests that in addition to the high degree of general cross-linguistic similarities in the order of acquisition of words, the dissimilarities between them likely reflect differences in the wordforms of the target languages being learned.

Because the languages we studied here are far from a reliable, representative sample of the world's languages, the correlation between linguistic similarity and acquisition order similarity is hard to interpret definitively (Naroll, 1965). Languages in which word forms are similar are also likely to have similar cultural beliefs around parenting, similar household organization and incomes, and generally share other non-linguistic features in common. Nonetheless, these analyses suggest that in addition to early communicative need—which may be quite similar cross-linguistically—language and culture-specific features govern the order of acquisition. In the following section, we take on the the relationship between early communicative need and linguistic variability directly, asking whether acquisition orders are equally cross-linguistically similar over development, or whether they instead diverge or converge as children learn more words.

8.4 Consistency across Development

In the next analysis, we ask whether similarities in ages of acquisition are constant over the course of acquisition, or whether the similarity across languages changes over development. If variability in acquisition trajectories across languages reflects variability in those languages, we might expect that children's trajectories diverge over the course of language acquisition as the structure of their target language or their cultural milieu play a stronger role in guiding which words are easy or important to learn. Put more simply: our analyses of the first 10 words above shows striking similarity in the earliest words. Does this similarity decrease for the next 300 words?

In order to measure change in cross-linguistic consistency over development, we extend the age of acquisition-correlation approach we have used throughout this chapter. For each concept that appeared in at least 8 languages, we computed its average age of acquisition across all languages in whose CDIs it appeared in both comprehension and production. We then ordered these words from the

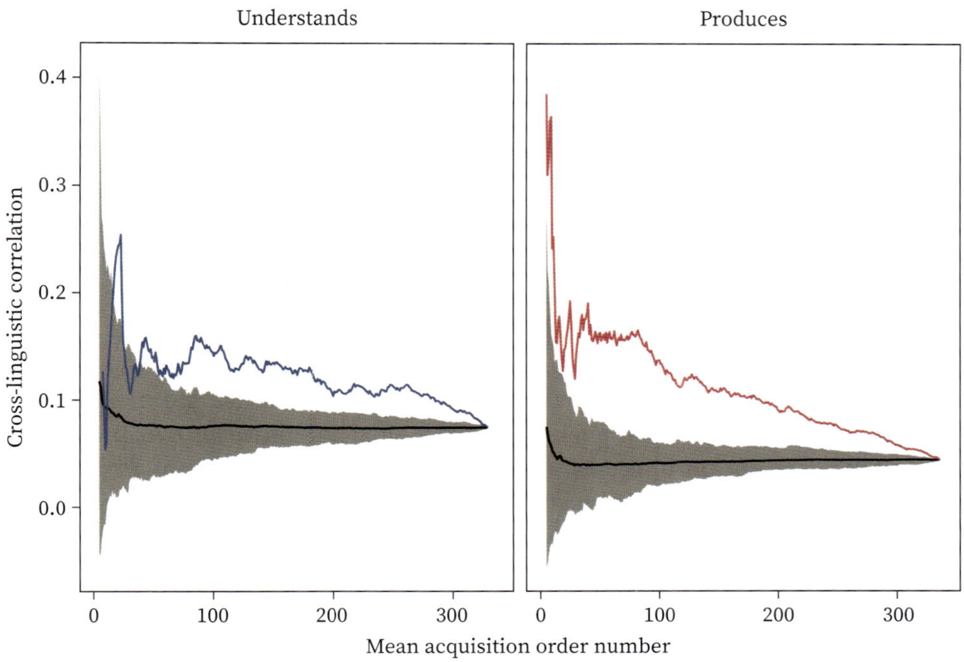

Figure 8.5
Cross-linguistic correlation ages of words' acquisition over the course of language development. Colored lines show empirical correlations, the gray area shows a 95 percent confidence interval for a randomly shuffled baseline. Especially in production, cross-linguistic similarity declines over the course of language development.

earliest learned word on average (*mommy*) to the latest learned word (*how*). We then computed the average cross-linguistic correlation in age of acquisition for the increasingly large sets of words starting with 5 words to 335 words. If the correlation increases over acquisition, we can infer that acquisition trajectories become more similar as more words are learned, that is, the hardest to learn words are learned more similarly across languages. In contrast, if the correlation decreases, we can infer that children start out learning similar concepts regardless of their native language, but that linguistic and cultural variability plays a greater role in the learning of later words.

Figure 8.5 shows these correlations for both comprehension and production over the course of acquisition. In addition, the gray shaded region shows a 95 percent confidence interval for a random baseline in which the concepts were ordered randomly, rather than in average acquisition order. This baseline is important to control for changes in measurement error that arise from changing numbers of concepts in the correlation. For both comprehension and production, the trajectories are reliably above the shuffled baseline. This trend is much more apparent for the earliest words in production, mirroring our qualitative sense from the analysis of the first 10 words above. Further, both trajectories clearly decrease over the course of acquisition.

These results confirm that there is substantially more similarity in the earliest learned words than in later learned words cross-linguistically, especially in production. This pattern of results is consistent with an account in which cross-linguistically shared communicative needs are a strong driver of the earliest acquired words. After these needs are met by the initial vocabulary, language-specific factors factors—variability in the forms, frequencies, and contexts of use for words—may play a larger role in the order of children's acquisition.

8.5 Conclusions

Children in all languages and culture learn language, but the languages they learn vary, and the cultures into which they are born may have quite different cultural practices around both language and cognitive development. Nonetheless, the order in which children learn the word for specific concepts in their language shows a substantial degree of cross-linguistic similarity. Further, dissimilarities are well-explained by measurable linguistic dissimilarity. This cross-linguistic similarity in concepts decreases over the course of acquisition. While the first ten words acquired in each language are highly consistent, later words are substantially more different.

As we noted in the introduction, the general observation of cross-linguistic similarity in early vocabulary has been taken as evidence for a wide variety of different theoretical claims. Our view is that these results indicate a shared core of concepts—for example, social routines, important people, and some early foods and household animals—that are perhaps especially important for communication independent of their linguistic realization.

We acknowledge, however, that there are likely many reasons for consistency of early words. One intriguing suggestion is that the phonological forms of words used with children actually evolve (or are adapted by parents) to be easier for children to say. One version of this hypothesis comes from Jakobson (1960), who hypothesized that parents adapt the word forms for *mother* and *father* to be easy for children to say or even to babble. Thus, the sound convergence across languages in the forms of words for these concepts (which is quite substantial) is due to convergence in what sounds are easy for children to say. This same mechanism could operate over other important early vocabulary as well, though note that this account already presupposes some notion of cognitive importance!

Regardless of the precise reason for this phenomenon, the similarity in early vocabulary is undeniable (ratifying suggestions by Clark, 1973, and others). As acquisition unfolds, however, the features that make languages (and cultures) different from one another play an ever increasing role in driving vocabulary development. In chapter 9, we explore demographic differences in acquisition that help to explain why two children learning the same language may acquire different words at different rates.

9 Demographic Variation in Individual Words

In chapter 6, we documented demographic differences in total vocabulary size. But where do these differences come from? Concretely, if girls say more words than boys, *which words* do they say more frequently? Is it the case that they are simply more likely to be producing each word with some relatively uniform probability, or are there individual words that they are much more likely to produce? In this chapter we consider the possibility that the probability of comprehending or producing individual words is related to demographic variables. We assess which words are learned differentially earlier or later by girls versus boys, by first-born versus later-born children, and by children with different levels of maternal education.

9.1 Methods

9.1.1 Data

As reviewed in chapter 3, subsets of the datasets in Wordbank are coded for one or more demographic variables. Here, as in previous analyses, we examine three: birth order, level of maternal education, and sex. For these analyses, we extract all of the instruments with demographically coded data and combine them into two datasets: comprehension from Words and Gestures (WG) forms and production across both WG and Words and Sentences (WS) forms (using the "by item stitching" approach described in appendix C). To avoid sparsity, we coded demographic variables into the values: First, Second, and Third+ for birth order; Below Secondary, Secondary, and College and Above for maternal education; and Female and Male for assigned sex at birth. This approach creates six different analyses, one for each

Authors' Note: The analyses below were presented at the 2019 Biennial Meeting of the Society for Research in Child Development; an earlier version of the sex analyses was presented at the Boston University Conference on Language Development in 2016.

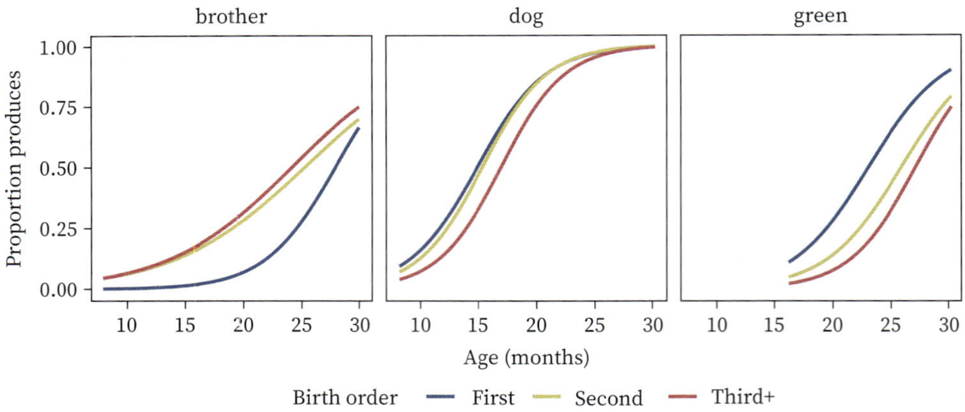

Figure 9.1
Developmental trajectories for the production of three example American English words by birth order.

combination of measure (comprehension/production) and demographic variable. We exclude a language from a given analysis if it has fewer than 50 children for each level of that demographic variable.

Each dataset yields a trajectory for each word, that is created by fitting a logistic curve to the proportion of children reported to understand or produce the word over age. These trajectories can be computed separately for each value of the demographic variable. For example, figure 9.1 illustrates the trajectories for some sample items in English for production data split by birth order. Note that the word *brother* is spoken much earlier by second-born and third-born children than by firstborn children while *green* is spoken much earlier by firstborn children. And the word *dog* is produced only slightly earlier by firstborn or second-born children than by later-born children. Averaging all of these trajectories together reproduces the demographic vocabulary curves reported in chapter 6.

9.1.2 Models

There are a number of complementary methods to estimate individual item effects of the type visualized in figure 9.1. In chapter 6, we explored relatively model-free approaches to estimating demographic effects across groups, mostly using descriptive statistics. Here we are interested in estimating these effects for individual items, and data are sparser for each individual item. Thus, estimating an independent statistic for each item would be noisier and more variable. In exploring this issue, we found that it was much more effective to reduce this variance by using a model in which demographic effects are estimated simultaneously at the level of all items and, specifically, for individual items.

In particular, in the analysis below we use mixed-effects logistic regression to predict whether children understand or produce each item based on their age and their level of a single demographic variable, with random effects of age and demographic variable by item. A model of this type is fit separately to the data for each language and demographic measure; for example, birth order is specified as:

$$\texttt{produces} \sim \texttt{age} + \texttt{birth\_order} + (\texttt{age} + \texttt{birth\_order} \mid \texttt{item}).$$

For each demographic variable, we specify the contrasts such that their coefficient compares each level of the variable to the previous level. For example, the coefficients for birth order reflect the overall difference between firstborn children as compared with second-born children and the overall difference between second-born children as compared with later-born children. Our key coefficients of interests are the fitted slopes for each demographic group and item. These indicate the size of the demographic differences for that item, over and above the main effect.

Since these effects are coefficients in a logistic regression, they represent the odds ratio between the demographic levels they are comparing. For example, the birth order model yields the odds ratio between firstborn and second-born children and the odds ratio between second-born children and later-born children. Thus, an effect of 1 indicates no difference (e.g., firstborn and second-born children are equally likely to know a word). An effect greater than 1 indicates an advantage for firstborn children (e.g., an effect of 2 would mean that the odds of firstborn children knowing a word are twice that of second-born children). Conversely, an effect lower than 1 indicates an advantage for second-born children (e.g., an effect of 0.5 would mean that the odds of firstborn children knowing a word are half that of second-born children).

9.2 Results

As discussed above, the primary targets of our analysis are the item random effects for each demographic variable, indicating our best estimate of the specific effect of a particular demographic on a particular item. The first question we address is the distribution of these random effects. Figure 9.2 shows the distribution of demographic random effects across all languages and measures, using a quantile-quantile (QQ) plot. Points on a diagonal line indicate conformity to the standard normal distribution while deviations suggest differences in distributional form.

The resulting plots show a broad, low-slope diagonal with skewed tails. The majority of coefficients are within a very tight range; 95 percent of the item effects

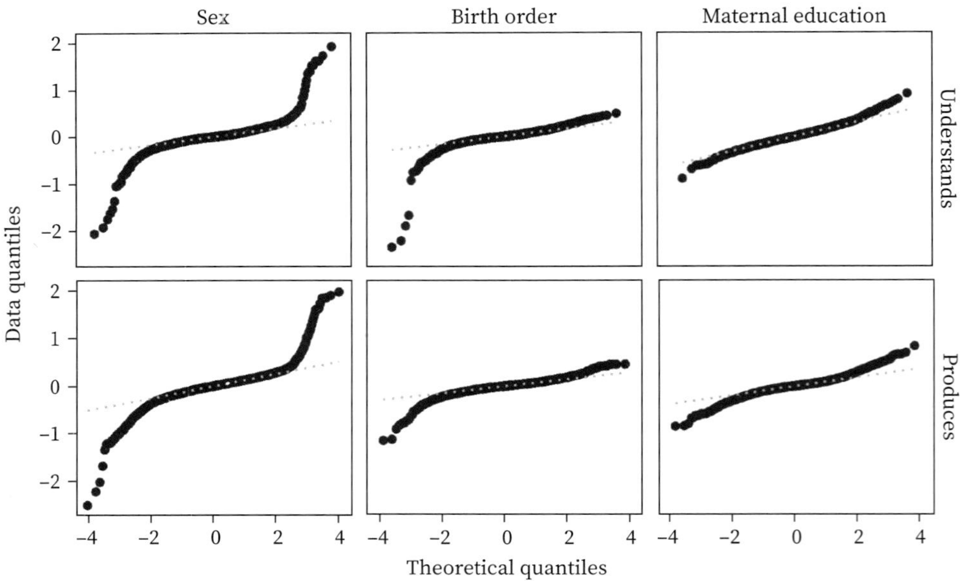

Figure 9.2
Quantiles of item random effects compared with theoretical quantiles of a normal distribution for both production and comprehension.

(across all languages, measures, and demographic factors) are within (0.74, 1.31), meaning that the odds for a child in one demographic group knowing the word are not more than 1.3 times higher than for another group for 95 percent of words. Most of the action is in the tails of the distribution: a few words vary much more in how often they are produced according to some demographic feature.

In the following subsections, we examine these coefficients. For each demographic factor, we first give the fixed effects of the demographic levels and then show the distribution of item random effects and the top 10 largest effects in each direction. It is these extreme items that we are most interested in.

9.2.1 Sex

As shown in chapter 6 and reflected here in figure 9.3, there is a highly consistent advantage for girls in language production. This advantage is slightly less pronounced for comprehension but is still present. Independently of this advantage, we also see specific items emerge as understood differentially for boys or girls.

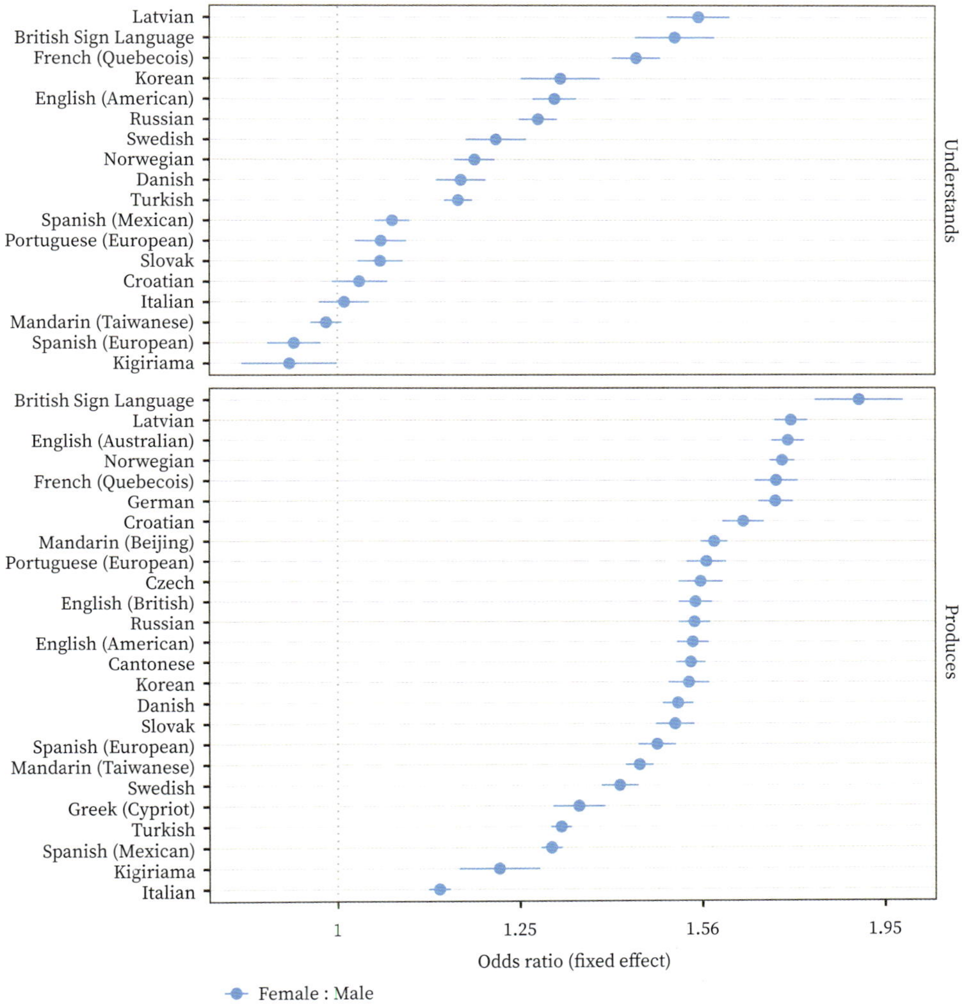

Figure 9.3
Main effect of sex for each language and measure.

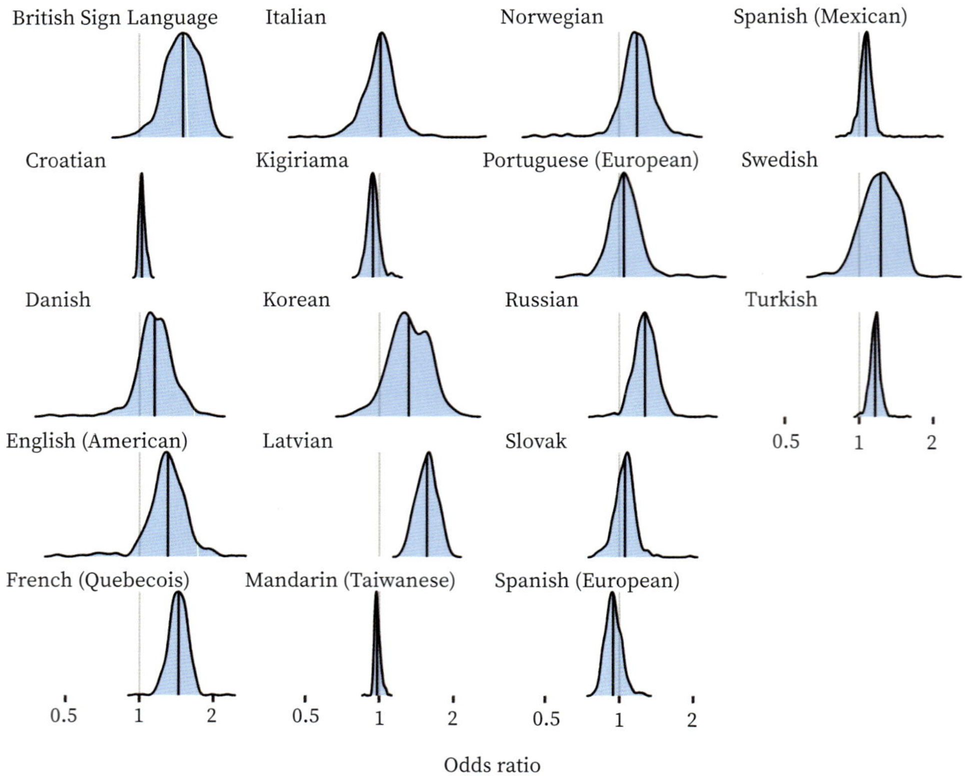

Figure 9.4
Distribution of sex item random effects for comprehension data in each language.

Figure 9.4 gives the full distribution of item effects for comprehension, and figure 9.5 shows the top 10 items most biased in each direction, in each language. These are almost exclusively traditionally gendered items of clothing and toys plus genital words. For English, for example, the words with a substantial male advantage are mostly vehicle- and tool-related while the female advantage words are mostly clothing-related. Thus, our first impression is that most of these tend to be specific content items associated with gendered play.

Figures 9.6 and 9.7 give the same measures for production. There are considerably more words per language with large sex biases (at least 1.5 times higher in either direction) for production (mean across languages is 2.62 percent) than for comprehension (mean across languages is 1.61 percent). The content of these sex-biased words is extremely similar across languages. For English, we again see the largest biases in each direction for genital terms, the largest male biases for vehicles and objects associated with traditionally male activities (e.g., sports),

British Sign Language

	Female > Male		Male > Female
1.4	hurry	1.7	peekaboo
1.4	pretty	1.6	daddy
1.3	zoo	1.6	no
1.3	before	1.6	toast
1.3	myself	1.5	toy
1.3	stuck	1.5	hello
1.3	coke	1.5	shower
1.3	tape	1.5	you
1.3	teacher	1.5	door
1.3	tights	1.5	light

Croatian

	Female > Male		Male > Female
1.1	pa-pa	1.1	brat
1.1	lopta	1.1	ga-ga
1.1	tata	1.1	proslava
1.1	nos	1.1	zoološki
1.1	mama	1.0	bazen
1.1	lutka	1.0	strašno
1.1	cipele	1.0	više
1.1	auto	1.0	crvena
1.1	knjiga	1.0	plava
1.1	baciti	1.0	sestra

Danish

	Female > Male		Male > Female
6.8	kjole	7.7	tissemand
5.0	dukke	2.7	lastbil
3.8	tissekone	2.6	hammer
1.7	halskæde	2.3	brandbil
1.6	baby	2.1	årnnn
1.6	sæbe	1.9	motorcykel
1.5	potte	1.7	kost
1.5	pige	1.7	tog
1.4	frakke	1.6	bil
1.4	knæ	1.5	flyvemaskine

English (American)

	Female > Male		Male > Female
4.6	dress	2.8	hammer
4.0	doll	2.8	truck
2.3	pretty	2.3	firetruck
2.0	necklace	2.2	broom
1.8	purse	2.0	shovel
1.6	when	2.0	motorcycle
1.6	baby	1.9	vroom
1.6	sweater	1.9	train
1.5	naughty	1.8	vacuum
1.5	girl	1.8	choo choo

French (Quebecois)

	Female > Male		Male > Female
1.6	poupée	1.5	camion
1.4	robe	1.4	vroum
1.2	poisson	1.3	balai
1.2	fromage	1.2	ballon
1.2	papillon	1.2	auto
1.2	(je) veux	1.2	fatigué
1.2	fille	1.2	bonjour
1.2	parler	1.2	train
1.2	quoi	1.2	tchou tchou
1.2	ton	1.2	bobo

Italian

	Female > Male		Male > Female
3.0	organo genitale femminile	3.9	organo genitale maschile
2.7	bambola	2.1	camion
1.7	gonna	1.8	motocicletta
1.6	cullare	1.6	automobile
1.6	ombelico	1.5	bicicletta
1.5	vestito	1.5	lavatrice
1.5	collana	1.5	martello
1.4	asciutto	1.4	treno
1.4	pettinare	1.4	coperchio
1.4	farfalla	1.4	scopa

Kigiriama

	Female > Male		Male > Female
1.3	Chakurya	1.2	Mupira
1.2	Madzi	1.1	Haha
1.2	Dzina ra ye mwana mwenye	1.1	Piga teke
1.2	Dzina ra murezi	1.1	Changu
1.2	Hombo/Nyonyo/Kopo ra mwana	1.1	haryahu
1.2	Lumira	1.1	Mimi
1.1	Mwana mutsanga	1.1	Kwenda
1.1	Kuha/ shiba	1.1	Lazha
1.1	Muhoho/ Dede	1.1	Agwa
1.1	Mudzini/ Nyumbani	1.1	Mwaga

Korean

	Female > Male		Male > Female
5.0	언니	6.8	형
4.5	오빠	5.7	누나
1.7	나비	5.0	고추
1.5	인형	1.7	총
1.5	예뻐	1.6	무/무우
1.5	시계	1.6	공
1.5	단추	1.5	밤
1.4	반지	1.5	차/자동차
1.4	치즈	1.5	다
1.4	꽥꽥	1.4	더

Latvian

	Female > Male		Male > Female
1.3	ļaudis	1.3	bumba
1.2	zils	1.3	suns
1.2	melns	1.2	kurpes
1.2	sarkans	1.2	mašīna
1.2	zooloģiskais dārzs	1.2	kaķis
1.2	noguris	1.2	dzert
1.2	rīts	1.2	maize
1.2	nakts	1.2	Atā!/Uz redzēšanos!
1.2	brālis	1.2	vau-vau
1.2	laimīgs	1.2	telefons

Mandarin (Taiwanese)

	Female > Male		Male > Female
1.1	抱(抱)	1.1	小ㄐㄐ/小鳥(生殖器)
1.1	牛奶(ㄋㄟㄋㄟ)	1.0	鍋子
1.1	謝謝	1.0	電鍋
1.1	爸爸/爹地	1.0	醫院(診所)
1.1	媽媽/媽咪	1.0	玩
1.1	小狗(汪汪)	1.0	公園
1.1	水(茶茶)	1.0	XX園(某遊樂園名稱)
1.1	吃	1.0	船
1.1	再見(byebye)	1.0	7-11(任何商店名稱)
1.1	弟弟	1.0	電梯

Figure 9.5
Top 10 most sex-biased words in each language for comprehension data.

Norwegian

Female > Male		Male > Female	
5.6	kjole	2.6	lastebil
3.9	dukke	2.6	brannbil
1.7	halskjede	2.2	hammer
1.6	søt	2.2	brrr
1.5	pen	2.1	motorsykkel
1.5	baby	1.9	tiss
1.5	børste	1.9	buss
1.5	fin	1.9	tog
1.4	mage	1.7	fly
1.4	navle	1.6	støvsuger

Portuguese (European)

Female > Male		Male > Female	
3.3	Boneca	4.6	Pilinha / Pila / Pilau
2.3	Vestido	1.7	Golo!
1.8	Pipi	1.6	Vrum-vrum
1.8	Beijinho!	1.6	Carro
1.6	Adeus / Xau!	1.5	Moto/a
1.6	Saia	1.5	Tinoni
1.5	Água	1.4	Comboio
1.4	Olá!	1.3	Todo(a) / todos(as)
1.4	Mãe	1.3	Pumba! / Pum!
1.4	Pai	1.3	Sol

Russian

Female > Male		Male > Female	
2.5	платье	1.6	бpp/ввв и т.п.
1.8	кукла	1.4	машина
1.4	туфли	1.4	молоток
1.3	красивый	1.4	штаны
1.3	кенгуру	1.3	дверь
1.3	мокро	1.3	трактор
1.3	яйцо	1.3	грузовик
1.2	баран/овечка	1.2	мяч
1.2	камень	1.2	холодильник
1.2	платок	1.2	залезть

Slovak

Female > Male		Male > Female	
1.8	bábika	1.3	motorka
1.4	nos	1.3	traktor
1.3	oko	1.3	vlak
1.3	lienka	1.3	garáž
1.3	moja	1.3	autobus
1.2	ústa / pusa	1.3	kostol
1.2	ucho	1.2	vú
1.2	klobúk	1.2	šiši / ššš
1.2	vlasy	1.2	Brm / tú; tí
1.2	pupok	1.2	metla

Spanish (European)

Female > Male		Male > Female	
1.3	nombre propio*	1.2	bici
1.3	cinco lobitos	1.2	mar
1.3	muñeca	1.2	fútbol
1.3	nombre del niño/a	1.2	avión
1.3	shh	1.2	gol
1.2	nena/niña	1.1	lavadora
1.2	mamá/mami	1.1	cerrar
1.2	papá/papi	1.1	camión
1.2	no	1.1	fuera/afuera
1.2	aúpa	1.1	apagar

Spanish (Mexican)

Female > Male		Male > Female	
2.0	muñeca	1.3	guapo
1.8	vestido	1.2	avión
1.4	bonita	1.2	dónde está
1.2	osito	1.2	tuyo
1.2	tortillitas	1.2	pantalón
1.2	collar	1.2	orejas
1.1	sopa	1.2	camión/troca
1.1	araña	1.1	doler
1.1	chupón/chupete	1.1	carro/coche
1.1	aretes	1.1	¡pum!

Swedish

Female > Male		Male > Female	
1.9	docka	1.7	bil
1.5	syster	1.7	lastbil
1.4	åsna	1.7	traktor
1.3	trädgård	1.6	köra
1.3	apa	1.5	hammare
1.3	söt	1.5	telefon
1.3	bort	1.4	öppna
1.3	elefant	1.4	badkar
1.3	halsband	1.4	smörgås
1.3	kalas	1.4	motorcykel

Turkish

Female > Male		Male > Female	
1.4	Bebek	1.2	Cee
1.2	Elbise	1.2	Gol
1.2	Tara	1.2	Kamyon
1.1	Bebek	1.1	Düt
1.1	Alkış	1.1	Bisiklet
1.1	El salla	1.1	Anne
1.1	Tarak	1.1	Baba
1.1	Yedir	1.1	Müzik
1.1	Saç	1.1	Muhallebi
1.1	Yastık	1.1	Tişört

Figure 9.5
(continued)

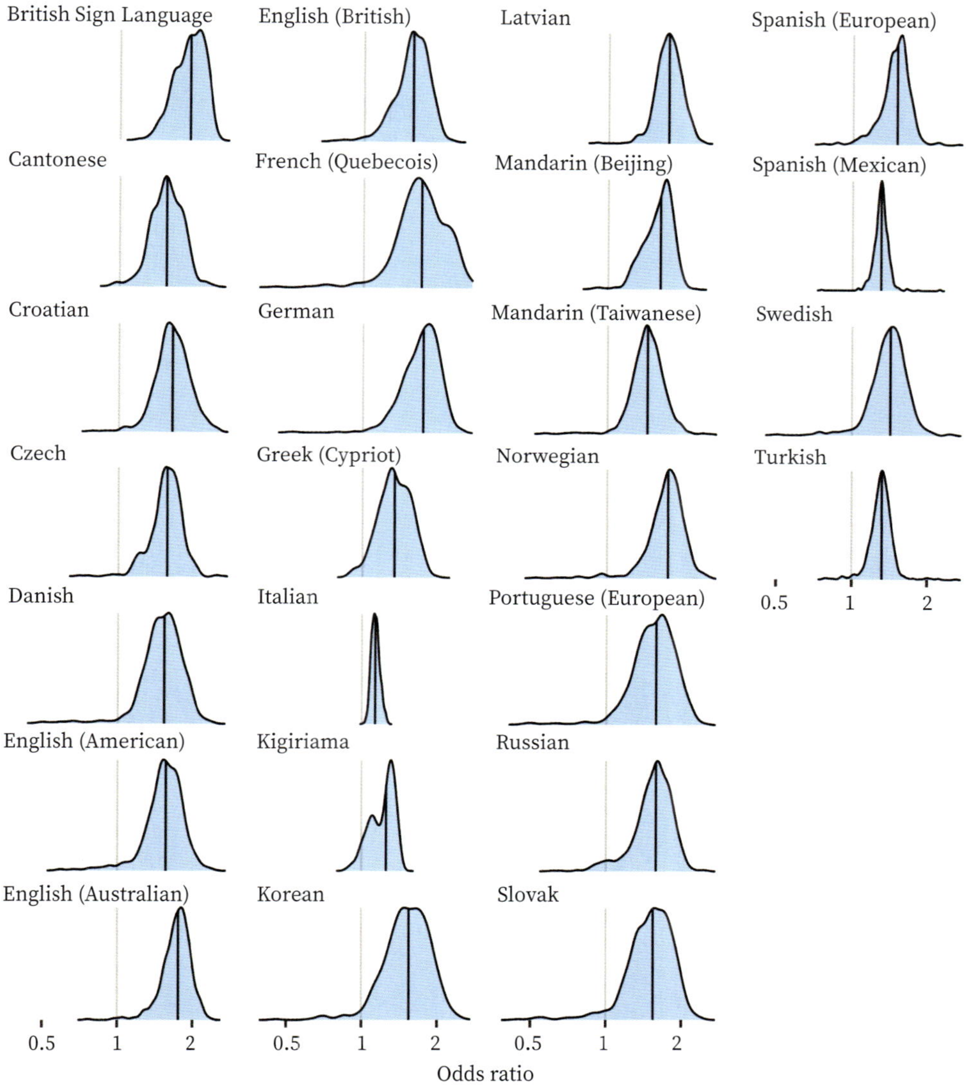

Figure 9.6
Distribution of sex item random effects for production data in each language.

British Sign Language

Female > Male		Male > Female	
1.3	zoo	1.6	peekaboo
1.3	room	1.5	hello
1.3	which	1.5	shower
1.3	pretty	1.5	mummy
1.2	wait	1.4	bye
1.2	garage	1.4	light
1.2	hamburger	1.4	you
1.2	will	1.4	eat
1.2	coke	1.4	tea
1.2	tonight	1.4	more

Croatian

Female > Male		Male > Female	
1.8	haljina	2.1	brr-brr
1.5	suknja	1.9	grr
1.5	zub	1.8	Ga-ga
1.5	beba	1.6	njemu/mu
1.5	papir	1.6	ku-ku
1.5	tanjur	1.6	penis
1.5	kiša	1.6	ono
1.5	meso	1.6	o
1.4	usta	1.5	ga-ga
1.4	jastuk	1.5	to/taj/ta

Danish

Female > Male		Male > Female	
6.4	kjole	3.1	årnnn
5.7	strømpebukser	3.0	tissemand
3.1	dukke	2.8	brandbil
2.6	tissekone	2.6	brandmand
2.4	halskæde	2.5	helikopter
1.9	trusser	2.4	politi
1.6	barnets eget navn	2.3	hammer
1.5	fin	2.3	traktor
1.5	navle	2.2	(benzin)tanken
1.5	baby	2.1	græsslåmaskine

English (Australian)

Female > Male		Male > Female	
2.5	dress	2.3	vroom/broom
2.0	doll	1.8	tractor
1.6	necklace	1.8	fire truck/engine
1.6	pretty	1.7	bat
1.4	hair	1.7	truck
1.3	butterfly	1.6	lawn mower
1.3	arm	1.6	choo choo
1.3	dry	1.5	ladder
1.3	thirsty	1.5	fireman
1.3	teddy bear	1.5	motorbike

French (Quebecois)

Female > Male		Male > Female	
4.0	robe	3.8	pénis
3.0	collants	2.9	vroum
2.8	poupée	2.7	tracteur
2.8	vagin vulve	2.5	marteau
1.8	collier	2.4	hélicoptère
1.6	fille	2.4	camion de pompier
1.6	joue	2.4	pompier
1.6	crayon	2.3	police
1.6	parce que	2.2	clou
1.6	pantalons	2.0	tondeuse

Cantonese

Female > Male		Male > Female	
1.9	裙	1.7	Bi Boo
1.5	BB	1.6	Bi Boo車
1.5	靚	1.6	貨車
1.5	拜拜/再見	1.6	車轆
1.5	麵包/包包	1.6	電車/叮叮
1.5	爸爸/爹哋	1.6	直升機
1.4	波	1.6	Mem Mem
1.4	屁股/Pek Pek	1.5	救傷車
1.4	媽媽/媽咪	1.5	Hello/Hi
1.4	俾	1.5	Bom Bang

Czech

Female > Male		Male > Female	
2.1	pipinka	2.2	vrn vrn
1.8	kalhoty	2.1	pindík
1.7	šaty	1.7	šššš hů
1.6	sukně	1.6	bagr
1.6	panenka	1.6	hasiči
1.4	umyvadlo	1.6	tik tak
1.3	bunda	1.5	ee [záchod]
1.3	večer	1.4	crrr
1.3	náušnice	1.4	íí áá
1.3	zmrzlina	1.4	náklaďák

English (American)

Female > Male		Male > Female	
4.6	dress	9.3	penis*
4.3	vagina*	2.6	vroom
4.1	tights	2.5	tractor
2.9	doll	2.3	truck
1.9	necklace	2.3	hammer
1.8	pretty	2.1	bat
1.8	underpants	2.0	dump
1.6	purse	2.0	firetruck
1.5	girl	2.0	police
1.5	sweater	1.9	motorcycle

English (British)

Female > Male		Male > Female	
2.2	doll	2.1	tyre
1.8	dress	1.9	vroom
1.5	eye	1.9	cowboy
1.4	slide	1.8	hose/hose-pipe
1.4	please	1.7	blade
1.4	baby	1.6	choo choo
1.3	draw	1.6	lorry / truck
1.3	nappy	1.6	microscope
1.3	pretty	1.5	fire engine
1.3	purse	1.5	ladder

German

Female > Male		Male > Female	
2.9	Kleid	3.3	Penis
1.9	Puppe	2.2	Bagger
1.8	Puppenwagen	1.9	Müllabfuhr
1.7	Scheide	1.8	Feuerwehrmann
1.5	schlafen	1.6	Hammer
1.5	malen	1.6	an
1.4	klatschen	1.6	aus
1.4	Handtuch	1.6	dies
1.4	kleckern	1.6	Feuerwehr
1.4	Strumpfhose	1.5	Trecker

Figure 9.7
Top 10 most sex-biased words in each language for production data.

Greek (Cypriot)

Female > Male		Male > Female	
1.5	μαμά*	1.5	δέκατος
1.4	μπάλα	1.5	έβδομος
1.4	γάλα	1.5	ένατος
1.4	γιαγιά	1.5	είντα
1.4	νερό	1.5	σαγιονάρες
1.3	παππούς	1.5	μυρμήγκι
1.3	μήλο	1.5	όγδοος
1.3	αυτοκίνητο	1.5	πέμπτος
1.3	αυτιά	1.5	σεσουάρ
1.3	μαλλιά	1.5	τέταρτος

Italian

Female > Male		Male > Female	
1.1	ciao	1.1	brum brum
1.1	mano	1.1	panettiere
1.1	mamma	1.1	barbiere
1.1	papà	1.1	jeep
1.1	nonna	1.1	soldato
1.1	naso	1.1	disubbidiente
1.1	capelli	1.1	vi
1.1	acqua	1.1	vostro/a
1.1	bocca	1.1	portiere
1.1	mio/a	1.1	babysitter/tata

Kigiriama

Female > Male		Male > Female	
1.2	Hehi na	1.4	Chakurya
1.2	Pia	1.4	Dzina ra ye mwana mwen
1.2	Angalau	1.4	Kelesi
1.2	Baadaye	1.4	Loha
1.2	Musichana	1.4	Ndugu
1.2	Bomani	1.3	Ima
1.2	Gazhirana	1.3	Tsukula
1.2	Gula	1.3	Gesa
1.2	Wao	1.3	Rira
1.2	Mukokoteni	1.3	Tsukirwa

Korean

Female > Male		Male > Female	
7.3	오빠	12.4	형
6.5	언니	3.4	고추
3.6	치마	2.5	아들
2.3	공주	2.3	총
2.2	인형	2.2	주유소
2.1	팬티	2.2	소방수
2.0	딸	2.2	트럭
1.8	아가	2.2	불자동차/소방차
1.7	예뻐	2.2	누나
1.6	손가락	2.0	(총을)쏴

Latvian

Female > Male		Male > Female	
1.7	kleita	1.9	br/brm/brum
1.6	svārki	1.5	motocikls
1.4	meita	1.4	āmurs
1.4	ikviens	1.4	traktors
1.3	mētelis	1.4	ņam-ņam
1.3	jauks/jauki	1.4	māte*
1.3	krelles	1.4	uz redzēšanos/atā
1.3	tādēļ	1.3	paldies
1.3	vai	1.3	nē
1.3	lelle	1.3	vau-vau

Mandarin (Beijing)

Female > Male		Male > Female	
1.7	裙子	1.8	吉普车
1.4	爸爸	1.8	老吊车
1.3	裤衩	1.7	卡车
1.3	躺	1.7	梆梆（打枪）
1.3	妈妈	1.6	救护车
1.3	再见/Byebye	1.5	小轿车
1.3	说	1.5	夹克
1.3	手绢	1.5	匹
1.3	拖鞋	1.4	粘糊
1.3	我的	1.4	棋子

Mandarin (Taiwanese)

Female > Male		Male > Female	
2.9	裙子	5.4	小ㄐㄐ/小鳥(生殖器)
1.8	內褲(三角褲)	2.6	吊車
1.7	娃娃(洋娃娃)	2.1	手槍
1.6	項鍊	2.0	怪手(挖土機)
1.5	好漂亮/漂亮的	1.9	球棒
1.5	梳子	1.9	卡車
1.5	頭髮	1.8	消防車
1.4	扣子	1.7	直昇機
1.4	鞋子	1.7	救護車
1.4	袋子	1.6	機器人

Norwegian

Female > Male		Male > Female	
5.1	kjole	3.3	brrr
3.3	dukke	2.7	brannbil
2.0	halskjede	2.5	brannmann
1.9	veske	2.3	traktor
1.5	børste	2.2	hammer
1.5	barnets navn	2.2	politi
1.4	søt	2.1	motorsykkel
1.4	strømpebukse	1.9	helikopter
1.4	knapper	1.9	lastebil
1.4	jente	1.9	tog

Portuguese (European)

Female > Male		Male > Female	
6.8	Vestido	7.6	Pilinha/Pila/Pilau
6.4	Totós	3.3	Helicóptero
5.2	Gancho	3.1	Trator
5.0	Saia	3.1	Camião
5.0	Pipi	2.8	Vrum-vrum
3.1	Boneca	2.5	Ambulância
2.3	Fita	2.4	Tinoni
2.0	Colar / Fio	2.2	Golo!
1.9	Fato de banho	2.0	Polícia
1.8	Sandálias	2.0	Moto/a

Russian

Female > Male		Male > Female	
3.3	платье	2.5	бpp/ввв и т.п.
2.1	дочка	2.3	пожарная машина
1.8	кукла	2.0	грузовик
1.8	туфли	2.0	гараж
1.6	мама	1.8	ту-ту/чух-чух
1.6	коляска	1.8	фу
1.5	бусы	1.8	тик-так
1.5	гулять	1.8	сын
1.4	кофта	1.7	биби
1.4	платок	1.7	трактор

Figure 9.7
(continued)

Slovak

Female > Male		Male > Female	
2.6	sukňa	3.4	formula
2.4	bábika	3.0	džíp
2.4	šaty	2.9	typy áut
2.2	cop / vrkoč	2.8	mercedes
2.0	mašľa / gumička	2.8	miešačka
1.6	papučky	2.6	bager
1.5	sedieť	2.6	vlečka
1.5	ponožky	2.3	lokomotíva / rušeň
1.5	moje	2.2	žeriav
1.5	vlastné mená blízkych osôb	2.1	kamión

Spanish (European)

Female > Male		Male > Female	
1.8	muñeca	1.9	brumbrum
1.8	vestido	1.7	tractor
1.5	pendientes	1.7	gol
1.5	labios	1.7	grúa
1.5	cocinitas	1.5	camión
1.5	pulsera	1.5	iiiiaaaa
1.4	falda	1.5	auuuu
1.3	tripa	1.5	tata
1.3	cuello	1.4	¡pumba!
1.3	estar	1.4	coche policía

Spanish (Mexican)

Female > Male		Male > Female	
1.7	muñeca	1.7	pene
1.7	vestido	1.2	bat
1.4	vagina	1.2	avión
1.3	falda	1.2	camión/troca
1.3	medias	1.2	helicóptero
1.3	mía	1.2	moto
1.3	aretes	1.2	piedra
1.2	collar	1.2	camión de bombero
1.2	bolsa	1.2	carro/coche
1.2	bebé	1.2	tren

Swedish

Female > Male		Male > Female	
3.7	klänning	2.8	penis
2.2	trosor	2.1	hammare
2.2	vagina	2.0	motorcykel
1.8	kjol	1.9	såg
1.7	docka	1.9	skruvmejsel
1.7	nattlinne	1.9	traktor
1.5	söt	1.9	brandbil
1.5	tights	1.9	lastbil
1.5	barnets eget namn	1.7	helikopter
1.4	vakna	1.7	brr

Turkish

Female > Male		Male > Female	
2.1	Etek	1.7	İtfaiye
2.0	Toka	1.7	Motosiklet
1.7	Taç	1.6	Vınn
1.7	Elbise	1.6	Helikopter
1.6	Küpe	1.5	Kamyon
1.6	Bebek	1.4	Traktör
1.5	Tayt	1.4	Ambulans
1.5	Bilezik	1.4	Tüfek
1.5	Prenses	1.4	Çekiç
1.4	Bebek	1.4	Robot

Figure 9.7
(continued)

and the largest female biases for female-coded clothing and toys. This pattern is replicated quite robustly across languages.

In sum, there appear to be two different processes at work in the sex effects we observe. The first is a general shift in the probability that *any* word will be produced or understood such that girls are slightly more likely to produce or understand it than boys. The average magnitude of this fixed effect across languages is 1.2 for comprehension and 1.5 for production. In other words, if a male child had a 50 percent chance of saying a word (odds 1:1), a female child would on average have 1.5 times higher odds of saying it—that is, a 60 percent chance. However, beyond this fixed effect, there are also variable effects for individual words. Most of these effects are small, but a few of them are quite large. For example, if an English-speaking male child has a 50 percent chance of saying the word *dress*, a female child would have a 82 percent chance; if a female child has a 50 percent chance of saying the word *hammer*, a male child would have a 70 percent chance.

9.2.2 Birth Order

We next consider individual items that are more or less likely in the vocabularies of firstborn versus later-born children. Here we consider both the contrast between second-born and firstborn children and between later-born and second-born children. As shown in figure 9.8, across languages, second-born children are advantaged over later-born children in both comprehension and production, while firstborn children are advantaged above second-born children in production. Somewhat surprisingly, across languages the reverse is true of comprehension—later-born children have somewhat bigger comprehension vocabulary. We can only speculate as to the source of this pattern, especially since it is not present in the production data. The number of languages for which we have birth order data is small, so conclusions are somewhat tentative.

Figures 9.9 and 9.10 again represent random effects coefficients for particular items in comprehension. In general, there are few surprises here: the words for *brother* and *sister* are much more likely to be understood by second-born children and even more likely for later-born children.[1] Additionally, several languages show a few other words that second-born and later-born children might be more likely to be exposed to via their siblings, such as *school* in English and Norwegian (*skole*).

The same general patterns are present in the production data (figures 9.11 and 9.12), with additional evidence that having older siblings appears to be related to exposure to sweets, at least in some cultures: *popsicle*, *donut*, and *candy* all appear in the English data and *tyggegummi* (gum) and several soda- and candy-related

1. Of course, children could be marked as understanding a term like *brother* even without a true understanding of its relational structure!

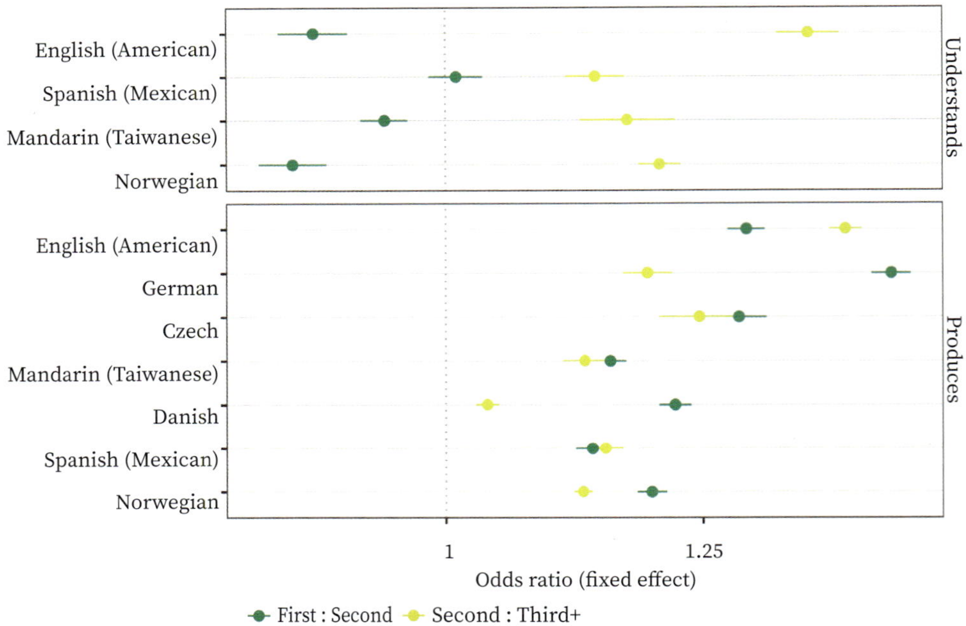

Figure 9.8
Main effect of birth order for each language and measure.

words appear in the Norwegian data. *Hate* also appears in the English data, suggesting some emotional expression due to having a sibling. We interpret this pattern with caution, however, as birth order is likely partially confounded with socioeconomic status (SES), in that families from lower SES populations tend to have more children (Huber et al., 2010). So, later-born children might also be more likely to come from low-SES families, who have more environmental exposure to "junk foods" like soda and candy (Ghosh-Dastidar et al., 2014; see below).

In sum, across languages, a given word has lower odds of being understood by a firstborn child than by a second-born child (by a factor of 0.9) and higher odds of being understood by a second-born child than a later-born child (by a factor of 1.2). It also has higher odds of being produced by both first-born compared to second-born (by a factor of 1.3) and second-born compared to later-born (by a factor of 1.3). Additionally, a handful of individual items show some substantial differences by birth order: in American English, a firstborn child having a 50 percent chance of saying the word *green* corresponds to a second-born child having a 61 percent chance of saying it; conversely, a second-born child having a 50 percent chance of saying the word *brother* corresponds to a firstborn child having a 76 percent chance of saying it.

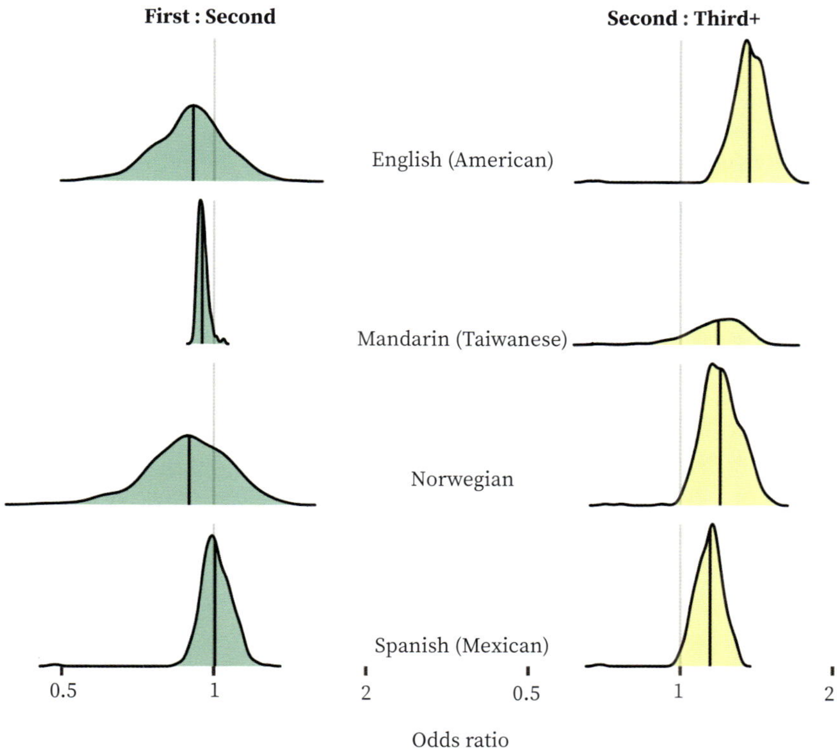

Figure 9.9
Distribution of birth order item random effects for comprehension data in each language.

A specific claim has been made in the literature regarding effects of birth order on language development—Oshima-Takane et al. (1996) reported that second-born children learn second-person personal pronouns (e.g., *you*) earlier, likely due to the disambiguating effect of having siblings addressed using such pronouns in overheard speech. We examined this pattern in our own American English data in figure 9.13 but did not find global support for it. The major trend that emerged was a smaller firstborn advantage for *me* and *mine* than average (and smaller than for *my*, *you*, and *your*). Prima facie, this finding is less consistent with a global increase in second-person disambiguation—which would have predicted a reversal of the global pattern for these terms—than with later-born children needing to assert "mine" to declare ownership.

Examining the data across languages in figure 9.14, we see a similar pattern. *Me*, *my*, and *mine* are all learned earlier for second-born children than firstborn. In contrast, *you* and *your*, while numerically positive, have confidence intervals that overlap with zero. These data thus support the hypothesis that first-person

English (American)
Later born biased

Second > First		Third+ > Second	
6.6	brother	2.1	brother
5.2	sister	2.0	sister
1.6	cry	1.2	smile
1.6	hit	1.2	school
1.5	backyard	1.2	babysitter
1.5	candy	1.2	some
1.5	doll	1.2	backyard
1.4	school	1.2	play pen
1.4	eat	1.2	pool
1.4	swing	1.2	pattycake

English (American)
Earlier born biased

First > Second		Second > Third+	
1.6	bunny	1.2	nose
1.5	blue	1.2	shoe
1.5	duck	1.2	duck
1.4	clock	1.2	eye
1.4	nose	1.2	toothbrush
1.4	read	1.2	cow
1.4	careful	1.2	ear
1.4	people	1.2	read
1.4	bear	1.2	airplane
1.4	cow	1.2	brush

Mandarin (Taiwanese)
Later born biased

Second > First		Third+ > Second	
1.1	XX園(某遊樂園名稱)	1.7	牛奶(ㄋㄟㄋㄟ)
1.0	電鍋	1.7	媽媽/媽咪
1.0	鍋子	1.7	抱(抱)
1.0	乾淨的	1.6	爸爸/爹地
1.0	汽水	1.5	再見(byebye)
1.0	翹翹板	1.4	吃
1.0	他	1.4	拍(手)
1.0	船	1.4	自己的名字
1.0	裙子	1.4	水(茶茶)
1.0	鹿	1.4	拍拍手

Mandarin (Taiwanese)
Earlier born biased

First > Second		Second > Third+	
1.1	牛奶(ㄋㄟㄋㄟ)	1.3	XX園(某遊樂園名稱)
1.1	媽媽/媽咪	1.3	電鍋
1.1	抱(抱)	1.2	鍋子
1.1	爸爸/爹地	1.2	乾淨的
1.1	再見(byebye)	1.2	翹翹板
1.1	拍(手)	1.2	汽水
1.1	吃	1.2	船
1.1	自己的名字	1.2	他
1.1	拍拍手	1.2	裙子
1.1	水(茶茶)	1.2	鹿

Norwegian
Later born biased

Second > First		Third+ > Second	
10.3	søster	1.7	bror
9.0	bror	1.6	søster
2.5	skole	1.3	skole
1.9	sukkertøy	1.2	pappa
1.8	corn flakes	1.2	legge
1.7	dytte	1.2	vondt
1.7	is	1.2	hysj
1.6	frokostblanding	1.2	munn
1.5	dyrehage	1.2	ha det
1.5	redd	1.2	fot

Norwegian
Earlier born biased

First > Second		Second > Third+	
1.6	voff voff	1.3	oldefar
1.5	hund	1.3	blå
1.5	der	1.2	rød
1.5	bok	1.2	tog
1.5	bilde	1.2	i morgen
1.4	Borte!	1.2	ugle
1.4	klokke	1.2	skilpadde
1.4	senere	1.2	medisin
1.4	mamma	1.2	oldemor
1.4	nei	1.2	hvorfor

Spanish (Mexican)
Later born biased

Second > First		Third+ > Second	
2.1	hermana	1.7	hermana
2.1	hermano	1.6	hermano
1.2	estar	1.2	papá/papi
1.2	bicicleta	1.2	cama
1.2	bacinica	1.1	cosquillitas
1.1	tijeras	1.1	pan
1.1	trapo	1.1	hacer la meme
1.1	mosca	1.1	estar
1.1	columpio	1.1	ojitos
1.1	zoológico	1.1	¡am!

Spanish (Mexican)
Earlier born biased

First > Second		Second > Third+	
1.3	gracias	1.1	diferente
1.2	espejo	1.1	hipopótamo
1.2	fotos	1.1	verde
1.2	abuela	1.1	difícil
1.2	alla/allí	1.1	tortuga
1.2	ahí	1.1	búho
1.2	¡ay!	1.1	azul
1.2	sacar	1.1	mono
1.2	zapato	1.1	rana
1.1	tía	1.1	trabajar

Figure 9.10
Top 10 most birth-order-biased words in each language for comprehension data.

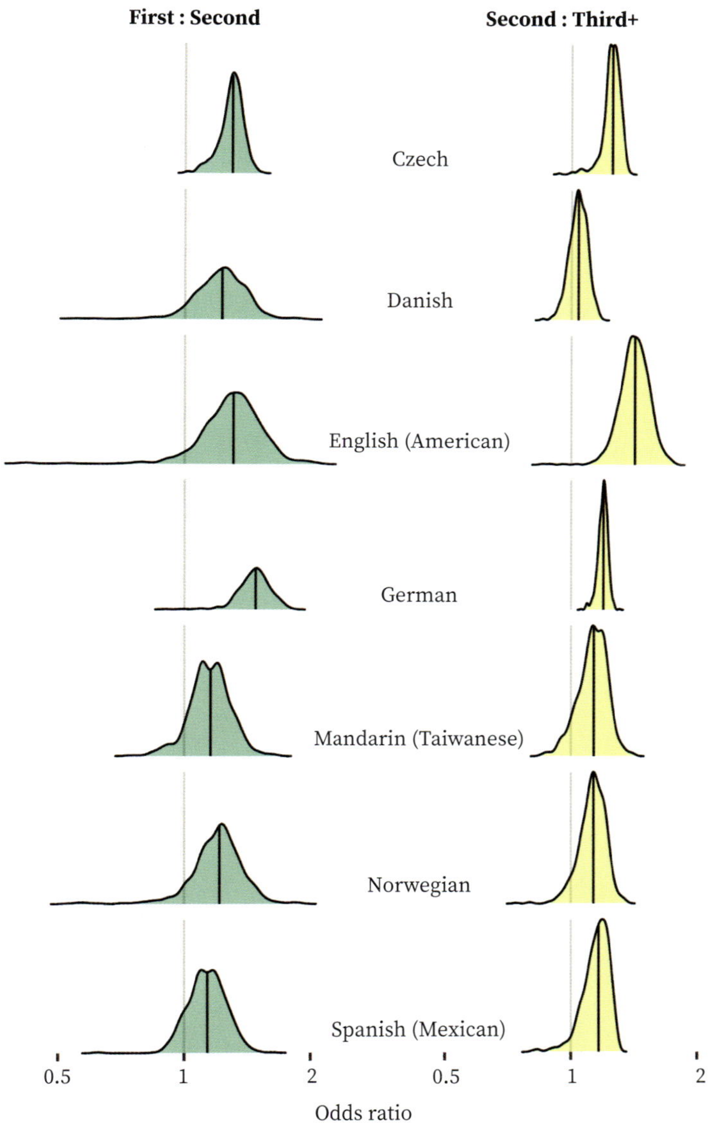

Figure 9.11
Distribution of birth order item random effects for production data in each language.

Czech
Later born biased

Second > First		Third+ > Second	
1.3	au	1.3	vrn vrn
1.3	máma	1.3	ee [záchod]
1.3	táta	1.2	tak
1.2	jméno jiné osoby	1.2	cákycáky
1.2	bác	1.2	ham
1.2	ham	1.2	tik tak
1.2	pití	1.2	crrr
1.2	haf	1.2	pšt!
1.2	vrn vrn	1.2	mňam
1.2	ahoj!	1.2	hači

Czech
Earlier born biased

First > Second		Second > Third+	
1.2	vaše	1.1	stůl
1.2	vy	1.1	noha
1.2	oni	1.1	nůž
1.1	aby	1.1	zmrzlina
1.1	černý	1.1	červený
1.1	včera	1.1	nůžky
1.1	vzadu	1.1	špinavý
1.1	ona	1.1	vajíčko
1.1	začít	1.1	břicho
1.1	den	1.1	zelený

Danish
Later born biased

Second > First		Third+ > Second	
2.2	skole	1.2	skole
2.0	bror	1.2	bror
1.7	søster	1.2	kinderæg
1.6	tyggegummi	1.2	biografen
1.5	spil	1.2	mælkesnitte
1.5	kinderæg	1.2	søster
1.5	slik	1.1	person
1.5	mælkesnitte	1.1	tyggegummi
1.4	biografen	1.1	camping
1.4	camping	1.1	lærer

Danish
Earlier born biased

First > Second		Second > Third+	
1.6	orange	1.1	nej
1.5	hvid	1.1	vov
1.5	grøn	1.1	ja
1.5	sort	1.1	muh
1.5	olde(mor)	1.1	grøn
1.5	blå	1.1	mad
1.5	gul	1.1	blå
1.4	æsel	1.1	gul
1.4	rød	1.1	rap
1.4	hvad	1.1	juice

English (American)
Later born biased

Second > First		Third+ > Second	
3.1	brother	1.7	brother
3.1	sister	1.6	sister
2.5	gum	1.5	gum
1.8	hate	1.4	hate
1.7	popsicle	1.3	candy
1.7	donut	1.3	because
1.7	candy	1.2	donut
1.6	school	1.2	so
1.6	lollipop	1.2	school
1.6	because	1.2	but

English (American)
Earlier born biased

First > Second		Second > Third+	
1.6	green	1.2	owl
1.6	blue	1.2	blue
1.5	sheep	1.2	green
1.5	red	1.2	duck
1.5	owl	1.2	peas
1.5	yellow	1.2	red
1.5	clock	1.2	clock
1.5	peas	1.2	sheep
1.5	white	1.2	tractor
1.5	brown	1.2	button

German
Later born biased

Second > First		Third+ > Second	
1.6	Bruder	1.1	Bruder
1.5	Schwester	1.1	Schule
1.5	Schule	1.1	da
1.4	ich	1.1	Bonbon
1.4	Bonbon	1.1	ich
1.3	Kakao	1.1	ab
1.3	Kindergarten	1.1	Papa
1.3	Knete	1.1	Kakao
1.3	mir	1.1	Mama
1.2	will	1.1	Schwester

German
Earlier born biased

First > Second		Second > Third+	
1.2	gestern	1.1	gestern
1.2	ein bisschen	1.1	Geschenk
1.2	Orange/Apfelsine	1.1	waren
1.2	Kaffee	1.1	ein bisschen
1.2	morgen	1.1	orange
1.2	Stadt	1.1	müssen
1.2	Geschenk	1.1	schwarz
1.2	Tee	1.1	können
1.2	neu	1.1	wenn
1.2	Bagger	1.1	welche

Mandarin (Taiwanese)
Later born biased

Second > First		Third+ > Second	
1.6	書包	1.3	姊姊
1.5	口香糖	1.3	爸爸/爹地
1.4	雞塊	1.3	媽媽/媽咪
1.4	冰棒	1.3	抱(抱)
1.4	彩色筆	1.3	書包
1.4	果凍	1.3	再見(byebye)
1.4	鉛筆	1.3	打
1.4	棒棒糖	1.2	小ㄐㄐ/小鳥(生殖器)
1.3	可樂	1.2	哥哥
1.3	姊姊	1.2	鞋子

Mandarin (Taiwanese)
Earlier born biased

First > Second		Second > Third+	
1.5	弟弟	1.3	藍藍的/藍色的
1.4	三角形的	1.2	綠色的
1.4	藍藍的/藍色的	1.2	明天
1.4	綠色的	1.2	三角形的
1.4	斑馬	1.2	黃色的
1.3	黃色的	1.2	白色的
1.3	牛(哞)	1.2	昨天
1.3	貓頭鷹	1.2	以後
1.3	白色的	1.2	傷心的/難過的
1.3	圓的/圓形的	1.2	先...然後...

Figure 9.12
Top 10 most birth-order-biased words in each language for production data.

Norwegian
Later born biased

Second > First		Third+ > Second	
2.3	nugatti	1.5	nugatti
2.2	popcorn	1.4	skole
2.2	godteri	1.4	søster
2.1	skole	1.3	corn flakes
2.1	tyggegummi	1.3	cola
2.0	bror	1.3	bror
2.0	søster	1.2	popcorn
1.9	cola	1.2	potetgull
1.7	brus	1.2	lærer
1.7	kino	1.2	mygg

Norwegian
Earlier born biased

First > Second		Second > Third+	
1.6	gul	1.2	tog
1.5	grønn	1.2	blå
1.5	oransje	1.2	pasta
1.5	rød	1.2	buss
1.5	hvit	1.2	oldemor
1.4	blå	1.2	sjiraff
1.4	brun	1.2	stein
1.4	svart	1.2	huske
1.4	i morgen	1.2	ape(katt)
1.3	mørkt	1.2	bolle

Spanish (Mexican)
Later born biased

Second > First		Third+ > Second	
1.8	hermana	1.4	mamá
1.8	hermano	1.4	papá
1.3	peanut butter	1.4	agua
1.3	entre	1.4	papá/papi
1.3	nana	1.4	mamá/mami
1.3	fábrica	1.3	pipí
1.3	trineo	1.3	hermana
1.3	aquellos	1.3	guaguá
1.3	ruidoso	1.3	¡ay!
1.2	tranquilo	1.3	hermano

Spanish (Mexican)
Earlier born biased

First > Second		Second > Third+	
1.4	agua	1.1	despacio
1.3	bebé	1.1	negro
1.3	sí	1.1	morado
1.3	guaguá	1.1	hipopótamo
1.3	mamá/mami	1.1	enfrente
1.3	mamá	1.1	luego
1.3	abuela	1.1	descompuesto
1.3	leche	1.1	nuestro
1.3	avión	1.1	mejor
1.3	shh	1.1	suyos

Figure 9.12
(continued)

possessives are learned slightly earlier for later-born children (perhaps related to property conflicts; e.g., "that's mine!").

9.2.3 Maternal Education

Our final set of analyses examines vocabulary items that are differentially present in the vocabulary of children with differing levels of maternal education. As noted in chapter 6, there are substantial cross-linguistic differences in how large the overall socioeconomic stratification is. For example, we observe large differences in children's vocabulary size in the American English data, with children of less educated mothers reporting substantially lower production vocabulary. Fixed effects from this analyses are shown in figure 9.15. Children with higher levels of maternal education generally have larger productive vocabularies but, perhaps surprisingly, smaller comprehension vocabularies. We discussed this finding in depth in chapter 6; here we reiterate that we believe it is plausibly due to reporting biases.

Figures 9.16 and 9.17 show item random effects for comprehension. Many more words are strongly affected by maternal education for American English than for other languages: 7 percent of American English words have effects of at least

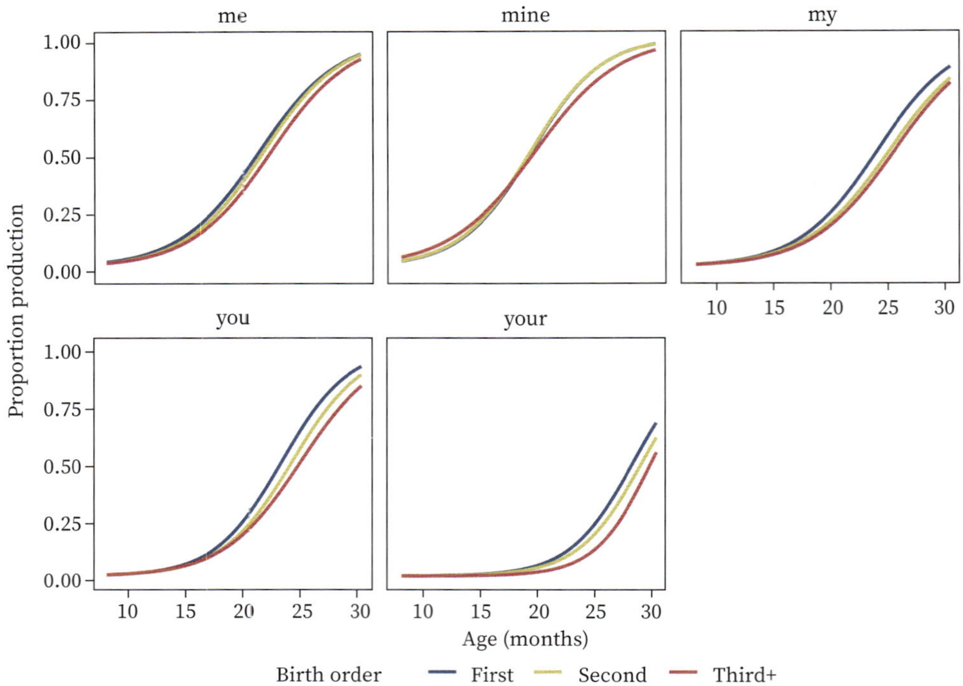

Figure 9.13
Developmental trajectories for personal pronouns in American English words by birth order. Note that lines are overplotted, especially in the case of *mine*.

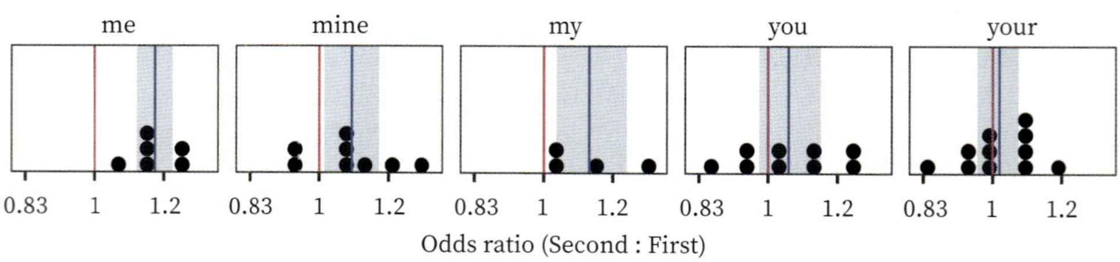

Figure 9.14
Histogram showing item random effects for personal pronouns by birth order, with each point representing the effect in a given language. Red lines marks no effect, blue lines marks the mean, and blue bands shows bootstrapped 95 percent confidence intervals.

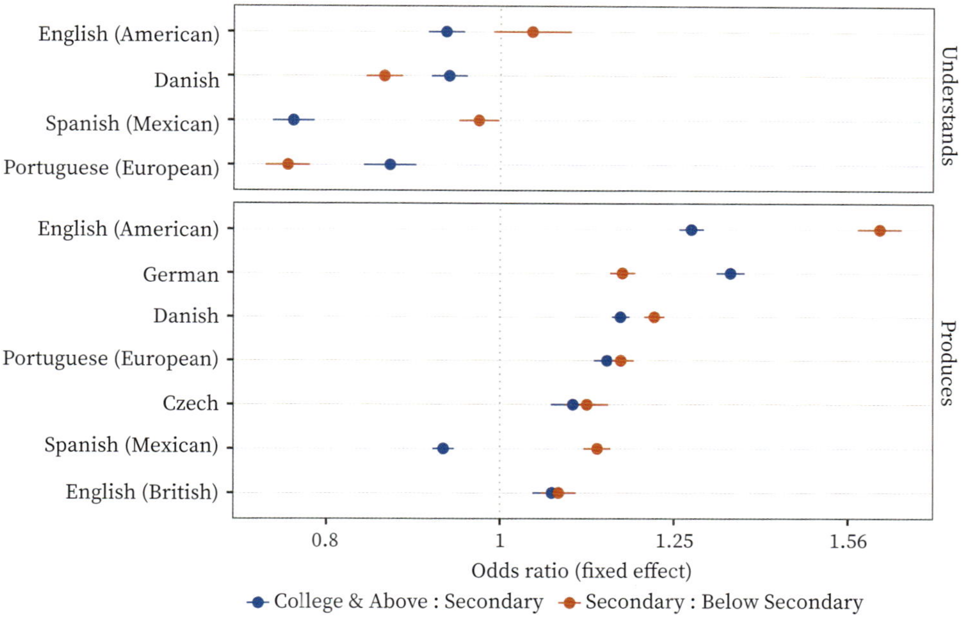

Figure 9.15
Main effect of maternal education for each language and measure.

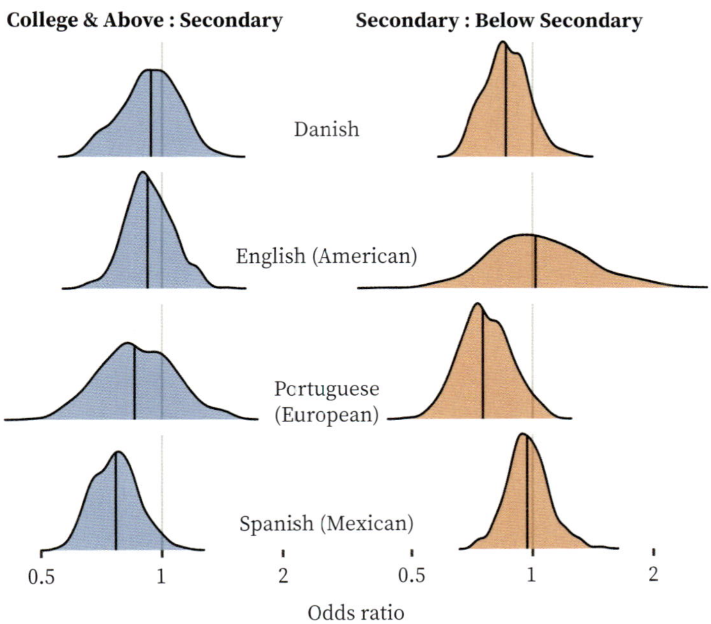

Figure 9.16
Distribution of maternal education item random effects for comprehension data in each language.

Danish
Lower maternal education biased

Below Secondary > Secondary		Secondary > College & Above	
1.3	slæde	1.5	smile
1.3	lærer	1.5	faster
1.3	kælk	1.5	onkel
1.3	småpenge	1.4	pony
1.3	person	1.4	morbror
1.3	mørk	1.4	farbror
1.3	hvornår	1.4	made
1.3	sår	1.4	moster
1.3	menneske	1.4	person
1.3	fest	1.4	lærer

Danish
Higher maternal education biased

College & Above > Secondary		Secondary > Below Secondary	
1.5	mæh	1.5	mor
1.5	muh	1.5	far
1.4	rosiner	1.4	nej
1.4	rap	1.4	borte tit/titte bøh
1.4	ko	1.4	barnets eget navn
1.4	får	1.4	hej
1.4	tiger	1.3	klappe kage
1.4	mus	1.3	hej hej
1.4	æble	1.3	mm mm
1.3	gris	1.3	av

English (American)
Lower maternal education biased

Below Secondary > Secondary		Secondary > College & Above	
2.4	candy	1.5	candy
2.0	radio	1.4	toy
1.8	aunt	1.4	aunt
1.8	toy	1.4	bad
1.8	bad	1.4	radio
1.8	money	1.4	baa baa
1.8	tonight	1.3	uncle
1.8	penny	1.3	TV
1.7	uncle	1.3	brother
1.7	pillow	1.3	pillow

English (American)
Higher maternal education biased

College & Above > Secondary		Secondary > Below Secondary	
1.6	read	2.5	read
1.4	duck	2.2	cheerios
1.4	cow	2.1	duck
1.3	slide	2.0	bird
1.3	cheerios	1.9	daddy*
1.3	bird	1.9	bunny
1.3	stairs	1.9	cow
1.3	gentle	1.9	quack quack
1.3	bunny	1.8	where
1.3	horse	1.8	gentle

Portuguese (European)
Lower maternal education biased

Below Secondary > Secondary		Secondary > College & Above	
1.5	Chupa-chupa	1.8	Chupa-chupa
1.4	Sumo	1.6	Fumo
1.4	Chocolate	1.6	Palhaço
1.4	Fumo	1.6	Chocolate
1.4	Palhaço	1.6	Sumo
1.4	Pouco	1.5	Pouco
1.3	Gordo	1.5	Gordo
1.3	Coca-cola	1.5	Coca-cola
1.3	Cuecas	1.5	Cuecas
1.3	Terra	1.5	Joelho

Portuguese (European)
Higher maternal education biased

College & Above > Secondary		Secondary > Below Secondary	
1.7	Mãe	1.4	Livro
1.7	Livro	1.4	Cu-cu
1.6	Palminhas!	1.4	Ah!
1.6	Adeus / Xau!	1.4	Olá!
1.6	Pai	1.4	Palminhas!
1.6	Cu-cu	1.4	Pai
1.6	Olá!	1.4	Mãe
1.6	Ah!	1.4	Quá-Quá
1.5	Pé(s)	1.3	Pé(s)
1.5	Anda!	1.3	Adeus / Xau!

Spanish (Mexican)
Lower maternal education biased

Below Secondary > Secondary		Secondary > College & Above	
1.4	tortilla	1.4	lo
1.3	soda/refresco	1.4	les
1.3	atole	1.3	se
1.3	dinero	1.3	diferente
1.3	miedo	1.3	le
1.3	lo	1.3	dinero
1.3	piedra	1.3	igual
1.3	sopa	1.3	difícil
1.3	estar	1.3	me
1.3	me	1.3	su

Spanish (Mexican)
Higher maternal education biased

College & Above > Secondary		Secondary > Below Secondary	
1.5	teléfono	1.5	teléfono
1.5	cuna	1.5	escuela
1.4	no	1.4	alberca/piscina
1.4	adíos/byebye	1.4	tengo manita
1.4	tengo manita	1.4	cuna
1.4	¡am!	1.4	caballo
1.3	hola	1.3	gracias
1.3	acerrín	1.3	león
1.3	babero	1.3	resbaladilla
1.3	escuela	1.3	rana

Figure 9.17
Top 10 most maternal-education-biased words in each language for comprehension data.

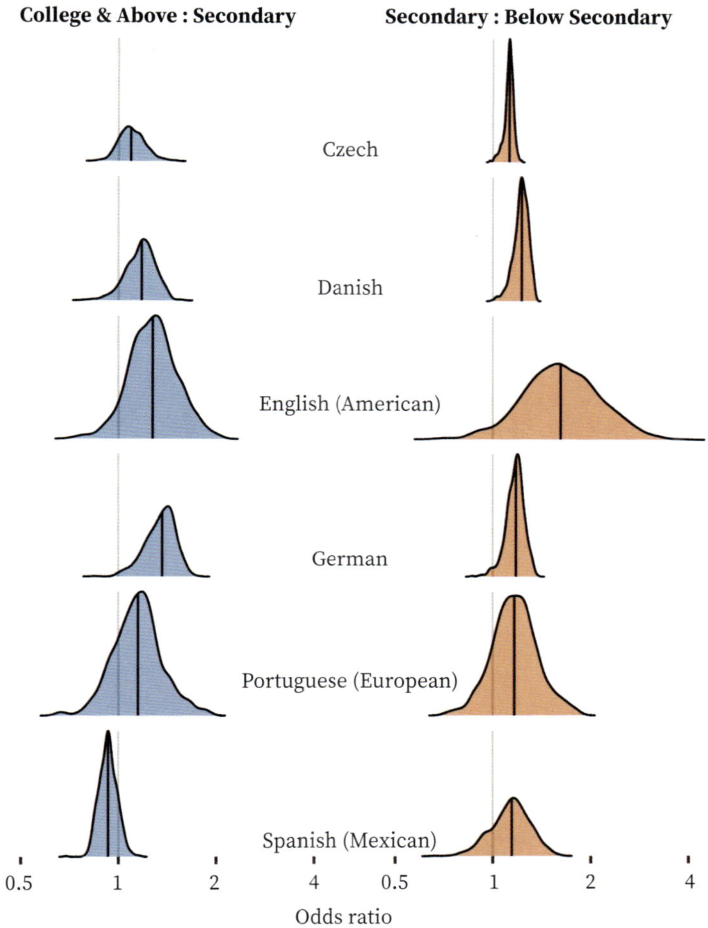

Figure 9.18
Distribution of maternal education item random effects for production data in each language

1.5 times in either direction as compared with 3 percent averaged over languages. This finding is consistent with the idea that maternal education shows a larger effect on total vocabulary size in the American English data than in other dataset—whether because of true cross-cultural differences in SES effects, the composition of the sample, or (most likely) both.

The words that are more likely to be understood by children of college-educated and secondary-educated mothers are often animal-related (e.g., Danish *får* [sheep], *tiger*; English *cow*, *quack quack*) and may speculatively be related to reading books about animals (since most of these animals are not prominent in most children's

Czech
Lower maternal education biased

Below Secondary > Secondary		Secondary > College & Above	
1.1	táta	1.3	párek
1.1	máma	1.3	jupí
1.1	ham	1.2	dej pusinku
1.1	bác	1.2	obličej
1.1	haf	1.2	žízeň
1.1	mňam	1.2	vaše
1.1	pápá	1.2	žvýkačka
1.1	vrn vrn	1.2	bráška
1.1	babička	1.2	sestřička
1.1	děda	1.2	knedlík

Czech
Higher maternal education biased

College & Above > Secondary		Secondary > Below Secondary	
1.4	máma	1.1	vaše
1.4	haf	1.1	vy
1.3	mňau	1.1	oni
1.3	ne	1.1	začít
1.3	pán	1.1	ona
1.3	babička	1.1	tvůj
1.3	táta	1.1	aby
1.3	pápá	1.1	ale
1.3	auto	1.1	on
1.2	bryndák	1.1	včera

Danish
Lower maternal education biased

Below Secondary > Secondary		Secondary > College & Above	
1.2	kæledyrets navn	1.5	mælkesnitte
1.2	dav	1.4	be om
1.2	så stor!	1.4	kinderæg
1.2	borte tit/titte bøh	1.4	klappe kage
1.2	mm mm	1.4	kæledyrets navn
1.2	årnnn	1.4	moster
1.2	klappe kage	1.3	oldefar
1.2	dagplejemors navn	1.3	dagplejemors navn
1.2	åh åhh	1.3	kondisko
1.2	grrrr	1.3	fester

Danish
Higher maternal education biased

College & Above > Secondary		Secondary > Below Secondary	
1.3	zebra	1.1	saks
1.3	også	1.1	pandekager
1.3	meget	1.1	papir
1.2	ugle	1.1	håndklæde
1.2	pandekager	1.1	tv/fjernsyn
1.2	bog	1.1	finger
1.2	i	1.1	hjælpe
1.2	lang	1.1	arm
1.2	et	1.1	ben
1.2	melon	1.1	mus

English (American)
Lower maternal education biased

Below Secondary > Secondary		Secondary > College & Above	
2.3	so	1.8	gum
2.3	walker	1.7	candy
2.2	gum	1.7	soda/pop
2.0	candy	1.6	give me five!
1.9	each	1.6	walker
1.9	could	1.6	so
1.9	wish	1.6	bed
1.9	but	1.5	penny
1.8	penny	1.5	drink
1.8	be	1.5	can

English (American)
Higher maternal education biased

College & Above > Secondary		Secondary > Below Secondary	
1.6	sheep	2.3	quack quack
1.6	giraffe	2.0	sheep
1.6	cockadoodledoo	2.0	babysitter's name
1.6	quack quack	2.0	cockadoodledoo
1.5	babysitter's name	2.0	duck
1.5	gentle	1.9	giraffe
1.5	owl	1.9	moo
1.5	zebra	1.9	owl
1.5	play dough	1.8	yogurt
1.5	mittens	1.8	zebra

German
Lower maternal education biased

Below Secondary > Secondary		Secondary > College & Above	
1.4	arm	1.6	arm
1.3	Name von Haustier	1.4	ihren
1.3	guck guck	1.4	Kino
1.3	Mama	1.3	sanft
1.2	Teddy	1.3	dasselbe
1.2	Papa	1.3	fieß
1.2	Tante	1.3	Name von Haustier
1.2	Bonbon	1.3	ihm
1.2	Onkel	1.3	fein
1.2	Opa	1.3	jeder

German
Higher maternal education biased

College & Above > Secondary		Secondary > Below Secondary	
1.3	Mama	1.2	müssen
1.3	Papa	1.1	passen
1.3	Ball	1.1	musst
1.2	Wasser	1.1	soll
1.2	Buch	1.1	sollst
1.2	Milch	1.1	in
1.2	nein	1.1	dürfen
1.2	tschüss	1.1	und
1.2	Baby	1.1	sind
1.2	Apfel	1.1	können

Portuguese (European)
Lower maternal education biased

Below Secondary > Secondary		Secondary > College & Above	
1.6	Ao	1.8	Aí
1.6	Chupa-chupa	1.7	Chupa-chupa
1.6	Aí	1.7	Da
1.6	Da	1.7	Ao
1.5	À	1.7	À
1.5	Coca-cola	1.5	Coca-cola
1.5	Lata	1.5	Manhã
1.5	Táxi	1.5	Lata
1.5	Manhã	1.5	Olé!
1.5	Olé!	1.4	Logo

Portuguese (European)
Higher maternal education biased

College & Above > Secondary		Secondary > Below Secondary	
1.7	Hipopótamo	1.6	Elefante
1.6	Elefante	1.6	Girafa
1.6	Girafa	1.5	Cocoró
1.6	Crocodilo	1.5	Hipopótamo
1.6	Zebra	1.5	Borboleta
1.6	Borboleta	1.5	Crocodilo
1.6	Azul	1.5	Tartaruga
1.6	Tartaruga	1.5	Zebra
1.6	Verde	1.5	Mu-Muuu
1.5	Tigre	1.5	Quá-Quá

Figure 9.19
Top 10 most maternal-education-biased words in each language for production data.

experience). Some of the largest differences are for *read* in English and *livro* [book] in Portuguese and are perhaps also related to reading practices (or the perception of the importance of these practices). Negatively linked words include some kinship terms (e.g., Danish *faster* and *moster* [aunt], *morbror* and *fabror* [uncle]; English *aunt, uncle, brother*), common sweets (e.g., English *candy*; Portuguese *chupa-chupa* [brand of lollipop]; Spanish *soda*), and money-related words (e.g., Danish *småpenge* [small change]; English *money, penny*).

Production data show a similar but more extreme picture (figures 9.18 and 9.19), with an even larger number of words linked to maternal education in American English (8 percent). Across languages, animal vocabulary is again more prevalent for children of more educated mothers (e.g., Danish *zebra*; English *sheep*; Portuguese *hipopótamo*) as is *babysitter's name*. Children of less educated mothers are again more likely to say words for relatives (e.g., Czech *bráška* [brother]; Danish *oldefar* [great-grandfather]), junk foods (e.g., English *gum, candy, soda*), and money (e.g. German *arm* [poor]; Spanish *dinero* [money]) along with other harder-to-categorize items.

9.3 Conclusions

Demographic factors like sex, birth order, and maternal education are related to children's vocabulary size. But, in addition to these global associations, they appear to be specifically associated with particular vocabulary items. Many of these are straightforwardly explicable in terms of differences in the environmental frequency (and importance) of particular lexical items for children in different circumstances. For example, there are many reasons why second-born children should say *brother* or *sister* more frequently than firstborn children!

More generally, item level variation relates to two issues of interest within the context of our project. The first is the validity of CDI-based measurement. From a psychometric perspective, the sort of variation reported here is known as "differential item function" (Hambleton et al., 1991) and is a negative characteristic of tests that impairs their validity. Thus, from a purely psychometric perspective, items like *babysitter's name* (or even *brother*) should probably not be included in global estimates of vocabulary size. (See chapter 4 for more details on this issue). However, in instruments with more than 500 items, this handful of items probably cause minimal decreases in reliability or validity.

The second broader issue is the question of mechanisms responsible for the demographic associations documented in chapter 6. Sex differences in vocabulary appear quite consistent across languages. Why is this? The analyses in this chapter allow us to gain one small piece of leverage on the issue by noticing that there appear to be two qualitatively different processes involved in the demographic effects we observed: first, girls have a small bump in their probability of producing almost every word and second, there are a small number of particular words for which their production probability is substantially different (higher or lower). To the extent these processes are separable and differ in magnitude, we might look for causal mechanisms that would provide a broader boost to language rather than trying to explain the small number of specifically sex-linked items identified above. Such hypotheses might appeal to dyadic factors like differences in amount and nature of language input directed to girls, to learner-internal factors like stronger social cognition, or to biological differences. A similar argument could be made for birth order and maternal education variables.

In sum, demographic differences in vocabulary likely have multiple sources. Average changes in vocabulary size could be a result of differences in the way children learn across groups, differences in the amount of language children in different groups hear, or an interaction between these two (for example, female children on average perhaps eliciting more language in interaction). However, specific differences in which words are part of the vocabulary are more probably due to differences in input environment, and thus we see in the prevalence of words like *brother* or *soda* the effects of specific changes in children's environment on their learning outcomes.

10 Predictive Models of the Acquisition of Individual Words

In this chapter, we take up the challenge posed in chapter 8—that is, to explain consistency and variability in the acquisition of individual words. Our approach is to define regression models that attempt to predict which words are learned earlier or later on the basis of a range of features drawn from different data sources. We fit these models to data across different languages and then interpret the resulting coefficients to draw conclusions about the potential contribution of different factors to children's learning.

10.1 Introduction

As discussed in chapter 1, one classic approach to word learning focuses on the specific mechanisms that children bring to bear on the learning problem. For example, across many laboratory experiments, a variety of mechanisms have been identified as plausible drivers of early word learning, including co-occurrence-based and cross-situational word learning (Schwartz and Terrell, 1983; Yu and Ballard, 2007), social cue use (Baldwin, 1993), and syntactic bootstrapping (Gleitman, 1990; Mintz, 2003). The individual contribution of each of these mechanisms has been difficult to assess, however.

Indeed, many theories of early word learning take multiplicity of cue types and mechanisms as a central feature (e.g., Hollich et al., 2000; Bloom, 2000). As important as this work is, though, these studies are typically aimed at understanding how a small handful of words are learned in the laboratory under precisely defined learning conditions. They do not directly address questions regarding the developmental composition and ordering of growth in the lexicon across many different children in their natural environments nor whether these patterns are consistent across different languages.

Authors' Note: The contents of this chapter are lightly adapted from Braginsky et al. (2019).

An alternate approach to word learning—one that we have been following throughout this book—asks why some words are learned so early and some much later. This question about the order of the acquisition of first words can provide a different window into the nature of children's language learning. In chapter 8, we approached this question by examining the consistency of acquisition order for children's earliest words. In this chapter, we advance this goal using quantitative models to understand acquisition ordering.

Posed as a statistical problem, the challenge is to find what set of variables best predicts the age at which different words are acquired. This approach was pioneered by Huttenlocher et al. (1991) and developed further by Goodman et al. (2008); it is now firmly established as an important method for understanding vocabulary learning at scale. This previous work has revealed that, within a lexical category (e.g., nouns, verbs) in English, words that are more frequent in speech to children are likely to be learned earlier (Goodman et al., 2008). Further studies in English have found evidence that the age of acquisition is likely to be earlier for words that have more phonological neighbors (e.g., Storkel, 2004; Stokes, 2010; Jones and Brandt, 2019; but see Swingley and Aslin, 2007; Stager and Werker, 1997); words that share more associations with other words in the learning environment (Hills et al., 2009); words that occur more often in isolation (Brent and Siskind, 2001; Swingley and Humphrey, 2018); words whose meanings are more concrete (Swingley and Humphrey, 2018); words that are rated more iconic and/or more associated with babies (such as "choo-choo" or "doggy"; Perry et al., 2015); and words that occur in more distinctive spacial, temporal, and linguistic contexts (Roy et al., 2015).

Each of these studies used a different dataset and focused on different predictors, however. In addition, nearly all analyzed data from English-learning children providing no opportunity for cross-linguistic comparison of the relative importance of the many relevant factors under consideration. In this chapter, we extend these approaches and assess the degree to which the predictors of word learning are consistent across different languages and cultures as well as whether there are similar patterns across different word types (e.g., nouns versus verbs).

We conduct cross-linguistic comparisons of the age of acquisition of particular words. We integrate estimates of words' acquisition trajectories from the Wordbank data with independently derived characterizations of the word learning environment from other datasets. The use of secondary datasets for these analyses is warranted because no currently available resource provides data on both children's language environments and their learning outcomes for more than a small handful of children. In particular, we derive our estimates of the language environment from transcripts of speech to children in the Child Language Data Exchange System (CHILDES) database (MacWhinney, 2000). This data-integration methodology was originated by Goodman et al. (2008); it relies on large samples to average out the (substantial) differences between children and care environments. This is a conservative strategy because it requires substantial commonalities across families. While introducing additional sources of variability, it also allows for analyses that cannot be performed on smaller datasets or datasets that measure only child or environment but not both.

As our particular measures of environmental input, we estimated each word's (a) frequency in parental speech to children, (b) mean length in words of the parental utterances containing that word (MLU-w), (c) frequency as a one-word utterance, and (d) frequency as the final word in an utterance. While these measures are crude, they are easy to compute and relatively comparable across the languages in our sample. To derive proxies for the meaning-based properties of each word, we accessed available psycholinguistic norms using adult ratings of each word's (a) concreteness, (b) valence, (c) arousal, and (d) association with babies. Integrating these environmental and meaning-based measures, which are based respectively on estimates of children's linguistic environment and words' meaning, we predict each word's acquisition trajectories. We assess the relative contributions of each predictor as well as how those predictors change over development and how they interact with the lexical category of the word being predicted.

These analyses address two questions. First, we ask about the degree of consistency across languages in the relative importance of each predictor. Consistency in the patterning of predictors would suggest that similar information sources are important for learners, regardless of language. Such evidence would suggest that superficial linguistic dissimilarities (e.g., greater morphological complexity in Russian and Turkish, greater phonological complexity in Danish) do not dramatically alter the course of acquisition. Conversely, variability would show the degree to which learners face different challenges in learning different languages, posing a challenge for more universalist accounts. Further, systematicity in the variability between languages would reveal which languages are more similar than others in the structure of these different challenges.

Second, we ask which lexical categories are most influenced by specific linguistic environment factors, like frequency and utterance length, compared with meaning-based factors like concreteness and valence. Division of dominance theory suggests that nouns might be more sensitive to meaning factors while predicates and closed-class words might be more sensitive to linguistic environment factors (Gentner and Boroditsky, 2001). Following syntactic bootstrapping theories (Gleitman, 1990), nouns are argued to be learned via frequent co-occurrence patterns in the input (operationalized by frequency) while verbs might be more sensitive to syntactic factors (operationalized here by utterance length; Snedeker et al.,

2007). Thus, examining the relative contribution of different predictors across lexical categories can help test the predictions of influential theories of acquisition.

10.2 Methods

10.2.1 Acquisition Trajectories

Since analyses in this chapter rely on "universal lemma" mappings (see section 3.1.5), the set of languages represented is smaller than in other chapters. We use data from the items on Words and Gestures (WG) forms for our comprehension measure and data from the items in common between WG and Words and Sentences (WS) forms for our production measure. Placeholder items, such as "child's own name," are excluded, as are longitudinal administrations. Table 10.1 gives an overview of our acquisition data.

See figure 10.1 for sample smoothed empirical item curves of the type being predicted in our subsequent analyses.

10.2.2 Word Properties

For each word that appears on the forms in each of our 10 languages, we used corpora of child-directed speech in that language from CHILDES to obtain an estimate of its frequency, the MLU-w of utterances in which it appears, its frequency as the sole constituent of utterance, and its frequency in utterance final position (with frequency residualized out of solo and final frequencies). Additionally, we computed each word's length in phonemes.

To capture meaning-based factors in acquisition, we included ratings of each word's concreteness, valence, arousal, and relatedness to babies. All of these

Table 10.1
Statistics for data from Wordbank and CHILDES. *N* indicates number of children.

		Production		Comprehension		CHILDES	
Language	Items	*N*	Ages	*N*	Ages	Types	Tokens
Croatian	388	627	8–30	250	8–16	12,064	218,775
Danish	381	6,112	8–36	2,398	8–20	4,956	195,658
English	393	7,312	8–30	1,792	8–18	45,597	7,679,042
French	396	1,364	8–30	537	8–16	28,819	2,551,113
Italian	392	1,400	7–36	648	7–24	7,544	188,879
Norwegian	380	7,466	8–36	2,374	8–20	10,670	231,763
Russian	410	1,805	8–36	768	8–18	5,191	32,398
Spanish	399	1,891	8–30	788	8–18	33,529	1,609,614
Swedish	371	1,367	8–28	467	8–16	8,815	359,155
Turkish	395	3,537	8–36	1,115	8–16	6,503	44,347

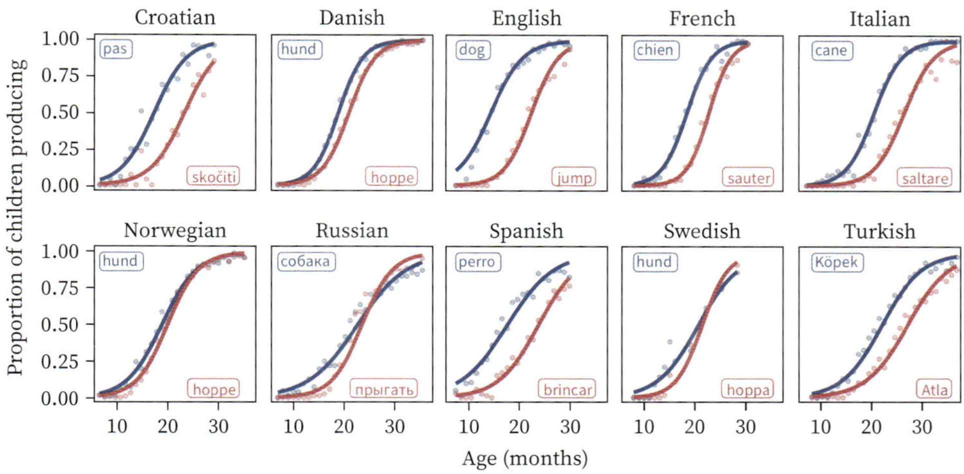

Figure 10.1
Example production trajectories for the words "dog" and "jump" across languages. Points show the proportion of children producing each word for each one-month age group. Lines show the best-fitting logistic curve. Labels show the forms of the words in each language.

ratings were compiled based on previous studies using adult raters. In addition, since existing datasets for all of these ratings are primarily available for English, we used the unilemma mappings (see section 3.1.5) to use the ratings for English words across languages. Example words for these predictors in English are shown in table 10.2.

Previous studies have shown robust consistency in the types of words that children learn very early (Tardif et al., 2008). These words seem to describe concepts that are important or exciting in the lives of infants in a way that standard psycholinguistic features like concreteness do not. Capturing this intuition quantitatively is difficult, but Perry et al. (2015) provides a proxy measure as a first step. This measure is simply the degree to which a particular word was "associated with babies." Intuitively, we expect this measure to capture the degree to which words like *ball* or *bottle* feature heavily in the environment (and presumably, the mental life) of many babies.

Each numeric predictor was centered and scaled so that all predictors would have comparable units.

Frequency. For each language, we estimated word frequency from unigram counts based on all corpora in CHILDES for that language (all corpora from a given language were included, regardless of dialect). Frequencies varied widely both within and across lexical categories. Each word's count includes the counts of words that share the same stem (so that *dogs* counts as *dog*) or are synonymous (so

Table 10.2
Items with the highest and lowest values for each predictor in English.

Predictor	Lowest	Highest
Arousal	today, asleep, shh	naughty, money, scared
Babiness	jeans, penny, donkey	baby, bib, bottle
Concreteness	that, now, how	apple, baby, ball
Final frequency	put, when, give	book, it, there
Frequency	babysitter, rocking chair, grrr	you, it, that
MLU	ouch, thank you, peekaboo	daddy, when, day
Number phonemes	i, eye, ear	refrigerator, cockadoodledoo, babysitter
Solo frequency	feed, bathroom, tooth	no, yes, thank you
Valence	ouch, hurt, sick	happy, hug, love

that *father* counts as *daddy*). For polysemous word pairs (e.g., *orange* as in color or fruit), occurrences of the word in the corpus were split uniformly between the senses on the Communicative Development Inventory (CDI; there were only between one and ten such word pairs in the various languages; in the absence of cross-linguistic corpus resources for polysemy sense disambiguation, this is a necessary simplification). Counts were normalized to the length of each corpus, Laplace smoothed (i.e., counts of 0 were replaced with counts of 1), and then log transformed.

Solo and final frequencies. Using the same dataset as for frequency, we estimated the frequency with which each of word occurs as the sole word in an utterance and the frequency with which it appears as the final word of an utterance (not counting single-word utterances). As with frequency, solo and final counts were normalized to the length of each corpus, Laplace smoothed, and log transformed. Since both of these estimates are by necessity highly correlated with frequency, we then residualized unigram frequency out of both of them, so that values reflect an estimate of the effects of solo frequency and final frequency over and above frequency itself.

MLU-w. MLU-w is only a rough proxy for syntactic complexity, but it is relatively straightforward to compute across languages (in contrast to other metrics). For each language, we estimated each word's MLU-w by calculating the mean length in words of the utterances in which that word appeared for all corpora in CHILDES for that language. For words that occurred fewer than ten times, MLU-w estimates were treated as missing.

Number of phonemes. In the absence of consistent resources for cross-linguistic pronunciation, we computed the number of phonemes in each word in each language based on phonemic transcriptions of each word obtained using the eSpeak tool (Duddington, 2012). We then spot-checked these transcriptions for accuracy.

Concreteness. We used previously collected norms for concreteness (Brysbaert et al., 2014), which were gathered by asking adult participants to rate how concrete the meaning of each word is on a 5-point scale from abstract to concrete.

Valence and arousal. We also used previously collected norms for valence and arousal (Warriner et al., 2013), for which adult participants were asked to rate words on a 1-to-9 happy-unhappy scale (valence) and 1-to-9 excited-calm scale (arousal).

Babiness. Lastly, we used previously collected norms of "babiness," a measure of association with infancy (Perry et al., 2015) for which adult participants were asked to judge a word's association with babies on a 1-to-10 scale.

Lexical category. Category was determined on the basis of the conceptual categories presented on the CDI form (e.g., "Animals," "Action Words"), such that the nouns category contains common nouns, predicates contains verbs and adjectives, and function words contains closed-class words (following Bates et al., 1994), and the remaining items are binned as Other.

Imputation. The resulting sets of predictor values for each language had varying numbers of missing values, depending on resource availability (number of phonemes 0 percent, concreteness 0–1 percent, arousal and valence 8–13 percent, [solo/final] frequency 2–4 percent, babiness 10–33 percent, MLU-w 2–53 percent). We used iterative regression imputation to fill in these missing values (separately within each language) by first replacing missing values with samples drawn randomly with replacement from the observed values and then iteratively imputing values for a predictor based on a linear regression fitting that predictor from all others.

Collinearity. A potential concern for comparing coefficient estimates is predictor collinearity. Fortunately, in every language, the only relatively large correlations are between MLU-w and solo frequency (mean over languages $r = -0.44$), as expected given the similarity of these factors, along with modest correlations between frequency and concreteness (mean over languages $r = -0.36$) and between frequency and number of phonemes (mean over languages $r = -0.33$), a reflection of Zipf's Law (Zipf, 1935). More importantly, the variance inflation factor for each of the predictors in each language is no greater than 2.3, indicating that multicollinearity among the predictors is low.

10.2.3 Analysis

We used mixed-effects logistic regression models (fit with the MixedModels.jl package in Julia; Bates et al., 2018) to predict whether each child understands/produces each word from the child's age, properties of the word, interactions between each property and age, and interactions between each property and lexical category (which was contrast coded). Each model was fit to all data from a particular language and included a random intercept for each word and a random slope of age for each word. Computational and technical limitations prevented us from including random effects for child or including data from all languages in one joint model.

The magnitude of the standardized coefficient on each property gives an estimate of its independent contribution to words being understood/produced by more children. Interactions between properties and age give estimates of how this effect is modulated for earlier-learned and later-learned words. For example, a positive effect of babiness means that words associated with babies are known by more children; a negative interaction with age means that high babiness leads to higher rates of production and comprehension for younger children compared with older children. Similarly, interactions between properties and lexical category give estimates of how the effect differs among nouns, predicates, and function words.

10.3 Results

English predictor effects. To illustrate the structure of our analysis, we first describe the results for English data, shown in figure 10.2 as the main effect and age interaction coefficient estimates and 95 percent confidence intervals, for comprehension and production. For main effects, words are more likely to be known by more children if they are higher in frequency or concreteness, as well as in babiness for comprehension and in sentence-final frequency or sole-constituent frequency for production. In contrast, words that appear in shorter sentences (MLU-w) are more likely to be reported as understood or produced. For age interactions, while most predictors have consistent effects over age, words that are higher in frequency or concreteness are more likely to be known more by older children, while words that are higher in valence have a greater effect on acquisition in younger children, with an additional negative interaction with babiness in comprehension and positive interaction with MLU-w in production.

Cross-linguistic predictor effects. Figure 10.3 shows the coefficient estimate for each predictor in each language and measure. We find that frequency is the strongest predictor of acquisition (mean across languages and measures $\beta = 0.23$). Other relatively strong overall predictors include concreteness ($\beta = 0.18$), solo frequency ($\beta = 0.17$), MLU-w ($\beta = -0.14$), and final frequency ($\beta = 0.13$). Number of phonemes is comparatively large for production ($\beta = -0.31$) but not comprehension ($\beta = -0.07$); conversely, babiness is comparatively large for comprehension ($\beta = 0.19$) but not production ($\beta = 0.08$). Finally, valence ($\beta = 0.06$) and arousal ($\beta = 0.003$) have much smaller effects.

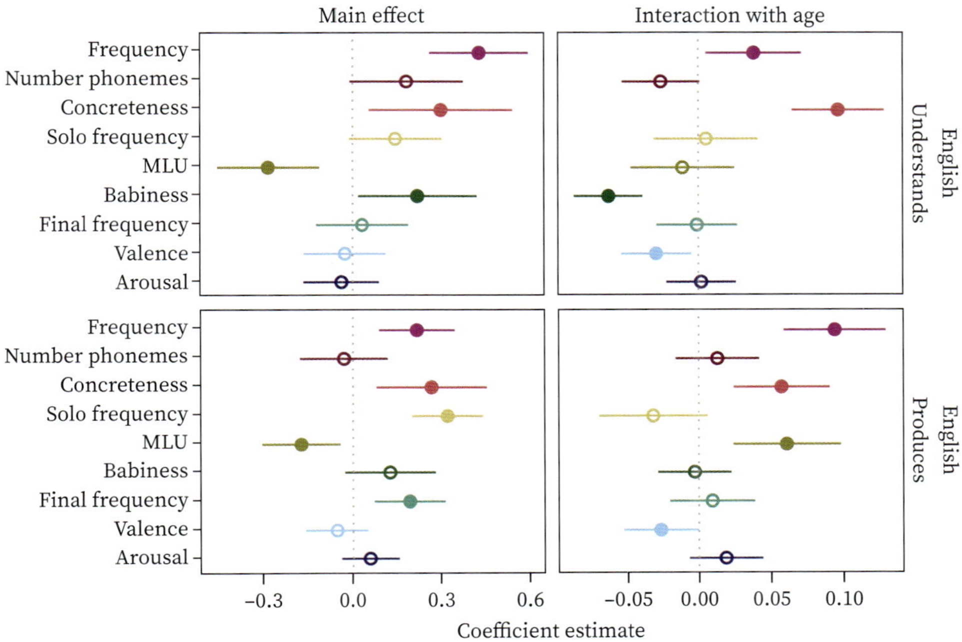

Figure 10.2
Estimates of coefficients in predicting words' developmental trajectories for English comprehension and production data. Error bars indicate 95 percent confidence intervals; filled in points indicate coefficients for which $p < 0.05$.

Given the emphasis on frequency effects in the literature (Ambridge et al., 2015), one might have expected frequency to dominate, but several other predictors are also quite strong. In addition, some factors previously argued to be important for word learning—namely, valence and arousal (Moors et al., 2013)—appear to have limited relevance when compared with other factors. These results provide a strong argument for our approach of including multiple predictors and languages in our analysis.

Consistency. Apart from valence and arousal, all other predictors have the same the direction of effect in all or almost all languages and measures (at least 17 of the 20). Thus, across languages, words are likely to be understood and produced by more children if they are more frequent, shorter, more concrete, more frequently the only word in an utterance, more associated with babies, more frequently the final word in an utterance, and tend to appear in shorter utterances.

Additionally, there is considerable consistency in the magnitudes of predictors across languages. A priori, it could have been the case that different languages have wildly different effects of various factors (due to linguistic or cultural differences),

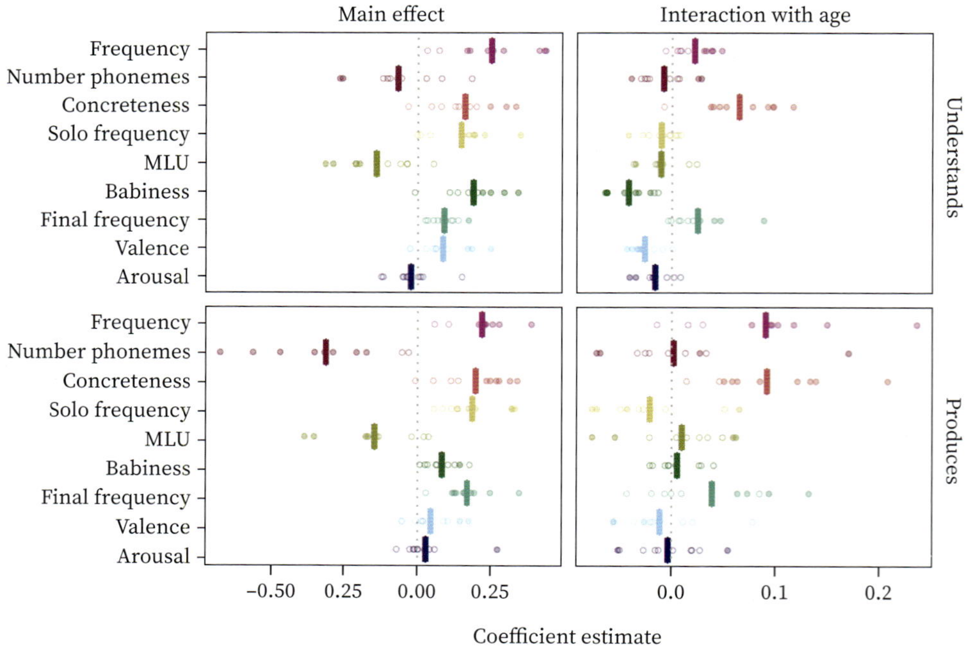

Figure 10.3
Estimates of coefficients in predicting words' developmental trajectories for all languages and measures. Each point represents a predictor's coefficient in one language, with the bar showing the mean across languages. Filled-in points indicate coefficients for which $p < 0.05$.

but this pattern is not what we observe. Instead, there is more consistency in the correlations among coefficients across languages than would be expected by chance. As shown in figure 10.4, each language's mean pairwise correlation with other languages' coefficients (i.e., the correlation of coefficients for English with coefficients for Russian, for Spanish, and so on) is outside of bootstrapped estimates in a randomized baseline created by shuffling predictor coefficients within language. The pairwise correlations are more consistent for production (mean $r = 0.72$) than for comprehension (mean $r = 0.56$), in which French and Russian effects are more idiosyncratic.

Variability. While some particular coefficients differ substantially from the trend across languages (e.g., the effect of frequency for comprehension in Spanish is near zero), these individual datapoints are difficult to interpret. Many unmeasurable factors could potentially account for these differences: Spanish frequency estimates could be less accurate due to corpus sparsity or idiosyncrasy, the samples of children in the Spanish CDI and CHILDES data could differ more demographically, or Spanish-learning children could in fact rely less on frequency.

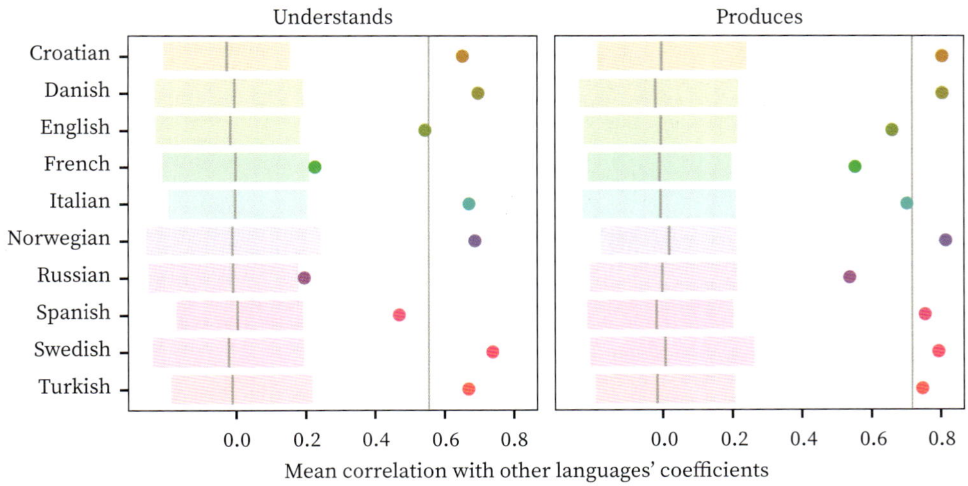

Figure 10.4
Correlations of coefficient estimates between languages. Each point represents the mean of one language's coefficients' correlation with each other language's coefficients, with the vertical line indicating the overall mean across languages. The shaded region and line show a bootstrapped 95 percent confidence interval of a randomized baseline where predictor coefficients are shuffled within language.

Rather than attempting to interpret individual coefficients, we instead ask how the patterns of difference among languages reflect systematic substructure in the variability of the effects.

To examine the substructure of predictor variability, we followed chapter 8 in using hierarchical clustering analysis to find the similarity structure among the pairwise correlations between languages' predictors. The resulting dendrograms are shown in figure 10.5, which broadly reflect language typology, especially for production data. This result suggests that some language-to-language similarity is captured by the profile of coefficient magnitudes our analysis returns.

Comprehension versus production. As mentioned above, word length is the one predictor of acquisition that varies substantially between measures: it is far more predictive for production than for comprehension. Although not all children will produce precisely the citation form of CDI words, words that have longer citation forms will also on average tend to have more difficult realizations in child language (e.g., some children will say "raf" for "giraffe" while others will attempt the full citation form). Given this, as measured here, length seems to reflect effects of production constraints (i.e., how difficult a word is to say) rather than comprehension constraints (i.e., how difficult it is to store or access). This result may explain why the hierarchical clustering analysis above appears more similar to linguistic typology in production than comprehension; that is, the role of production

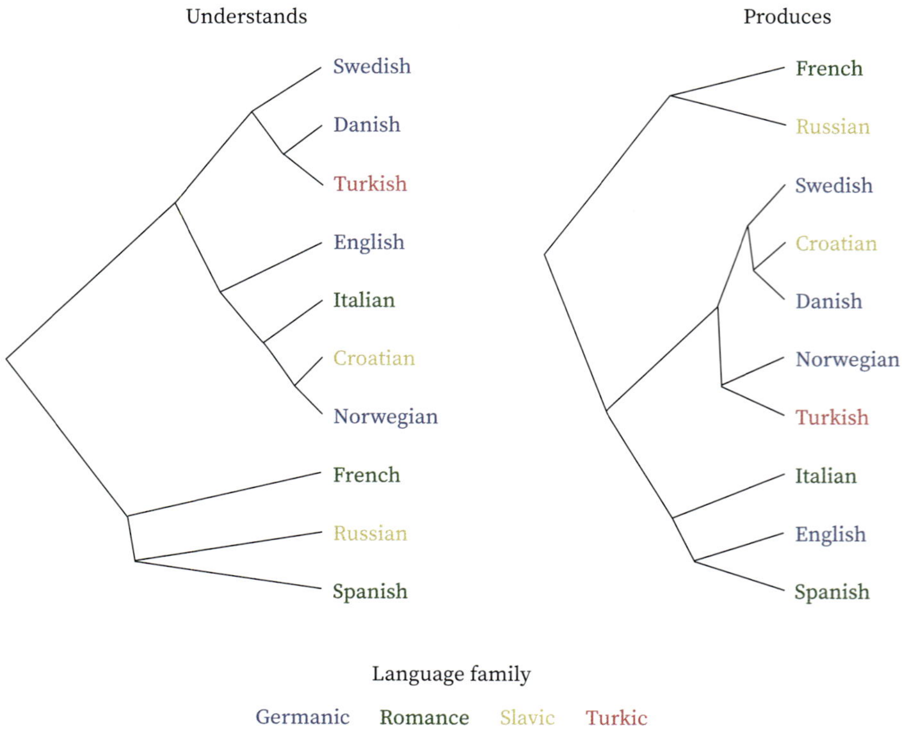

Figure 10.5
Dendrograms of the similarity structure among languages' coefficients.

difficulty may be more similar for more typologically related languages. Another possibility is that since the measures are confounded with age (comprehension is only measured for younger children), word length may play a larger role later in acquisition. Similarly, the stronger effect of babiness in comprehension over production could be due to its larger prominence earlier in development.

Developmental change. For both comprehension and production, positive age interactions can be seen in at least 9 out of 10 languages for concreteness and frequency. Conversely, there are negative age interactions for babiness and valence for comprehension in at least 9 out of 10 languages. This suggests that concreteness and frequency facilitate learning more so later in development, while babiness and valence facilitate learning earlier in development. This result is consistent with the speculation above that the babiness predictor captures meanings that have special salience to very young infants.

Lexical categories. Previous work suggests that predictors' relationship with age of acquisition differs among lexical categories (Goodman et al., 2008). We

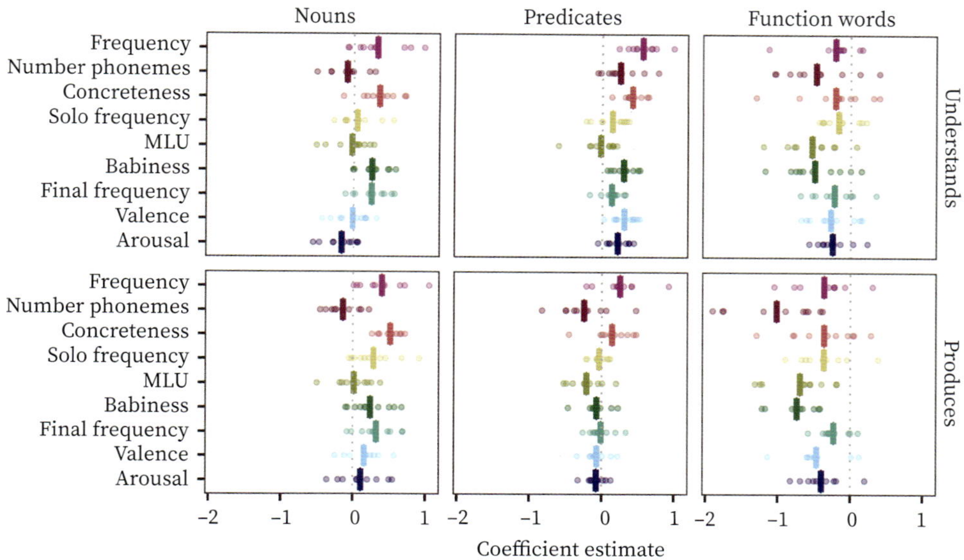

Figure 10.6
Estimates of effects in predicting words' developmental trajectories for each language, measure, and lexical category (main effect of predictor + main effect of lexical category + interaction between predictor and lexical category). Each point represents a predictor's effect in one language, with the bar showing the mean across languages.

investigate these differences by including lexical category interaction terms in our model. Figure 10.6 shows the resulting effects for each lexical category, combining the main effect of a given predictor with the main effect of the lexical category and the interaction between that predictor and that lexical category.

Across languages, the strongest predictors of acquisition for both nouns and predicates are concreteness (nouns β = 0.44; predicates β = 0.28) and frequency (nouns β = 0.36; predicates β = 0.41). Thus content words are most likely to be known by more children if they are more frequent or more concrete. Conversely, function words are most influenced by number of phonemes (β = –0.74), babiness (β = –0.61), and MLU-w (β = –0.61), meaning that function words are most likely to be known by more children if they are shorter, are less associated with babies, or appear in shorter sentences. These patterns are supportive of the hypothesis that different word classes are learned in different ways or at least that the bottleneck on learning tends to be different, with different information sources having varying degrees of relevance across categories.

Additionally, the mean pairwise correlation of coefficients among languages is much larger for nouns (r = 0.68) and predicates (r = 0.54) than for function words (r = 0.29). The higher between-language variability for function words suggests the learning processes differ substantially more across languages for function words than they do for content words.

10.4 Discussion

What makes words easier or harder for young children to learn? Previous experimental work has largely addressed this question using small-scale lab studies. While such experiments can identify sources of variation, they typically do not allow for different sources to be compared directly. In contrast, observational studies allow the effects of individual factors to be measured across ages and lexical categories (e.g., Goodman et al., 2008; Hills et al., 2009; Swingley and Humphrey, 2018), but they are limited in the size and scope of the datasets and languages that can be directly compared. We derived several new findings from our analyses, in part due to the larger set of languages and predictors we were able to include.

First, we found consistency in the patterning of predictors across languages at a level substantially greater than the predictions of a chance model. This consistency supports the idea that differences in culture or language structure do not lead to fundamentally different acquisition strategies, at least at the level of detail we were able to examine. Instead, they are likely produced by processes that are similar across populations and languages. We return to a discussion of these "process universals" in chapter 17.

Second, predictors varied substantially in their weights across lexical categories. Frequent concrete nouns were learned earlier, consistent with theories that emphasize the importance of early referential speech (e.g., Baldwin, 1995). For predicates, concreteness was somewhat less important and frequency somewhat more important. And for function words, length and MLU-w was more predictive, perhaps because it is easiest to decode the meanings of function words that are used in short sentences (or because such words have meanings that are easiest to decode). Overall, these findings are consistent with some predictions of both division of dominance theory, which highlights the role of conceptual structure in noun acquisition (Gentner and Boroditsky, 2001), and syntactic bootstrapping theory, which emphasizes linguistic structure over conceptual complexity in the acquisition of lexical categories other than nouns (Snedeker et al., 2007). More generally, our methods here provide a way forward for testing the predictions of these theories across languages and at the level of the entire lexicon rather than individual words.

In addition to these new insights, several findings emerged that confirm and expand previous reports. Environmental frequency was an important predictor of learning, with more frequently heard words learned earlier (Goodman et al., 2008; Swingley and Humphrey, 2018). Predictors also changed in relative importance across development. For example, certain words whose meanings were more strongly associated with babies appeared to be learned early for children across the languages in our sample, perhaps explaining our findings in chapter 8 (see also Tardif et al., 2008). Finally, word length showed a dissociation between comprehension and production, suggesting that challenges in production do not carry over to comprehension (at least in parent-report data).

11 Vocabulary Composition: Syntactic Categories

We have previously examined vocabulary with a wide angle lens, focusing on total vocabulary size (chapter 5), and with a narrower lens, examining individual words (chapter 10). Here we take a middle perspective, splitting vocabulary by syntactic category and analyzing consistency and variability across languages in the acquisition of these categories. A primary goal of this analysis is to quantify the "noun bias" across languages. In addition, we quantify the degree of bias for or against verbs and closed-class words. This chapter deals primarily with the aggregate biases for or against syntactic categories across entire datasets, but in chapter 15, we consider variation of this sort within individuals.

11.1 Introduction

11.1.1 The Composition of Early Vocabulary

As we reviewed in chapter 8, the first words children utter are quite consistent and primarily composed of names for people and things and words related to social routines (see also Tardif et al., 2008; Schneider et al., 2015). Soon after, however, they begin to add predicates, such as verbs (*go*) and adjectives (*pretty*), in greater proportions than earlier in development and may even begin to use closed-class forms, such as determiners (*the*). These patterns seem to suggest a developmental course that follows distinct "waves" of learning for words from different classes. That is, along with early social routines, nouns tend to predominate early vocabularies while other types of words, such as predicates and closed class forms, are learned later. This pattern may be further qualified by differences in the types of words learned in comprehension versus production (Benedict, 1979).

Examination of the composition of early vocabulary is complicated by the fact that we categorize words by their adult syntactic category. We do so in

Authors' Note: An earlier version of this work was presented at the Boston University Conference on Language Development in 2015.

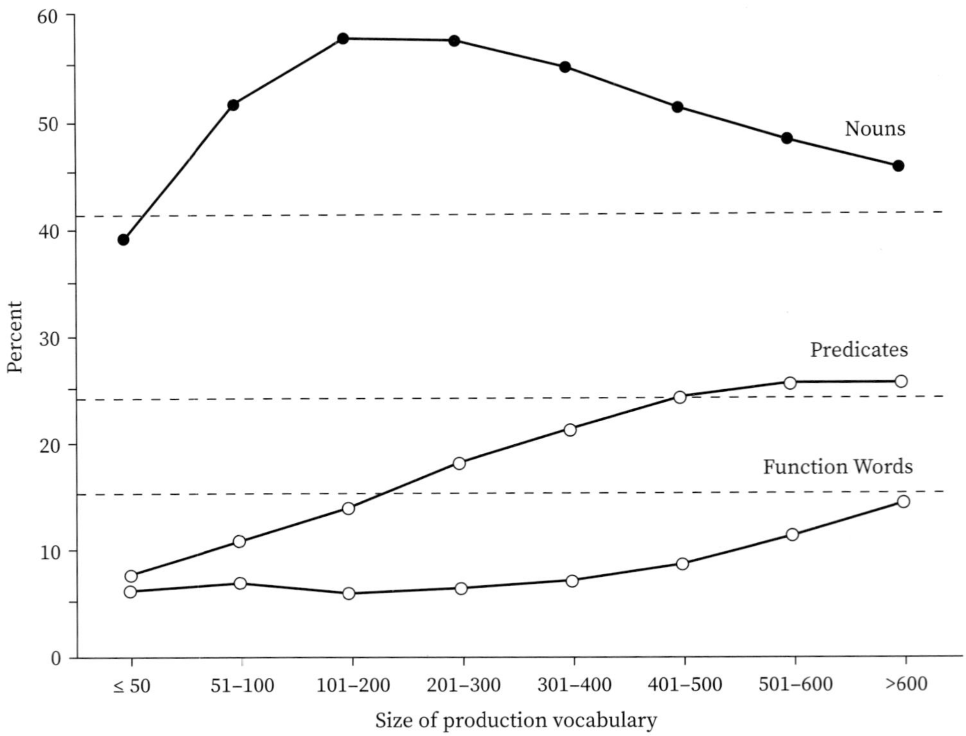

Figure 11.1
Developmental trends in the categorical composition of early vocabulary as number of items. Horizontal lines represent the number of opportunities on the vocabulary checklist for each category. *Source*: Figure 1 from Bates et al. (1994).

the discussion below without presupposing that children themselves do this categorization, however (Tomasello, 2000). Children may be sensitive to these categories very early in development (Valian, 1986; Yang, 2013), or they may discover them either gradually (Pine and Lieven, 1997) or more quickly (Meylan et al., 2017). Importantly, though, we treat adult syntactic categories as an analytic convenience that describes certain regularities in how groups of words are distributed in speech by adults (as captured, for example, in language samples) and how they function in different contexts rather than as an ontological fact about children's knowledge.

Bates et al. (1994) characterized these patterns of vocabulary composition in the following way. Figure 11.1 shows average vocabulary composition of nominals, predicates, and closed class forms as a function of children's vocabulary size for English-speaking children from the original Communicative Development Inventory (CDI) Words and Sentences (WS) norming study (Fenson et al., 1994). Note that when children only know a few words (i.e., fewer than 50),

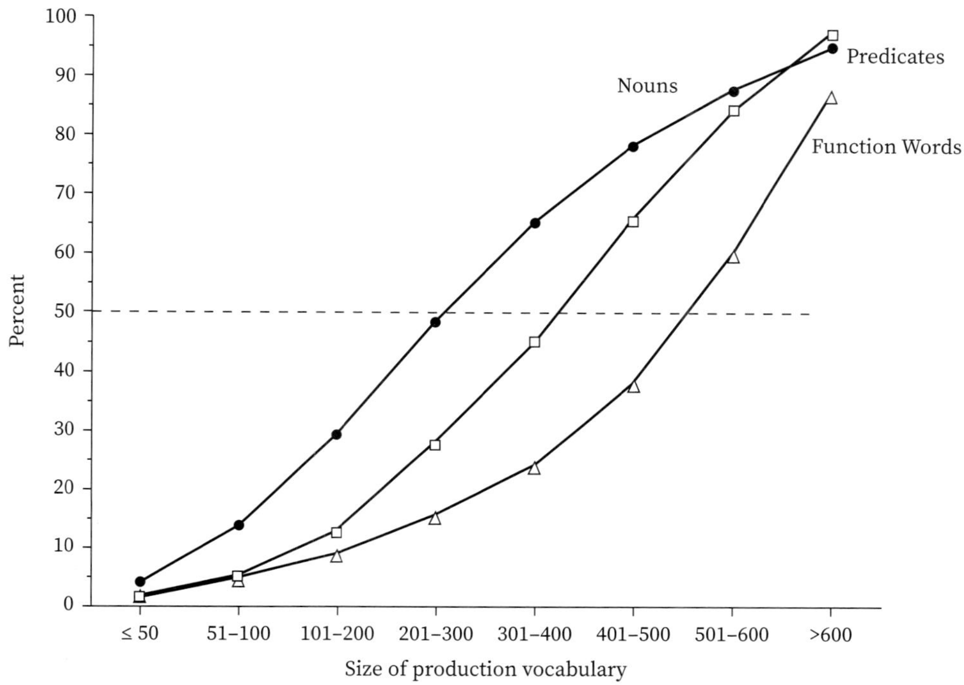

Figure 11.2
Developmental trends in the categorical composition of early vocabulary, using proportions of total opportunities in each category. Horizontal line represents the point at which 50 percent of the items are chosen on average for each category. *Source*: Figure 3 from Bates et al. (1994).

nominals comprise the greatest proportion of the words that children are reported to produce with very few predicates or closed class forms (< 10 percent). As children learn the next hundred words or so, the proportion of nominals increases even more dramatically with a gradual increase in the proportion of children's vocabularies that are predicates. Closed class forms remain a much smaller proportion over this period. Yet, after about 300 words, children do not appear to add nouns to their vocabularies at the same pace that they did earlier in development; consequently, the proportion of nominals tends to decrease.[1] During this developmental period, the proportion of predicates in the vocabulary tends to increase, followed by growth in the proportion of closed class forms.

A different way to capture these same trends is shown in figure 11.2. Here, each of the categories of words is plotted as a proportion of the opportunities on the

1. This effect may also reflect aspects of CDI form design—for example, "running out of nouns" to learn: children may increasingly be learning nouns that are not on the forms.

checklist, again as a function of total vocabulary size. Shown in this way, the curves reflect when, in development, children are reported to produce half of the words in each of the categories, as represented by the solid horizontal line. For example, it is easy to see that 50 percent of the nouns have been checked (on average) when total vocabulary is between 200 to 300 words, whereas 50 percent of the predicates are reported (on average) when overall vocabulary size falls between 300 and 450 words. Finally, closed-class opportunities do not reach the 50 percent mark until total vocabulary falls between 500 and 600 words.

11.1.2 The Noun Bias

Why do children learn nouns before verbs and other types of words? This question has received a great deal of attention in the literature; we briefly summarize some of the major issues here. Before doing so, we note that Bates et al. (1994) considers the contrast between nouns and predicates while other literature considers verbs specifically rather than grouping predicates together (e.g., Gentner, 1982). In the analysis below, we split the difference by beginning with predicates (to increase the amount of data we consider) and later breaking down that category into verbs and adjectives.

One reason for the widely reported "noun bias" (overrepresentation compared with verbs and predicates more generally) could be that nouns are simply more frequent in the talk to young children. It is well established that children learn the words that they hear more often (e.g., Hart and Risley, 1995). Many observational studies of English-speaking caregivers have demonstrated that caregivers use more nouns than verbs (in both types and tokens) with their children (e.g., Fernald and Morikawa, 1993; Goldfield, 1993; Gopnik et al., 1996; Kim et al., 2000; Poulin-Dubois et al., 1995; Tardif et al., 1997). Gentner (1982) considered and rejected this explanation, however: predicates and function words are often *more* frequent than nouns, not less; thus pure frequency is not a sufficient explanation of the phenomenon. We investigated the role of word frequency in more detail in chapter 10; our findings are consistent with Gentner's.

Other researchers have framed the "noun bias" in terms of universals about what and how different words "partition" things (e.g., objects, people, relations, qualities, and so on) in the world. For example, Gentner (1982) argued that children learn nouns before verbs because the meanings of nouns are easier to encode since they identify things that can be more easily differentiated in the world (e.g., common everyday objects). Verbs and other predicates, in contrast, express *relations* among things in the world. Hence, the meanings of verbs are less accessible to children through common, everyday experiences and thus are more difficult to map onto word forms without additional linguistic or social support. In addition, verbs and predicates may be more likely to vary in their precise meaning across languages.

Another reason that nouns might be easier than verbs for young children is that nouns tend to be less morphologically complex than verbs (e.g., Tardif et al., 1997). For example, in many languages, nouns are typically marked for number, case, and gender, while verbs are likely to additionally carry tense information. In English, at least, verbs might also be harder to learn because they tend to occur in sentence-medial position (rather than sentence final), which make verbs less salient in the input that children hear (Slobin, 1985; Caselli et al., 1995).

Finally, differences in children's acquisition of nouns versus verbs might result from differences in the contexts in which children hear nouns versus verbs in the speech from caregivers (e.g., Choi and Gopnik, 1995; Tardif et al., 1999). Several researchers have examined what caregivers talk about using naturalistic data of caregiver-child interactions. For example, caregivers in some cultures tend to emphasize the names for people or things in the world and spend a great deal of time providing labels and "names for things" for their children. In other cultures, caregivers do so much less frequently, instead focusing on the actions in which those objects or people engage (e.g., Fernald and Morikawa, 1993; Gopnik et al., 1996). These differences in input to children could influence which words are salient for children and hence which words they are most likely to learn.

What is the evidence that a noun bias is a universal property of children's vocabularies? Documenting the extent to which the noun bias is universal is relevant to understanding mechanisms of language learning—in particular, the presence of conceptual biases in early acquisition and the role of cross-cultural variability in the input that children receive from caregivers. Some studies find consistent evidence for a noun bias in English as well as in Korean and Italian (Bates et al., 1994; Au et al., 1994; Caselli et al., 1995; Kim et al., 2000). The literature is mixed, however, and other studies do not find evidence of a noun bias in languages as varied as French, German, Chinese, Estonian, and Korean (Bassano, 2000; Bloom et al., 1993; Choi and Gopnik, 1995; Kauschke and Hofmeister, 2002; Tardif, 1996; Tardif et al., 1999; Schults and Tulviste, 2016).

Identifying the extent of cross-linguistic variation versus consistency with respect to the noun bias has been difficult, however, since variation across studies may be due to the different methodologies that are used. Even within a single language—Korean—parent reports of children's first words find a noun bias (e.g., Au et al., 1994), whereas studies using direct observational methods find less evidence for this pattern (e.g., Gopnik et al., 1996). Few studies have had the scope to directly compare the extent of the noun bias across multiple languages using a common methodology.

One notable exception in a literature where samples have been small—in terms of both languages and children—is Bornstein et al. (2004a), in which the researchers compared vocabulary composition in seven languages. In this chapter, we follow their comparative approach (see also Tardif et al., 2008). Since we have access to many more observations, our approach offers a more comprehensive approach than these earlier studies. Moreover, we attempt to estimate the extent to which languages show a noun bias; we develop a statistical method for quantifying the extent of the noun bias across the entire developmental range in which a particular form is used.

11.2 Methods and Data

Each CDI form contains a range of words from different syntactic categories. As in chapter 10, we adopt the categorization of Bates et al. (1994), categorizing words into nouns, predicates (verbs, adjectives, and adverbs), and function words (also referred to as "closed class" words). For each child's vocabulary, we compute the proportion of the words in each of these categories that they are reported to produce. Following the approach developed by Bates et al. (1994), for each of the languages in our sample, we plot these proportions against total vocabulary. We use a style similar to figure 11.2, with some modifications; our version of this plot is shown schematically in figure 11.3. Taking nouns as our example, if a child learns words irrespective of syntactic category, then the proportion of nouns in the child's vocabulary should be the same on average as the proportion of total vocabulary that the child knows (the diagonal). In contrast, if the child learns more nouns than expected, their datapoint would be above the diagonal; if they learn fewer, their datapoint would fall below the diagonal. By averaging these datapoints, we can then assess the average bias for or against nouns (and other categories as well).

In this chapter, we limit our analysis to traditional WS and Words and Gestures (WG) forms (along with variants in these classes) because short forms, like the one used in the British English Twins Early Development Study (TEDS) study, do not typically include category information. We also exclude longitudinal administrations so as not to over-weight particular children in our estimates of the extent of category biases in the population.

Figure 11.4 shows this analysis for the American English WS data. Each point shows an individual child's vocabulary, and each panel shows a different lexical class (thus each child is represented once in each panel). The curves show the relationship between a class and the whole vocabulary. We capture the overall trend in this plot by estimating a generalized linear model over the data, predicting category proportion as a function of total production (shown by the thick lines). This model is fit with third-order polynomials to allow both concave and convex

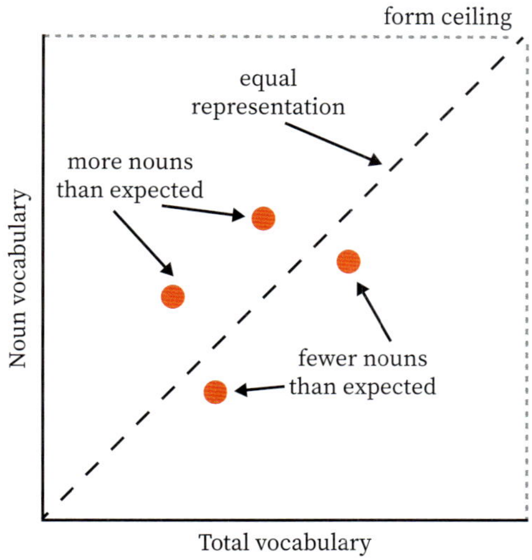

Figure 11.3
Schematic of our vocabulary composition analysis.

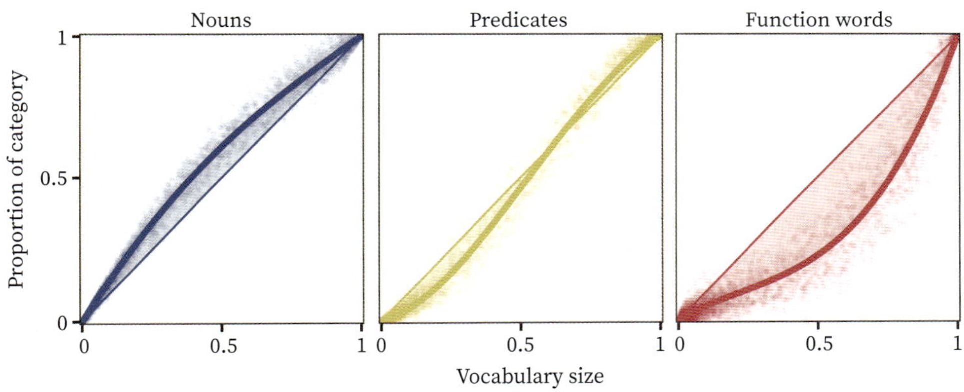

Figure 11.4
For American English WS data, proportion of each lexical category produced by each child as a function of the proportion of all vocabulary items produced by that child. Lines show model fits.

functions and changes in convexity. Using quadratic programming, we fit these models with the constraint that they must predict the point ($x=1, y=1$) so that they are guaranteed to arrive at the diagonal point in the special case that all words on a form are checked.[2]

The final step in our method is to capture the overall bias in a particular sample by estimating the difference in area between the curve and the diagonal. If the curve is substantially above the diagonal, this difference will be positive (indicating, e.g., a positive noun bias). In contrast, if the curve is below the diagonal, the difference will be negative. To capture uncertainty in this area estimate, we conduct a resampling analysis where we randomly resample each population of children 1,000 times with replacement, then recompute the area measurement. Confidence intervals displayed below are based on this resampling procedure.

Critically, this analysis controls for a number of confounds in previous analyses. First, because our interest is in the shape of the overall curve, underrepresentation of children in some age bands should add uncertainty but not bias. Of course, if data are too sparse, estimates will be unconstrained (visible in wide confidence intervals), but particulars of age sampling should not bias our estimates. Second, in principle, the analysis should not be biased by the number of items in a particular category as the analysis is relative to the numerical representation of a particular class on the form. Thus we should be able to compare across forms with larger or smaller numbers of items in particular sections.

11.3 Results

We present results of this analysis across languages, beginning with comprehension for WG-type forms and moving to production for WS-type forms. We do not analyze WG production data here, however. For the most part, production estimates on WG forms are quite low, and hence curves are relatively unconstrained (or determined by a small number of children who are reported to have very large early vocabulary sizes).

11.3.1 Comprehension (WG)

Comprehension results are shown in figure 11.5. This representation of the data is the most complete, but it can be somewhat overwhelming. The largest trend visible in these plots is the highly consistent underrepresentation of function words. In contrast, though most theoretical discussion has centered on nouns and

2. We experimented with a number of model classes and found that these polynomial models seemed flexible enough to provide good fit to a wide variety of patterns of data. Because our concern is not making statistical inferences about the shape of these curves but rather estimating the area under them, we are not specifically worried about overfitting.

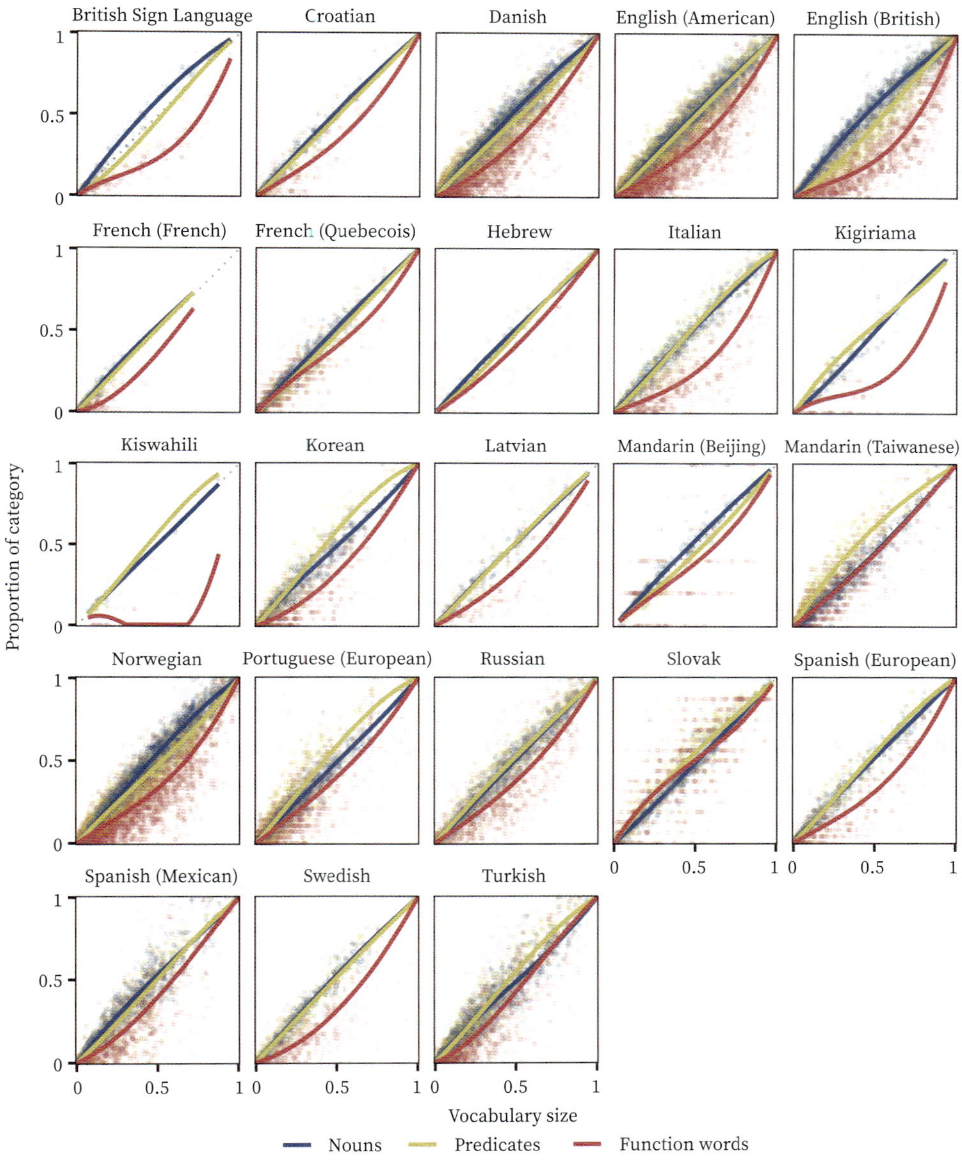

Figure 11.5
For each language's comprehension data, proportion of each lexical category produced by each child as a function of the proportion of all vocabulary items produced by that child. Lines show model fits.

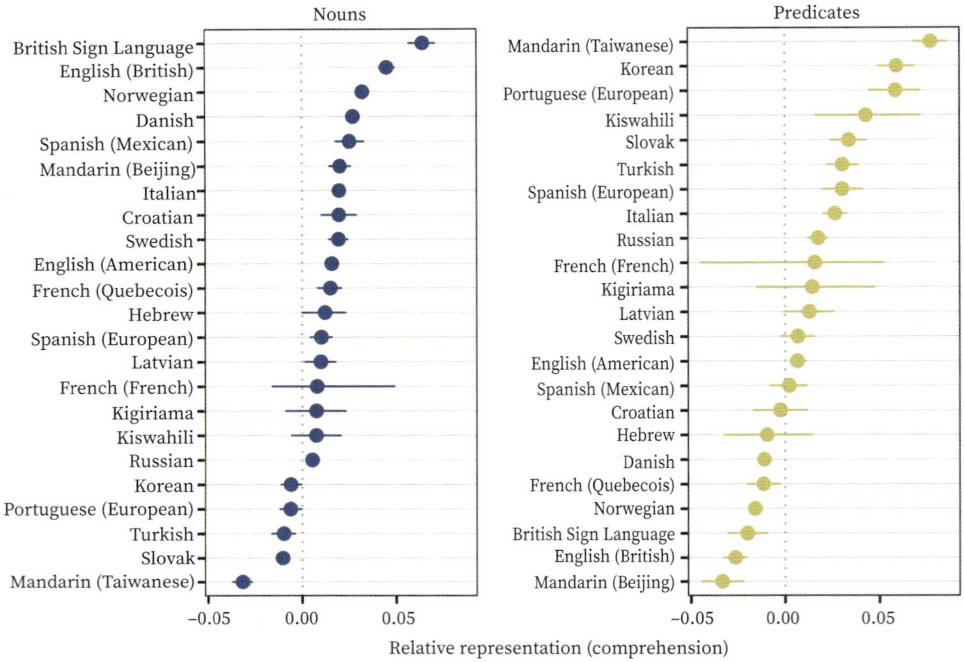

Figure 11.6
Relative representation in vocabulary compared with chance for nouns and predicates for comprehension data in each language (line ranges indicate bootstrapped 95 percent confidence intervals).

predicates, these two categories appear quite close to one another in most languages. For further detailed comparison, we show summaries of curve areas for each language in figure 11.6 and figure 11.7.

Nouns are overrepresented in many—but not all—languages. Mandarin, Slovak, Turkish, Portuguese, and Korean, a set of typologically and culturally distinct languages, show slight underrepresentation of nouns, while British Sign Language and British English show the largest overrepresentation of nouns. Predicates are underrepresented in some datasets and overrepresented in others. Even datasets that are very closely related (e.g., Taiwanese Mandarin and Beijing Mandarin) show substantial differences in the degree of predicate bias. Confirming the visual impression of a trade-off between nouns and predicates, there is a very strong negative correlation between noun and predicate bias measures ($r(21) = -0.82$); this correlation should be interpreted with some caution as nouns + predicates + function words are constrained to sum to one, so some degree of correlation is assured.

Function words are substantially underrepresented across nearly every language and dataset in our sample (except Slovak). The scale difference on this

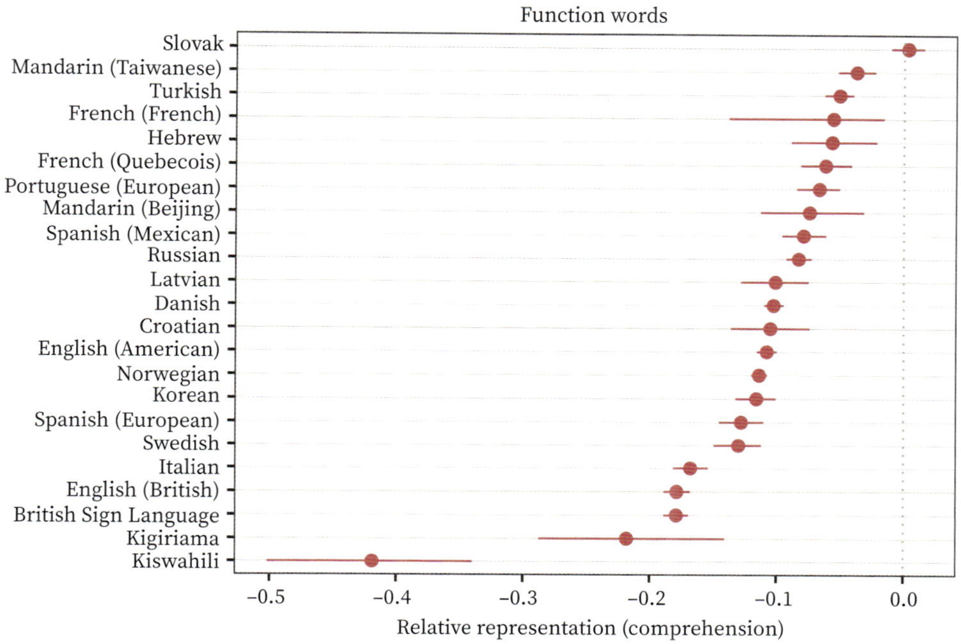

Figure 11.7
Relative representation in vocabulary compared with chance for function words for comprehension data in each language (line ranges indicate bootstrapped 95 percent confidence intervals).

plot should also be noted—function word bias values are far more extreme than noun and predicate bias values. These results likely reflect some combination of true underrepresentation of function words as well as the difficulty of reporting on function-word comprehension in very early language (see chapter 4 for more details on this issue). Such issues may also vary across cultures, languages, and administration methods. Function word representation might plausibly differ due to linguistic factors such as morphological complexity, pronoun dropping, agreement, and so on. However, it is also notable that the two lowest function-word scores come from Kiswahili and Kigiriama (Alcock et al., 2015), a study in which the predominantly rural, low-education parents may have had substantially less metalinguistic awareness.

11.3.2 Production (WS)

We next turn to production data from WS-type forms (figure 11.8). We can immediately see the same trend in the underrepresentation of function words as we observed in comprehension. In addition, however, in many but not all languages, a noun bias is more evident than it was in comprehension.

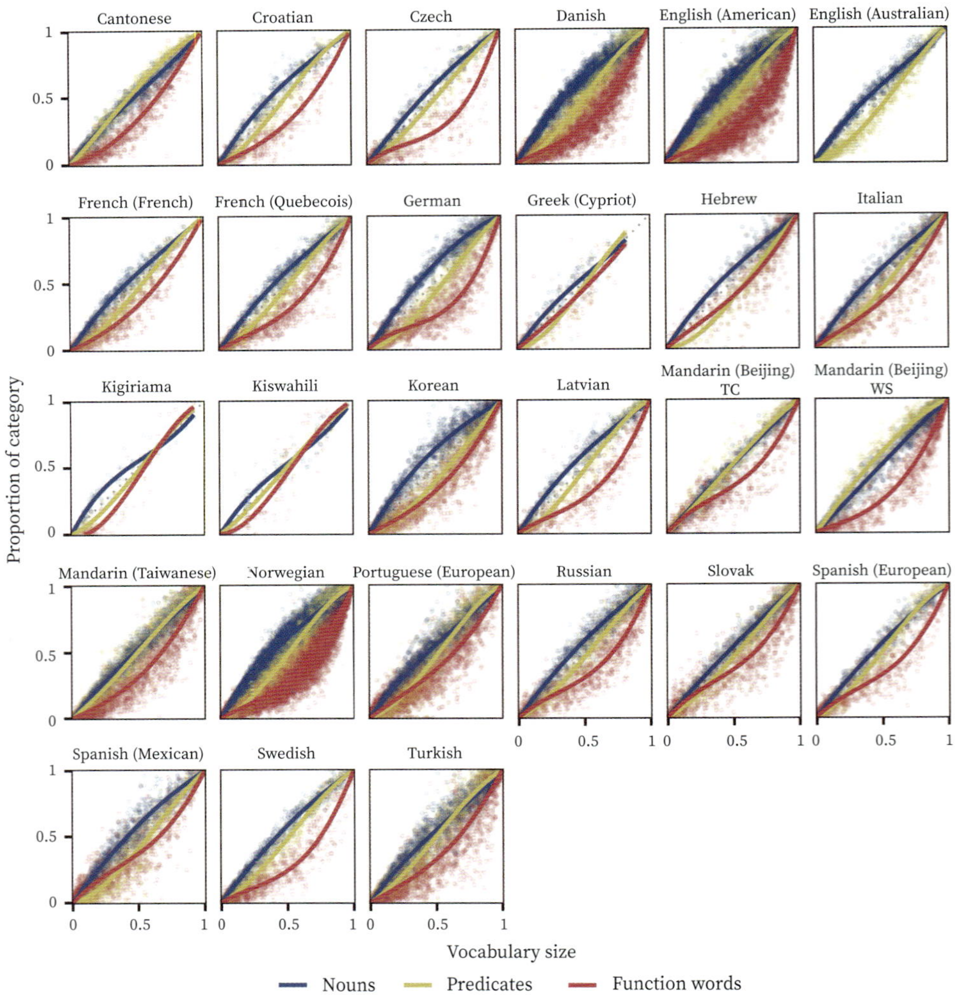

Figure 11.8
For each language's production data, proportion of each lexical category produced by each child as a function of the proportion of all vocabulary items produced by that child. Lines show model fits.

Turning to the language summaries (figure 11.9), we see a larger pattern of variation in nouns and predicate representation. Every language has a relative overrepresentation of nouns, though the degree of this overrepresentation varies. German and Korean have especially large noun biases; the Mandarin and Cantonese datasets have especially low biases (we return to this trend below). Overall the noun bias is both more extreme and more consistent in production data than comprehension. In contrast, predicate representation is both more variable and more negative than we observed for comprehension. Mandarin and Cantonese

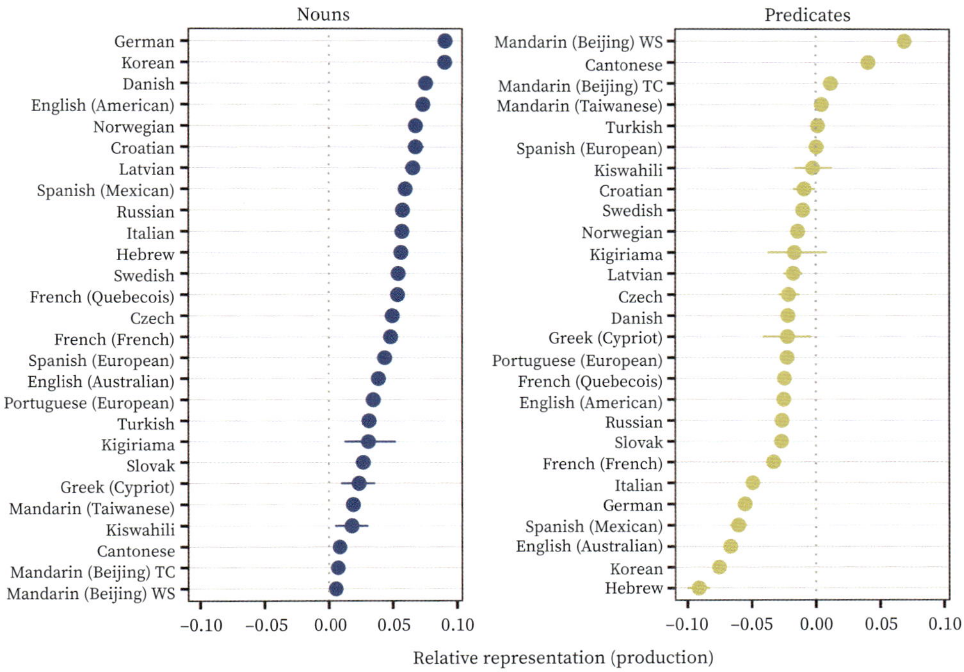

Figure 11.9
Relative representation in vocabulary compared with chance for nouns and predicates for production data in each language (line ranges indicate bootstrapped 95 percent confidence intervals).

are the only languages with overrepresentation of predicates in early vocabulary (and in some datasets, this overrepresentation is substantial). Further, noun and predicate representation is again negatively correlated ($r(25) = -0.65$).

The negative representation of function words (figure 11.10) is generally consistent in magnitude with that seen in comprehension. Across all languages, children are reported to produce fewer function words than would be expected by chance sampling.

11.3.3 Predicates

In contrast to Bates et al. (1994), much of the writing about the noun bias has emphasized the contrast between nouns and verbs rather than predicates more generally. Thus, as an auxiliary analysis to those above, we break down the predicates category further and examine the bias estimates in production for verbs and adjectives, as shown in figure 11.11. A handful of languages also have adverbs as part of their predicate set; for simplicity and because of the smaller sample size, we omit adverbs from this analysis. We also focus here on production rather than comprehension.

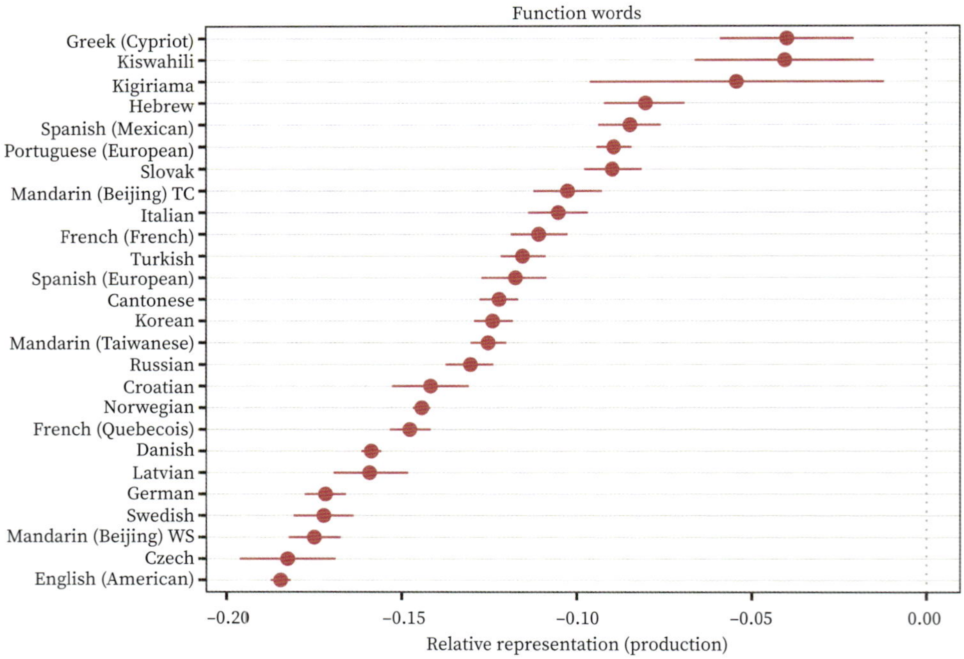

Figure 11.10
Relative representation in vocabulary compared with chance for function words for production data in each language (line ranges indicate bootstrapped 95 percent confidence intervals).

Overall, this analysis suggests substantial variability in verb bias with more languages with positive verb biases than languages with a positive predicate bias as a whole. The largest verb biases can be seen in Mandarin, Cantonese, and Kiswahili datasets, but we also see positive biases in Swedish, Turkish, Kigiriama, Russian, Croatian, and Norwegian, a typologically unrelated set of languages. In contrast, to verbs, adjectives tend to be underrepresented in almost all languages, with Beijing Mandarin varieties being the sole exception.

11.3.4 Reliability of Bias Estimates

Given the seeming differences between comprehension and production noted above, one natural question for further investigation is how consistent estimates of bias are across measures. Figure 11.12 shows—for the sample of languages in which we have data from both WG-type and WS-type instruments—the relative bias we recovered in the analysis above. Somewhat surprisingly, correlations between these different instruments are quite low. Function word bias is negatively correlated between production and comprehension ($r(19) = -0.39$, $p = 0.08$). This result is likely due to Kiswahili and Kigiriama, which, as discussed above, have the lowest

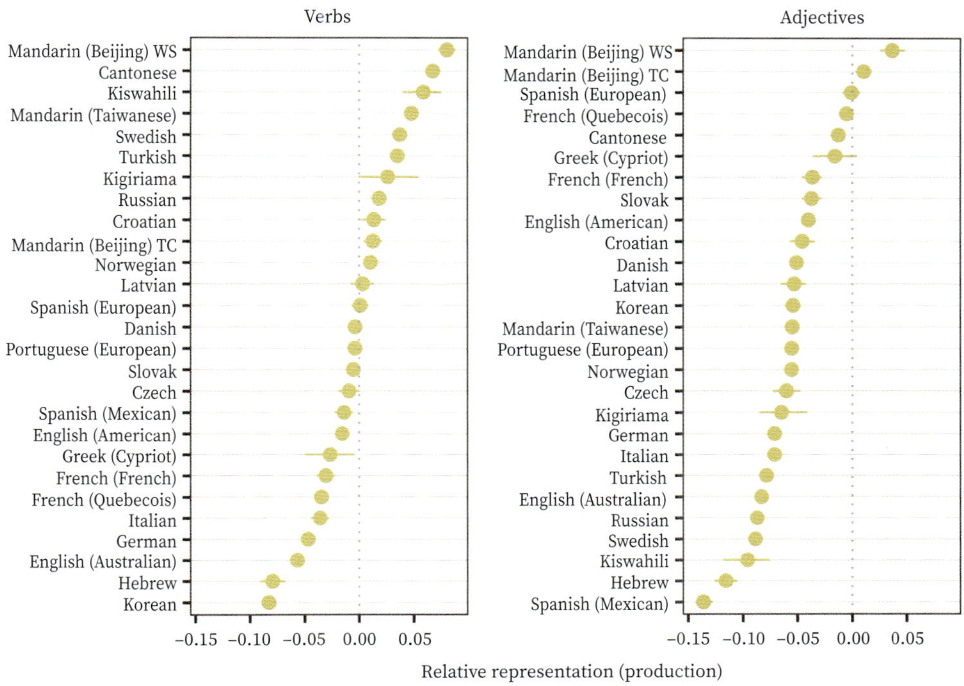

Figure 11.11
Relative representation in vocabulary compared with chance for verbs and adjectives for production data in each language (line ranges indicate bootstrapped 95 percent confidence intervals).

values for function word comprehension. But predicate bias estimates are close to uncorrelated with one another across measures ($r(19)=0.02$, $p=0.94$), and the correlation between noun bias estimates is modest though positive ($r(19)=0.43$, $p=0.05$).

This analysis is conducted with only 19 languages and hence has relatively low power (despite the many thousands of children necessary to carry it out). Nevertheless, it raises some important questions. A number of explanations come to mind and appear consistent with the data:

1. Lexical category bias differs between comprehension and production, perhaps because of different mechanisms at work acquiring words for comprehension versus production.
2. Estimates of bias are influenced by the composition of specific forms, so much so that WS- and WG-type forms yield radically different estimates of bias.
3. Estimates of bias are influenced by the stability of parent report for different measures such that comprehension estimates are less meaningful.

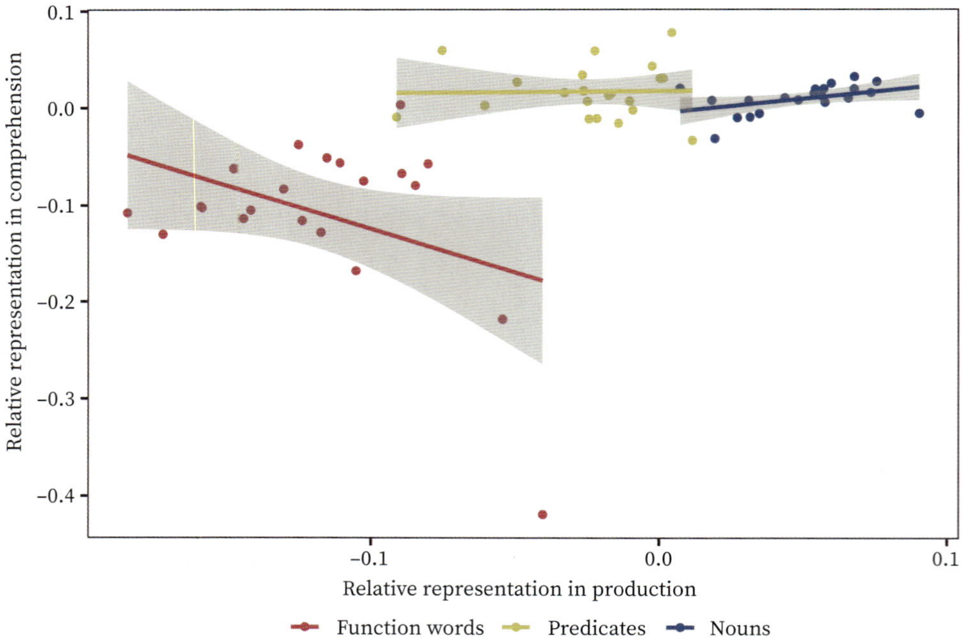

Figure 11.12
Relative representation in vocabulary for each lexical category for comprehension data compared with production data in each language (lines indicates linear regression fits).

4. Bias differs developmentally. Perhaps different biases are evident earlier versus later in acquisition.

We assess each of these explanations in turn.

Our first potential explanation is that category bias is simply different between production and comprehension because vocabulary is different. To test this hypothesis, we need data on a sample of children who are matched in both age and language on which we can compare production with comprehension. Unfortunately, as noted above, data on production from standard WG forms is simply too sparse to perform our bias assessment method; since most children do not produce half of the words on the form, the shape of the bias curves is driven primarily by older children. Thus, to assess comprehension/production differences as a source of bias, we examine the Oxford CDI (shown in figure 11.13). The Oxford CDI is relatively unique in that it includes comprehension questions even later in development, so we can compare bias estimates directly across younger and older children.

In the Oxford CDI data, the measured noun bias for comprehension is 0.05; for production it is 0.07. These values for predicates are −0.03 and −0.08, respectively. These values are somewhat similar to one another, but they do vary beyond

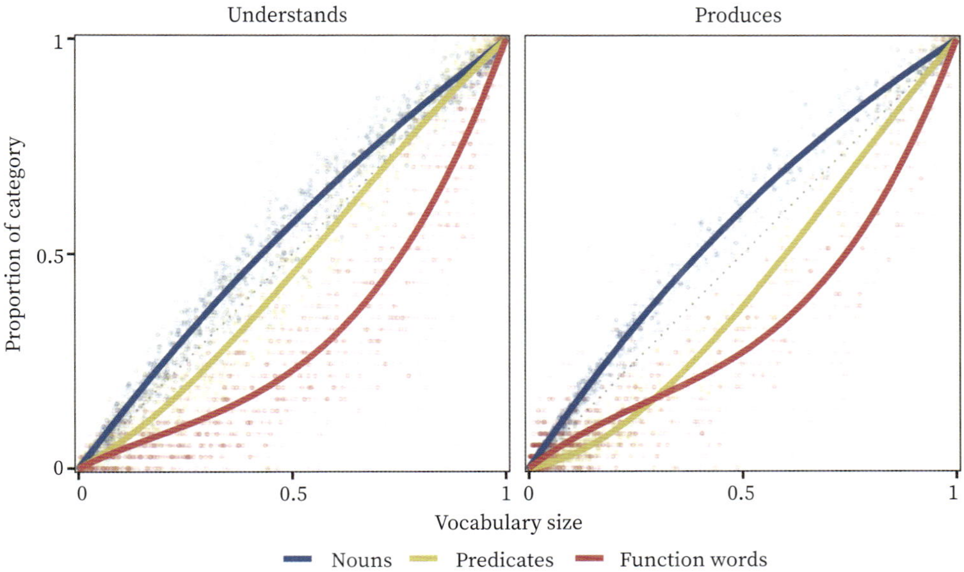

Figure 11.13
For Oxford CDI data, proportion of each lexical category produced by each child as a function of the proportion of all vocabulary items produced by that child. Lines show model fits.

the average confidence interval on each (± 0.01). Thus, there is some evidence for comprehension/production asymmetries.

What might be the mechanism for these asymmetries? At least for languages that do not allow argument dropping, producing a predicate typically requires producing other words as well. For example, English-speaking children do not often produce bare predicates. Further, felicitous predicate production requires some ability to combine words syntactically; in contrast, comprehension of predicates (especially verbs) can often be accomplished by guessing based on known arguments (e.g., Gillette et al., 1999). For these reasons, there may be a greater bias against predicates in production compared with comprehension. This explanation is consistent with our data (in which the average production predicate bias is -0.021, while the average for comprehension is 0.013). Thus, comprehension/production asymmetries likely explain some part of the differences we observed above.

The second potential explanation is that bias estimates are related to form composition. Although our method for calculating bias corrects for the number of items from a particular category on a form, it does not correct for the relative difficulty of these items. For example, a form with more predicates might actually show a *lower* degree of predicate bias—more predicates on the form would imply that some of those predicates are relatively more difficult (because the form designer

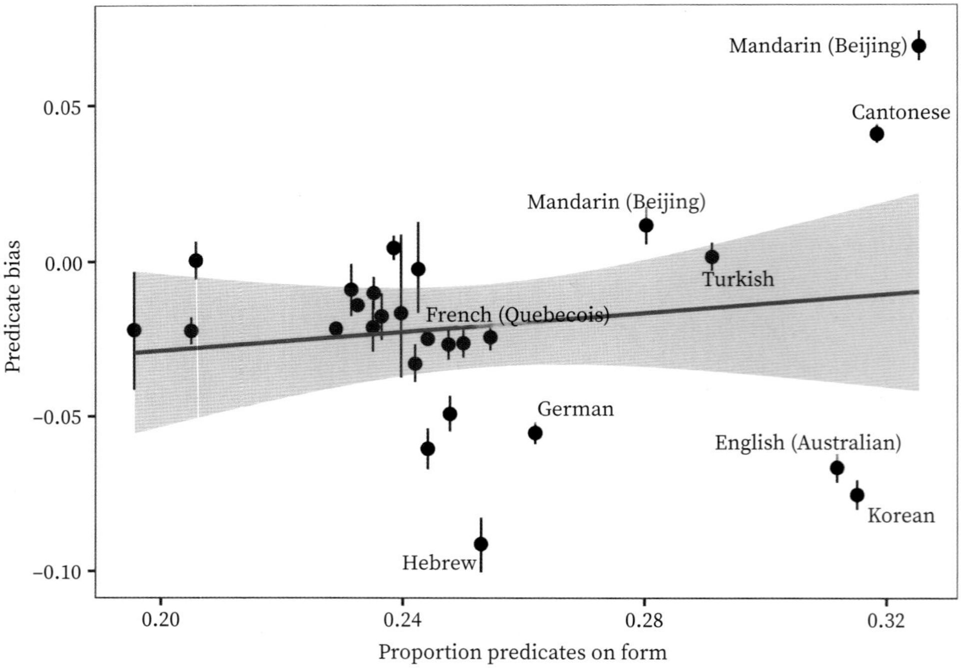

Figure 11.14
Relative representation in vocabulary for predicates as a function of proportion of predicates on form for comprehension data in each language, with languages labeled whose proportion of predicates is greater than 0.25 (line ranges indicate bootstrapped 95 percent confidence intervals). Blue line shows linear model fit.

had "run out" of easy predicates), and hence these predicates would not be checked as frequently by parents. Thus, the lack of correlation between comprehension and production bias measures might be a function of differences in composition across the forms.

We assess this hypothesis in two ways. We examine the relationship between predicate bias and predicate representation on forms (focusing on predicates because they are a minority on the form). Then we consider the case of Mandarin, where we have data from two forms with different compositions.

As shown in figure 11.14, there is no reliable relation between the proportion of predicates on a form and the predicate bias that is demonstrated ($r(25)=0.15$, $p=0.45$). Thus, a simple relation between form composition and bias is not supported. On the other hand, it does appear that there is greater variance in areas for languages with larger numbers of predicates on the form, and languages with the highest predicate representation do have the highest number of predicates on the form as well. Perhaps the causality is reversed: greater numbers of predicates

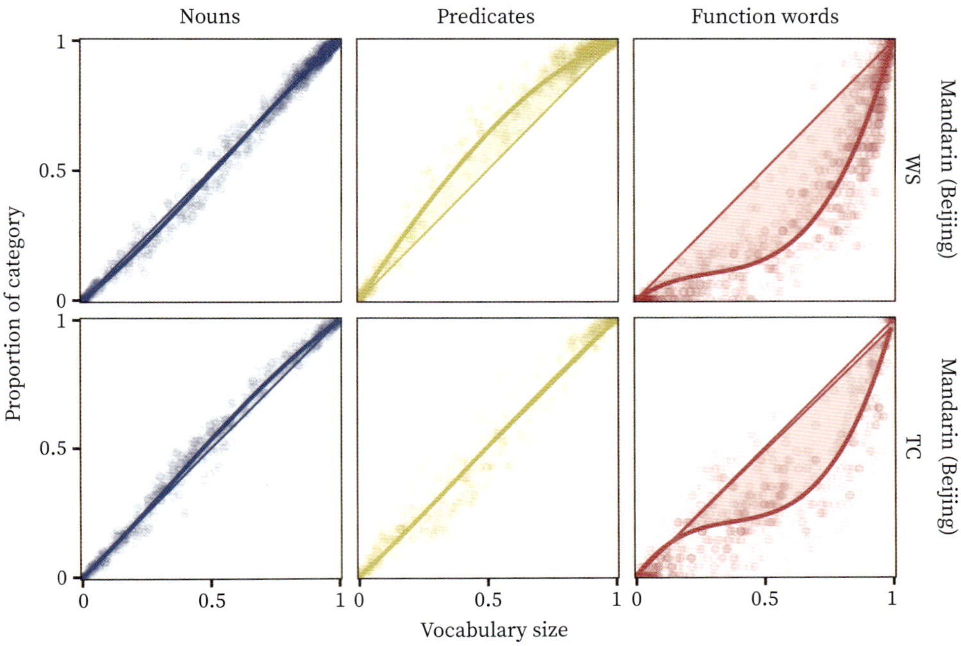

Figure 11.15
For Mandarin WS and Mandarin TC data, proportion of each lexical category produced by each child as a function of the proportion of all vocabulary items produced by that child. Lines show model fits.

have been included in forms for languages like Cantonese, Mandarin, and Korean where the predicate bias is an open theoretical question (or where the acquisition of predicates is of special interest).

The existence of two different forms for Mandarin opens the possibility of a further, more direct test of this issue. The Mandarin WS form (Tardif et al., 2009) and the Mandarin Toddler Checklist (TC) form (Hao et al., 2008) are completely independent forms but represent large datasets collected on Beijing Mandarin specifically. The lexical category representation for these two instruments is shown in figure 11.15. The Mandarin WS form is 33 percent predicates and shows overall a 0.069 predicate preference while the TC form is 28 percent predicates and shows overall a 0.012 predicate preference. The intersection of these forms yields 330 items, with 30 percent predicates. Interestingly, when analyzing only shared predicates, the predicate representation for the TC and WS samples still differs, if anything, more substantially: for WS it is 0.094 and for TC, 0.004.

This result is worrisome—these are samples from the same city and using the same items. They even include very similar age ranges: 16- to 30- month-olds and 17- to 30- month-olds, respectively, with approximately uniform sampling. The suggestion is then that differences in predicate bias can be substantial based on relatively minor details, such as specifics of administration or form context or specifics of sample composition. Extrapolating outward, despite the apparent stability of these estimates under resampling (confidence intervals for bias estimates are around ±0.005 in the analysis above), we should be cautious in overestimating our degree of certainty in particular bias estimates. Further, these data provide more evidence against the notion that form composition (or at least the specific sample of predicates being assessed) is the primary determinant of bias.

The third hypothesis that we examine is the possibility that comprehension and production bias estimates differ because of differences in parent report processes. This explanation is related to our first but is distinct—here, we appeal to the relatively lower psychometric stability of comprehension reports. Yet, we presented evidence in chapter 4 that comprehension, while apparently less reliable, is still fairly reliable and does not differ too much by syntactic category. Further, in other chapters (e.g., chapter 10) we saw strong consistency between comprehension and production results. Thus we do not believe that pure measure unreliability is a good explanation of bias.

The final hypothesis that we examine is that there are developmental differences in bias. Such differences would help explain the observed differences between bias in early comprehension and later production. To address this question, we split data from each language and form into older and younger groups at the median of the data for that sample. We then recomputed our bias estimates, shown in figure 11.16. There was a developmental difference such that older children showed less of a negative function word bias, but differences in noun and predicate bias were very slight for most languages. Thus, overall we do not see evidence that bias estimates for nouns and verbs are globally different for older versus younger children.

In sum, we did not find strong support for effects of form composition, measure reliability, or age on bias estimates. Each of these factors, of course, could contribute in part to the mismatch between WG comprehension and WS production estimates, but none are strongly supported. On the other hand, data from the Oxford CDI suggested some differences in bias estimates between comprehension and production on the same form, with the bias against predicates and for nouns being substantially more pronounced for production than for comprehension. Further, data from two different Beijing Mandarin datasets suggest possible factors relating to population and administration.

11.4 Discussion

This chapter presents a comprehensive examination of the issue of biases for and against particular syntactic categories in acquisition. Building on earlier work by

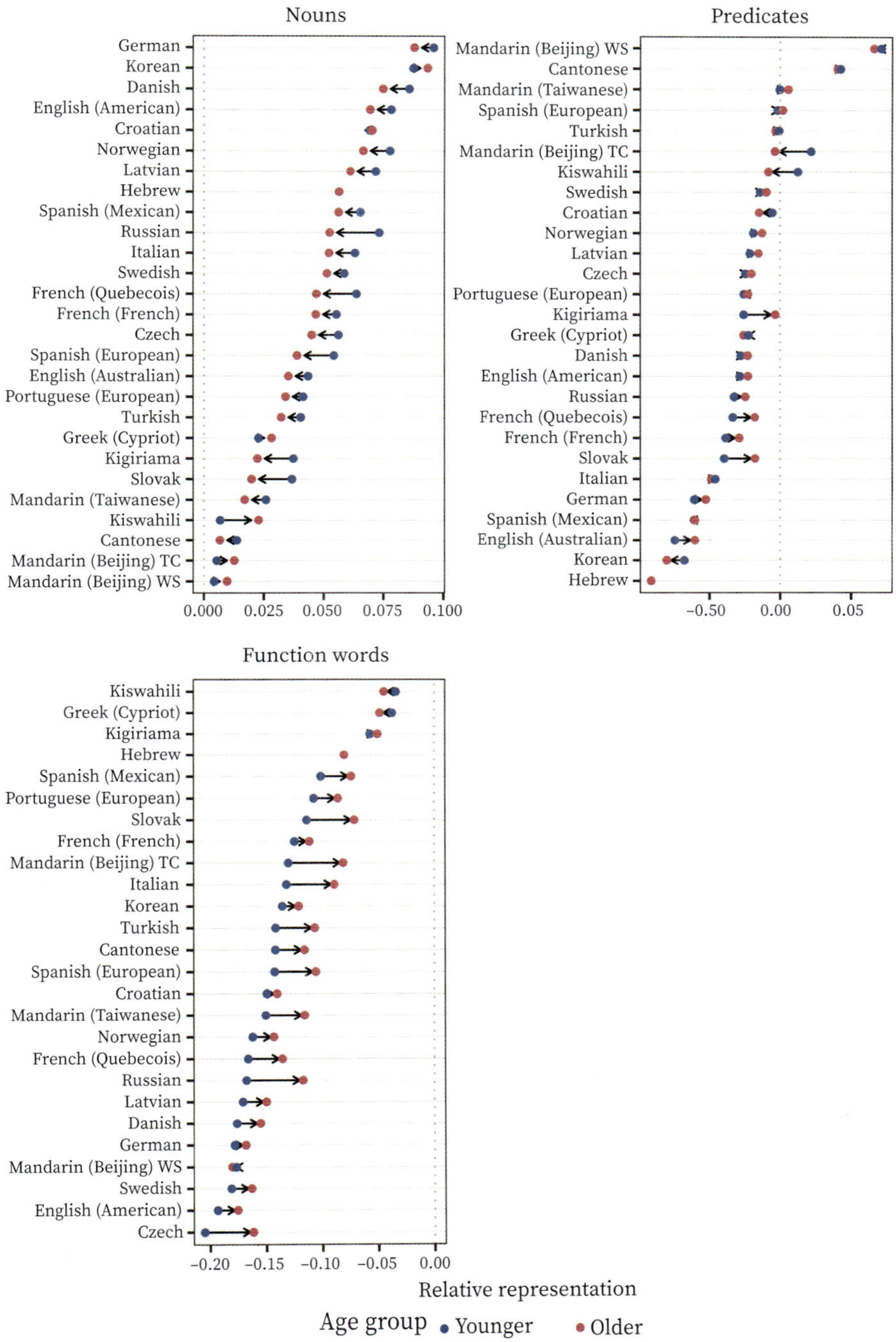

Figure 11.16
Relative representation in vocabulary for each lexical category per age group for production data in each language.

Bates et al. (1994), we created a quantitative measure of noun, predicate, and function word bias and examined variability in these measures across languages. Overall, a number of generalizations emerge regarding the composition of early vocabulary development.

Nearly every language showed a positive bias for nouns in children's early vocabularies, though the degree of this bias varied and was more pronounced in production. And further, every language showed a substantial bias against the early acquisition of function words, supporting the generalization that these are acquired much later than content words, despite their typically higher frequency in the ambient language (see chapter 10). This bias was larger, on average, than noun and predicate biases.

In children's early comprehension vocabularies, there was variability in the degree of predicate representation; in early production vocabularies, as has previously been reported, languages were mostly biased against predicates. There were a few notable exceptions, however, including Mandarin and Cantonese. This conclusion largely supports previous work on these languages (e.g., Tardif, 1996; Tardif et al., 1997). Our Mandarin findings contrast with our finding of no positive predicate bias in Korean, and this differs from the predicate bias found by Choi and Gopnik (1995) in a smaller study. Interestingly, predicate biases in production were in part driven by a relatively consistent bias against adjectives with more variability in the bias for verbs.

Finally, measures of bias in production and comprehension were not highly correlated with one another, especially for predicates and function words. There are likely many causes of this pattern, but one possible explanation appears to be greater predicate comprehension compared with production. However, within two different Mandarin production samples (even subset to the same items), we saw some differences. Thus, one important caveat to the work presented here is that—despite the huge amount of data going into our bias estimates—the resulting measurements may not be as stable or accurate for a particular language as we had hoped. Leaving aside this caveat, a suggestive conclusion of our investigation is that constraints on the production of predicates—that is, that verbs, adjectives, and adverbs are rarely uttered alone—might lead to their relative underrepresentation in production reports. Put simply, if you need to say nouns along with other words but you can also say nouns alone, on average the other words will look less frequent and less well represented in the vocabulary.

In sum, we see clear evidence for a positive noun bias and a negative function word bias with more cross-linguistic variability present for predicates. The precise measurement of predicate bias and an explanation of its variability are important topics for future work.

12 Vocabulary Composition: Semantic Categories

Following the approach in the previous chapter, we next investigate the consistency of semantic content categories across languages. By analogy with the "noun bias," we ask whether early vocabulary development is biased for or against items from a range of semantic categories. For example, are some languages more "vehicle biased" or "animal biased" than others? Consistent biases across languages can provide hints regarding attentional or conceptual factors influencing early word learning. We begin by exploring general patterns in semantic category bias and then focus specifically on a few theoretically interesting conceptual domains, such as words for time, color, body parts, and logical operators.

12.1 Introduction and Methods

In contrast to the "noun bias" literature, where a wide variety of hypotheses have been articulated over the preceding decades, analyses of differences in early vocabulary content have been less frequent. Thus, our current analyses are more exploratory than those presented in chapter 11. One notable piece of prior work is an analysis of cognitive biases in the early language of international adoptees by Snedeker et al. (2012), and we discuss those results further here and in chapter 17, but those data are limited to English learners; our goal is to measure cross-linguistic/cross-cultural variability.

Given the exploratory nature of this chapter's analyses, we focus on Words and Sentences (WS) forms and production measures. As discussed in previous chapters, we have reason to believe these will be most reliable; further, we can take advantage of the larger sets of items and categories available on most WS-type forms.

Rather than taking on the daunting task of creating novel semantic categorizations across languages, we make use of the fact that Communicative Development

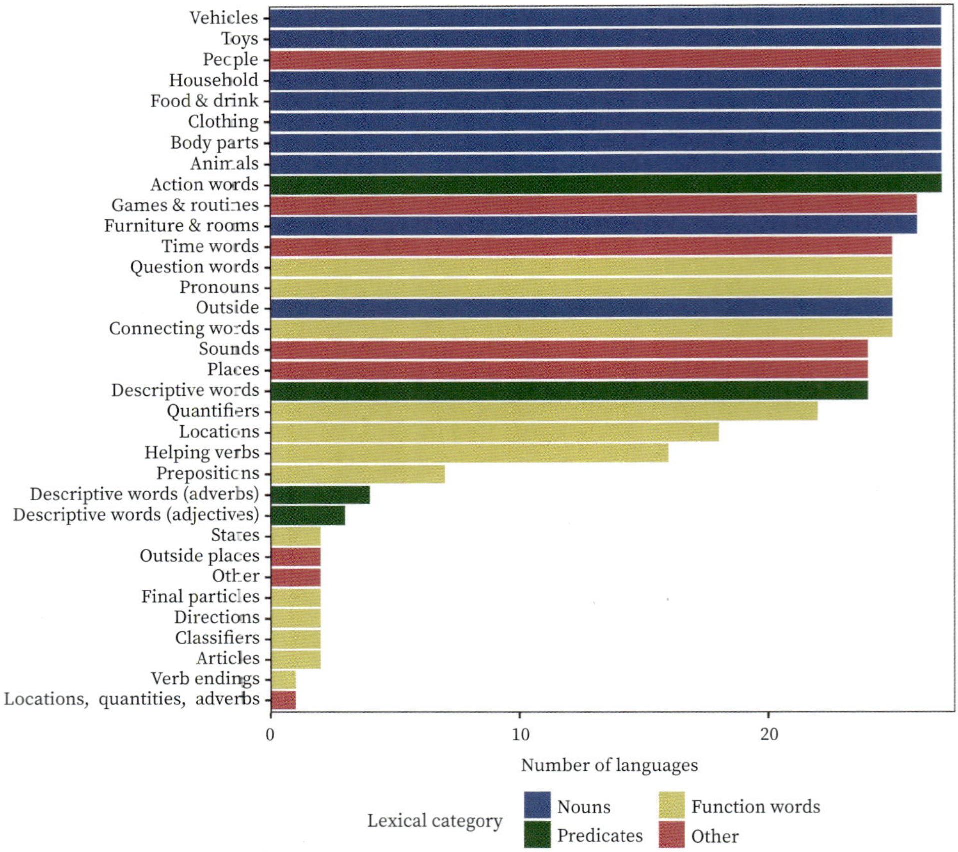

Figure 12.1
Number of languages whose forms contain each semantic category.

Inventory (CDI) forms are typically structured into semantic categories (e.g., Animals or Body Parts). As figure 12.1 shows, while some of these semantic categories are shared across many instruments, there are others that are quite rare (often corresponding to specific syntactic or semantic categories that are of interest in particular languages). We focus on those semantic categories with greater representation in the data. Further, to avoid duplicating our analysis in chapter 11, we focus on those semantic categories that fall into "nouns" and "other" lexical classes. Typically most or all of the predicates and function words we analyzed in that chapter are grouped into a small number of categories; thus, adding categories like Action Words (i.e., verbs) or Descriptive Words (i.e., adjectives) would simply repeat the prior analysis. This filtering step leaves 14 categories: Animals,

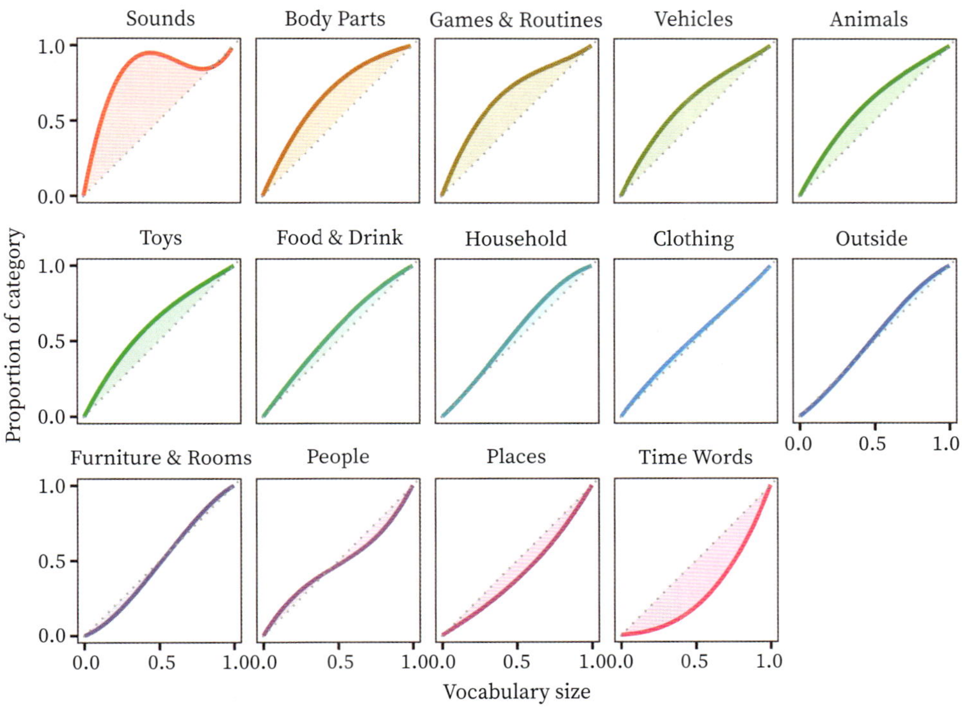

Figure 12.2
For American English WS data, model fit curves for proportion of each semantic category produced by each child as a function of the proportion of all vocabulary items produced by that child.

Body Parts, Clothing, Food and Drink, Furniture and Rooms, Games and Routines, Household, Outside, People, Places, Sounds, Time Words, Toys, and Vehicles.

We first illustrate our approach using data from the English WS form alone. Analogous to the plots in chapter 11, figure 12.2 shows areas where the data deviate from the pattern of category acquisition predicted by random item sampling. The size of the shaded region above versus below the diagonal gives evidence of oversampling versus undersampling for a particular semantic category (panels are ordered by the size of the bias). We omit the full distribution of datapoints as the visual impression is clearer when only shaded regions are shown.

Many of the results of this analysis for English are expected. Sounds items (onomatopoeia) are heavily overrepresented, as are Body Parts, Games and Routines and, to a slightly lesser extent, Toys, Animals, and Vehicles. These particular biases are likely related to particular parenting practices, cultural emphases (for example, on animal names), and young children's' idiosyncratic interests (DeLoache et al., 2007). The largest underrepresentation across categories is Time Words. This

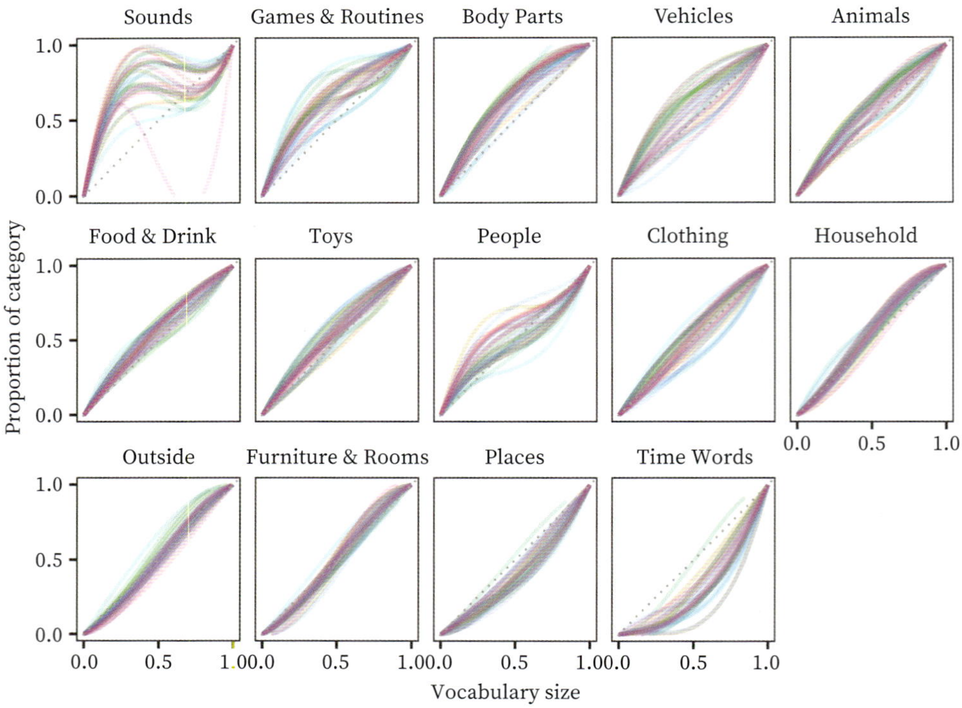

Figure 12.3
Model fit curves for each semantic category as a function of vocabulary size for each language.

pattern is consistent with the body of work on children's acquisition of the semantics of time words that suggests that children struggle with understanding these complex terms through age five (Harner, 1975; Clark, 1971; Tillman and Barner, 2015; Tillman et al., 2017). We next turn to how this pattern varies across languages.

12.2 General Results

12.2.1 Individual Domains

Because there are so many different languages represented in this analysis, the simplest analysis examines the spread of languages across categories (figure 12.3). Somewhat surprisingly, the ordering of categories looks quite similar to what was observed in English. Sounds, Games and Routines, and Body Parts are all over represented. Vehicles, Animals, Food and Drink, Toy, People, and Clothing are all more variable across languages. Small Household Items, Outside Things, and Furniture and Rooms show some variability but overall smaller bias. Finally, Places and Time Words are both underrepresented systematically across all languages.

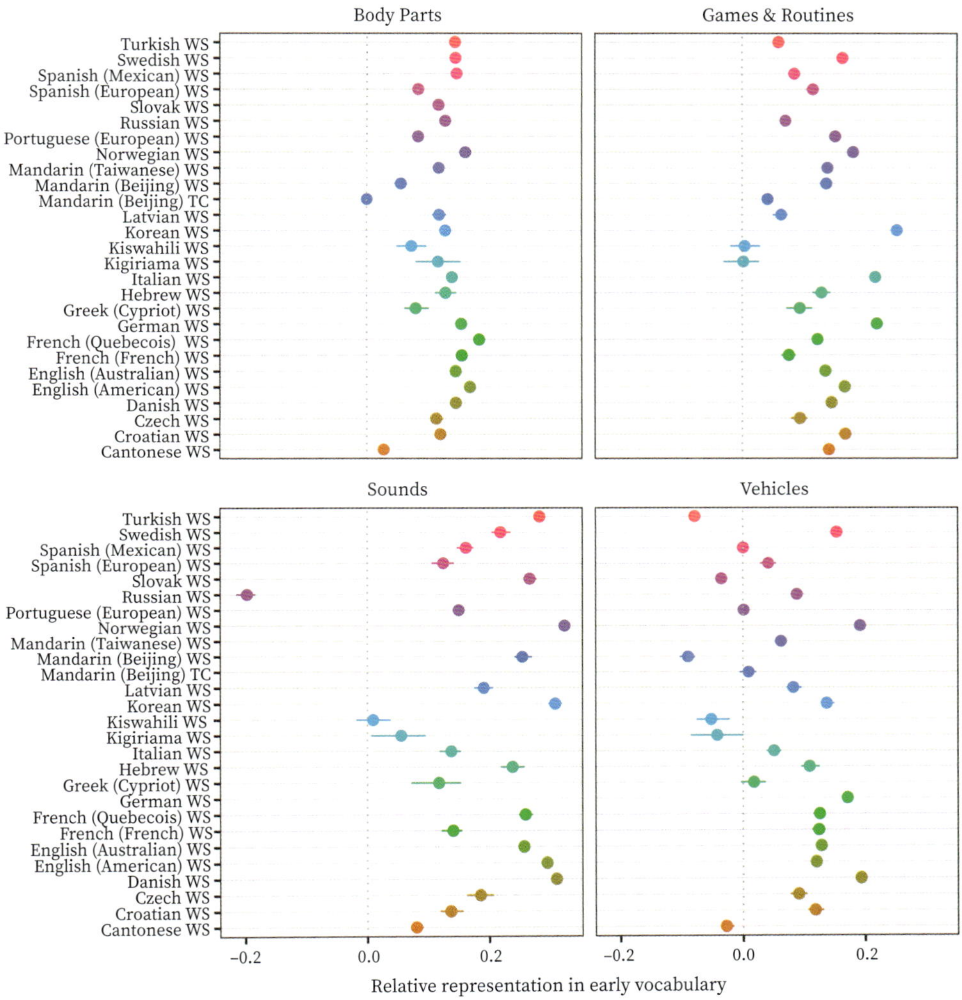

Figure 12.4
Relative representation in vocabulary compared to chance for categories that tend to be overrepresented across languages (line ranges indicate bootstrapped 95 percent confidence intervals).

We next zoom in on the most highly overrepresented categories (figure 12.4). The highest mean comes from Body Parts, which are overrepresented in just about every language (Andersen, 1978). Interestingly, the three datasets with the lowest representation of Body Parts are the two Mandarin datasets (WS and TC) and the Cantonese WS data. Games and Routines are generally overrepresented but somewhat more variable, with Kiswahili, Kigiriama, and Mandarin TC data being the lowest. Sounds are quite highly variable but almost all positive, with Russian being the outlier. Inspection of these items shows *negative* developmental trajectories

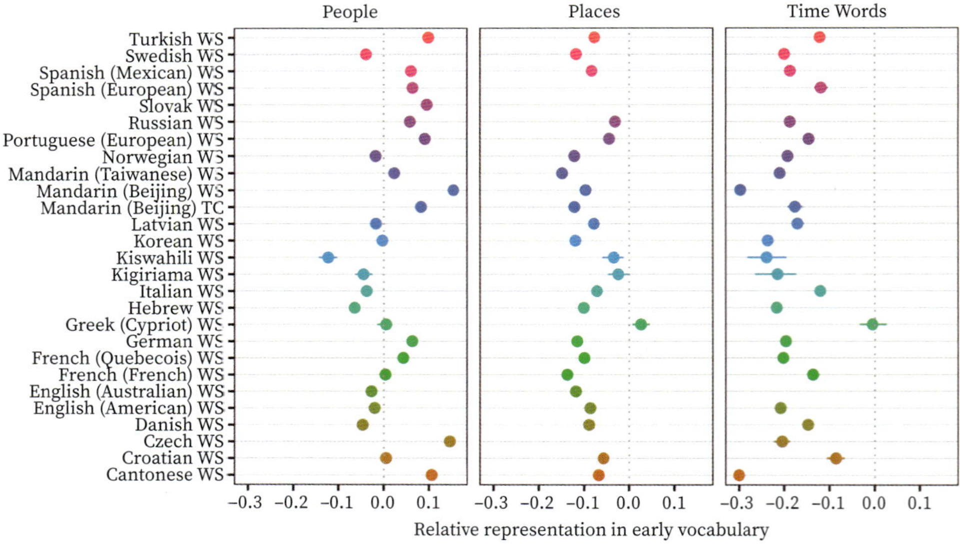

Figure 12.5
Relative representation in vocabulary compared to chance for categories that tend to be underrepresented or highly variable across languages (line ranges indicate bootstrapped 95 percent confidence intervals).

for a number of words in the Sounds category. We believe these data are likely an artifact of parents feeling that they should "trade off" with noun labels in the Animals category and hence that items in the Sounds category are "baby words" and should be discounted. Finally, words in the Vehicles category appear more variable but have positive bias across most languages.

We next consider People, Places, and Time Words (figure 12.5). The category people has highly variable bias, with some languages underrepresenting and others overrepresenting. Tardif et al. (2008) speculated that names for people were a substantial part of children's earliest words, but that may reflect that study's use of Mandarin and Cantonese data where people terms are very overrepresented due to cultural emphasis on family connections. Surprisingly, despite the relatively multigenerational and family-centric nature of children's experience in Kenya (Alcock and Alibhai, 2013), people words were relatively underrepresented in Kiswahili and Kigiriama.

In contrast to the heterogeneity in bias for People words, words in Places and, especially, Time Words were almost uniformly underrepresented in children's vocabulary. As noted above, time is known to be conceptually difficult for children. Time words offer a number of conceptual challenges in terms of mapping an ordered set of durations (second < minute < hour < day, etc.) to a set of concepts that do not map cleanly onto perceptual experience. Less has been written about children's understanding of geographical vocabulary, however. Many of the same conceptual difficulties that are true of time also may hold true for larger locational/geographical hierarchies (neighborhood < city < state < country). Or alternatively, the underrepresentation of places in children's early vocabulary may simply reflect the relative lack of diversity of their experiences with some of the items that traditionally populate this section (e.g., *beach*, *camping*, *church*, and *circus*, to name the first four). See chapter 4 for some evidence that *camping* especially may be a poor reflection of children's overall language experience.

12.2.2 Dimensionality Reduction

Our next analysis takes an exploratory dimensionality-reduction approach. Rather than examining each semantic category individually, we consider the space defined by variation in semantic preferences by running principal components analysis (PCA) on these data. PCA is a dimensionality reduction technique that projects high-dimensional data (e.g., bias by semantic category for each language) into a set of orthogonal dimensions where lower dimensions capture as much of the variance as possible.

Standard PCA requires datasets without missing data, so we removed languages with missing categories. This analysis thus includes 23 language/form combinations and 13 categories (we exclude words from sounds because of the issue with Russian in this category and with other missing data). The total variance explained by the first principal component is 44 percent, and by the second, 16 percent.

Figures 12.6 and 12.7 show the loadings of semantic categories on these two components and the data projected onto the space of the first two principal components, respectively. We examine the loadings of categories on components first. We see that the first principal component (PC1) appears primarily to capture increases in Vehicles, Animals, and Clothing relative to Places and People. Intuitively, this dimension seems like it might be distinguishing the tendency of children learning a language to name small objects versus other entities. In contrast, PC2 appears to capture variance related to differences between People and Games and Routines (both of which categories are overrepresented in the earliest words) compared with Places, Outside Words, and Time Words (most of which are quite conceptually difficult).

These dimensions can be clarified by examining the projection of languages into the reduced-dimensionality space. PC1 appears to be distinguishing languages in which children name objects and animals. English, many Northern European datasets, and Korean are clustered at the far right, with high scores on Vehicles, Clothing, and Animals. In contrast, for those datasets that do not show

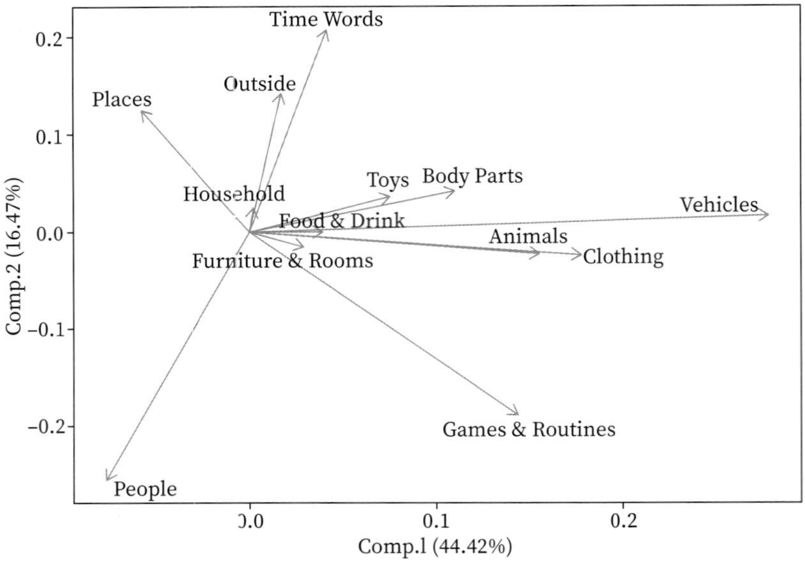

Figure 12.6
Loadings of each semantic category.

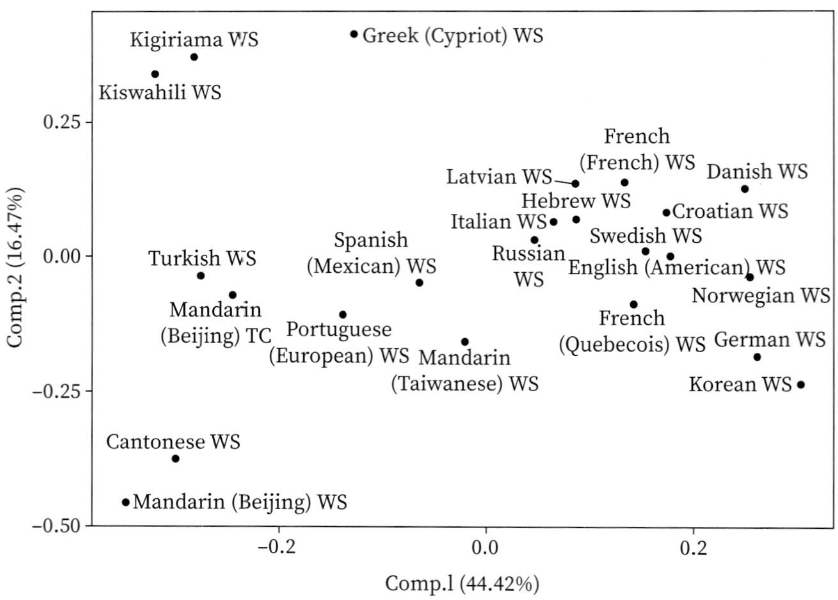

Figure 12.7
First two principal components for each language.

overrepresentation in these categories, we see PC2 being more diagnostic: Mandarin and Cantonese WS data are very far toward the direction of the categories People and Games and Routines. In contrast, Kiswahili, Kigiriama, and Cypriot Greek are especially far in the direction of Outside and Places, perhaps consistent with the datasets being collected in rural and semi-rural areas.

Overall, this analysis reveals some interesting structure that could be tested in future studies. However, we caution that care should be taken not to overinterpret. In particular, as we saw in chapter 11, within-culture differences (e.g., Mandarin TC versus Mandarin WS) are as large in size as between-culture differences. Thus, understanding factors underlying category bias variability should be an important goal for future work.

12.3 Individual Conceptual Domains

In this section we isolate individual items from specific domains of interest. Our approach is to use the "universal lemma" mappings (see section 3.1.5) to find matching lexical items across languages. The specific domains we consider are time, color, body parts, and logical words. We also investigated spatial prepositions and number words but do not include them here. Spatial prepositions present a wide variety of mapping issues since lexical items "cut up" space differently across languages (see, e.g., Bowerman, 1996). And number words are not found on enough CDI forms to have sufficient data for inclusion. Thus, although these categories are of major theoretical interest, other approaches beyond the relatively crude mapping approach we followed here will be necessary.

12.3.1 Time

As discussed above, the semantics of time words are very challenging for children through middle childhood (Tillman and Barner, 2015; Tillman et al., 2017). Despite this, parents report that children do produce some of them by age 2 1/2. The set of words with sufficient translation equivalents for inclusion is *after, day, later, morning, night, now, today, tomorrow, yesterday*.

Figure 12.8 shows trajectories for time lexical items across languages, sorted by difficulty. Because *night* is typically signaled by darkness, it is perceptually very concrete and likely easier than other time words. Similarly, *now* seems relatively more straightforward given that it has a common imperative meaning in sentences like "give me that right now." In contrast, the latest acquired is *yesterday*; this word is highly abstract and requires a sort of "mental time travel" in thinking retrospectively beyond the "here and now" (Busby and Suddendorf, 2005). While *tomorrow* shares those same features, it appears to be learned earlier than *yesterday* on average, whether due to frequency or to other factors.

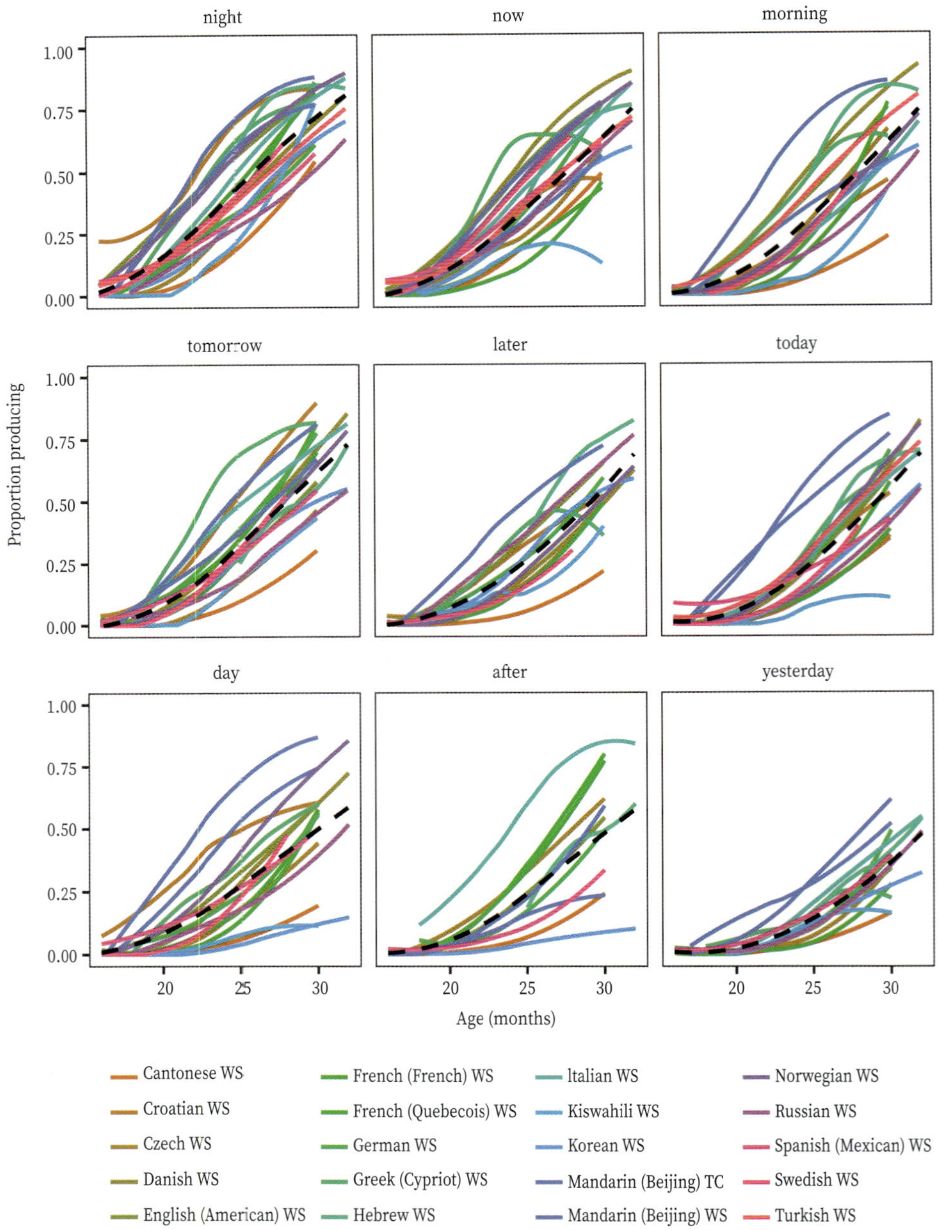

Figure 12.8
Developmental trajectory of each time word in each language.

12.3.2 Color

Color word acquisition has been a focus of interest at least since Carey (1978)'s influential study of "fast mapping." Although early work suggested that color words were learned almost simultaneously (Bartlett, 1977), recent studies have described a more protracted trajectory of partial knowledge. Many children learn some color words and overextend these to cover the rest of color space (Wagner et al., 2013). Adding to the complexity of this issue are the substantial cohort changes in the age at which colors are learned: while school-aged children struggled with their colors 50 to 100 years ago, more recently children learn colors in the age range spanned by the CDI forms (Bornstein, 1985).

There is tremendous cross-linguistic variation in the overall level of color vocabulary (Kay et al., 2009). We take advantage of the fact that most of the languages in our dataset have relatively large color vocabularies, and we can assume this means that individual colors probably have relatively similar extensions.[1] Despite this, most CDI forms do not include all the basic level color words. The set of color words with sufficient translation equivalents for inclusion is *black, blue, green, red, white, yellow*.

In contrast to the variability in overall level of production of color terms across languages, the relative ordering is fairly consistent. In this set of words (figure 12.9), we see that *red* is typically the first learned, although there is substantial variability in when it is first learned. It is followed by *yellow, blue,* and *green,* with *black* and *white* coming last, consistent with reports by Wagner et al. (2013)—see Yurovsky et al. (2015) for an account of factors involved in color word learning.

Additionally, we see a ranking of languages with respect to total rates of color word production reported (figure 12.10). As in other analyses (see chapter 5), Mandarin WS has the highest level of production. American and Australian English also tend to have high levels of color word production. Interestingly, Kiswahili has by far the lowest level of color word production; perhaps it is related to the limited availability of manufactured toys of contrastive colors (Bornstein, 1985).

12.3.3 Body Parts

Across languages, body parts have been claimed as lexical universals; that is, nearly every language provides terms for naming the body (Andersen, 1978). Despite this consistency, there is substantial cross-linguistic variation in exactly which body parts are named, reflecting different segmentation of body forms (e.g., some

1. Such an assumption would not be warranted if we were considering languages with just a handful of color terms where the extension of a term like *red* would be much larger than in English.

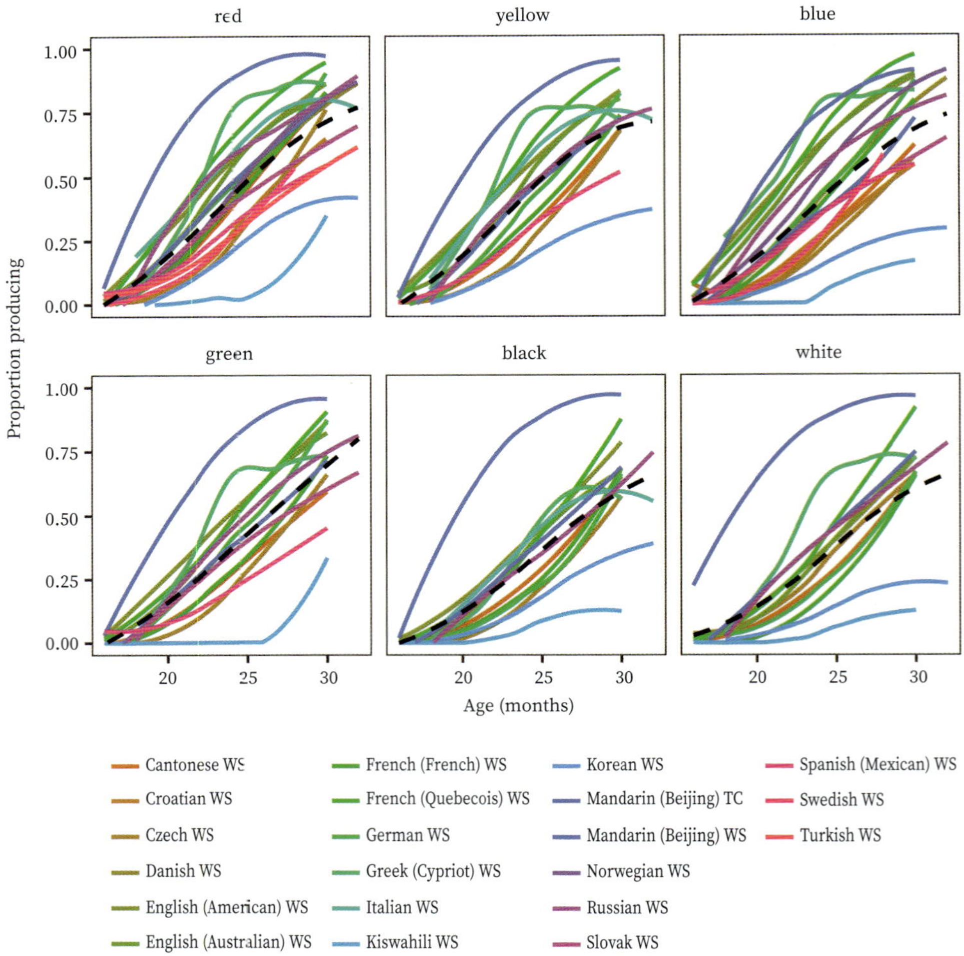

Figure 12.9
Developmental trajectory of each color word in each language.

languages name both *hand* and *arm* while others have a single word for both; Majid 2010). In our data, words for Body Parts (figure 12.11) are produced very early by most children, and the variance is quite low across languages (with the exception of a few terms in Cantonese and Cypriot Greek). One interesting pattern that is visible in these data is the ordering of *hand* and *foot* before *leg* and *arm*. Cantonese is probably an exception here because *hand* and *leg* are monomorphemic (and written with single characters) while *foot* and *arm* in Cantonese are morphological compounds.

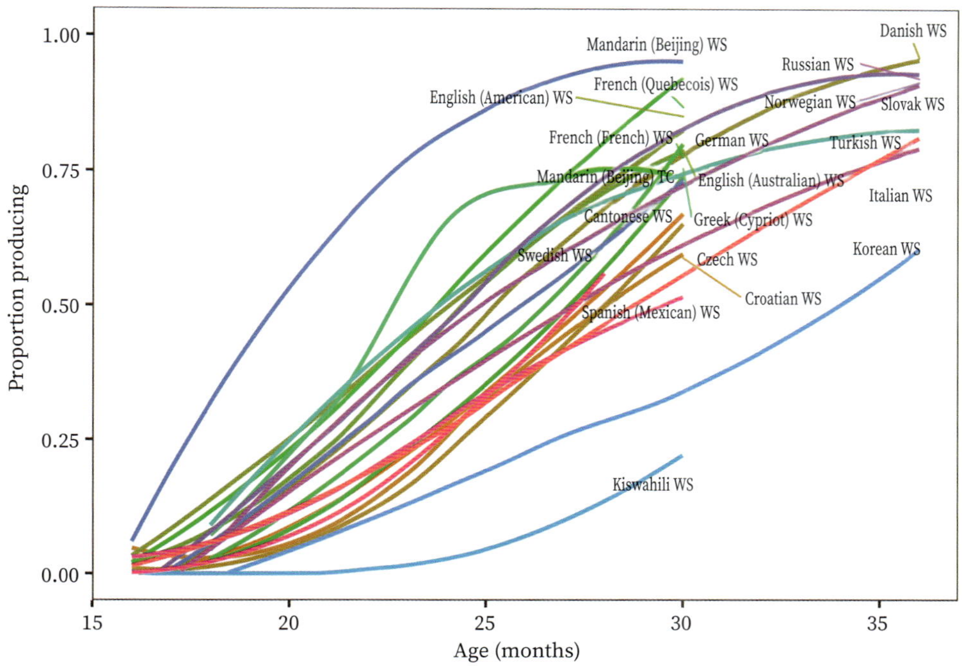

Figure 12.10
Mean developmental trajectory of color words in each language.

12.3.4 Logic

Finally, we examine words for logical operators (figure 12.12). The only items that are available across a substantial number of languages are *all, and, because, none, some, no,* and *not*. One important caveat that applies especially here (although of course to other words as well) is that children may produce many of these words with meanings that are quite different from those they take in adult language. To take just one example, there is a large literature investigating the semantics and pragmatics of "some" in children's language (see, e.g., Barner et al., 2011).

Keeping this caveat in mind, negative words are learned early, with an ordering consistent with Bellugi (1967) and Pea (1982). *No* is very early, and *not* is later. Interestingly, the quantifiers are not ordered as shown by Katsos et al. (2016) in their massive cross-linguistic study. In that study—as well as in work in English (Horowitz et al., 2017)—*all* was found to be understood better than *none*. In contrast, here we tend to find *none* is learned earlier than *all* and definitely learned earlier than *some*. One possibility is that these uses are only found in a restricted set of cases. Another is that contextualized production of negation is simpler than decontextualized comprehension, as found in work on the comprehension of negation in context (Nordmeyer and Frank, 2014, 2018).

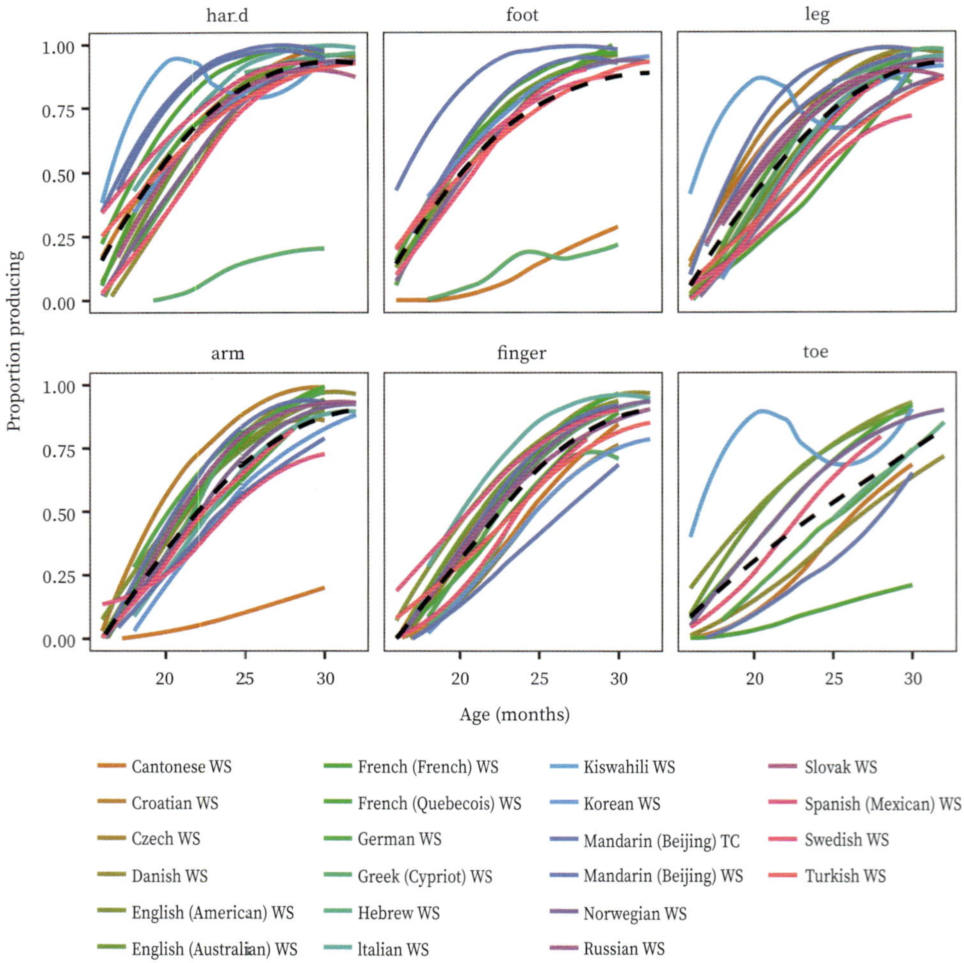

Figure 12.11
Developmental trajectory of each body part word in each language.

12.3.5 Category Variability

To end this section, we quantify the variability across languages for each of these restricted sets of lexical items using the coefficient of variation (CV; see chapter 5).[2]

2. We experimented with applying this approach to the broader set of lexical categories we investigated in the first part of this chapter but were confounded by a particular property of the CV. For those categories with very small mean bias, even if they had small variance, their CV was very high (simply because $CV = \sigma/\mu$, and hence, as μ goes down, CV must go up). For this reason, we choose only categories with some bias. We expand on this limitation of the CV analysis in chapter 16.

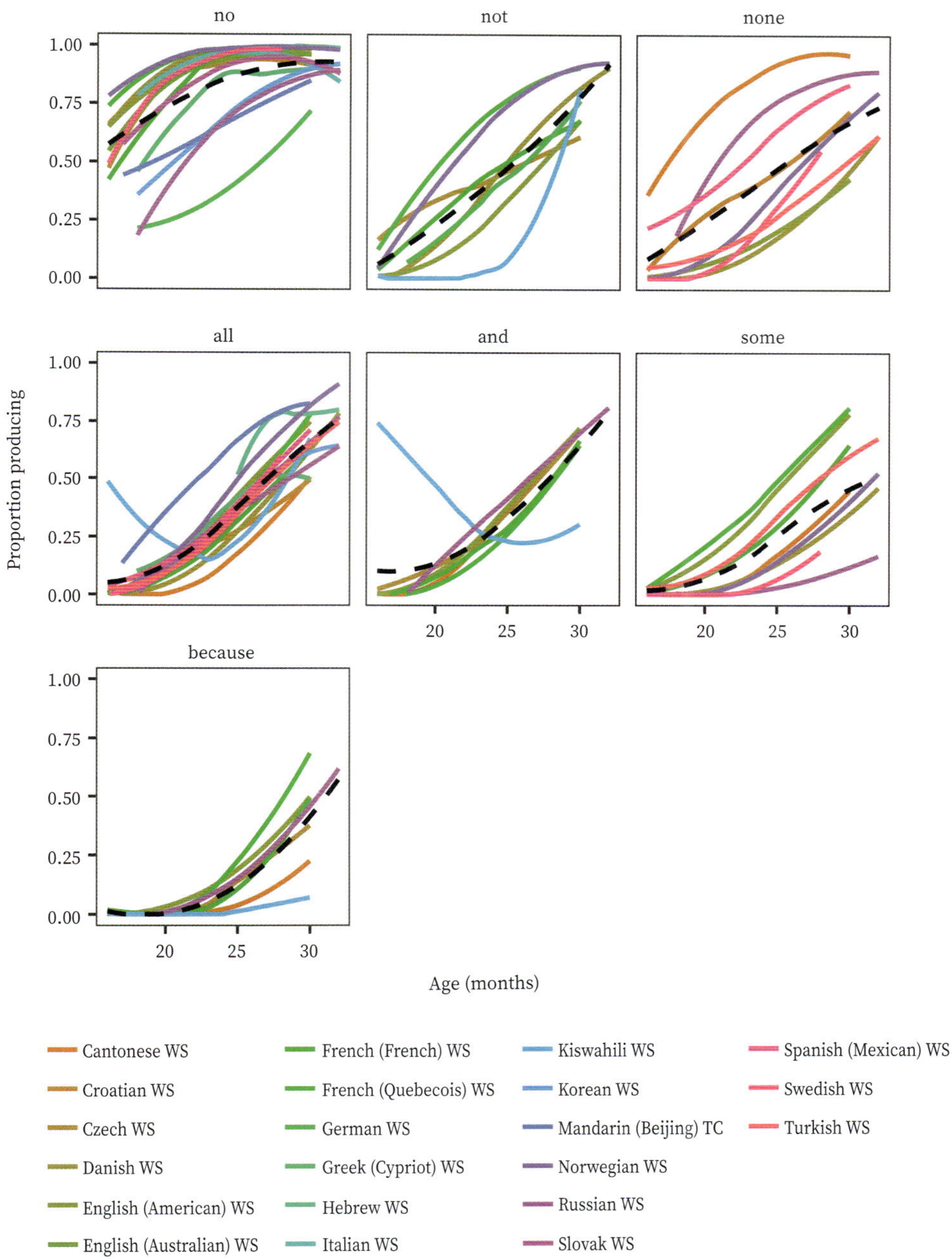

Figure 12.12
Developmental trajectory of each logic word in each language.

Table 12.1
Mean coefficient of variation for each semantic category.

Category	CV	SEM	*N*
Animals	0.32	0.06	17.78
Body	0.27	0.05	16.50
Color	0.47	0.09	14.83
Logic	0.47	0.11	10.57
Time	0.53	0.09	16.44

For 22- to 26-month-olds (chosen somewhat arbitrarily to be an age range of high coverage across forms that does not encompass too much developmental change), we compute the coefficient of variation for children at each age on each lexical item (see chapter 5 for the details of this analysis). We first average across ages and then across lexical items; reported *N*s are for the average number of contributing languages. We additionally add words from the Animals category for the sake of comparison. Table 12.1 gives the coefficient of variation for each category.

This analysis shows that the acquisition of animal and body words is highly consistent across languages. In contrast, color words, logic words, and time words are substantially less consistent cross-linguistically. These effects are likely somewhat affected by floor and ceiling effects, but inspection of individual items confirms the robustness of the general conclusion.

12.4 Discussion

In these exploratory analyses, we considered representation of different semantic categories across the different languages in our dataset. We found some surprising consistencies. Words from the Places and Time Words categories were underrepresented, while words from the Sounds, Games and Routines, and Body Parts categories were overrepresented. These results converge with our analyses in chapter 8 in suggesting that there are certain semantic categories that are quite common in children's earliest language and with our analyses in chapter 10 in suggesting that there may be specific domains that parents associate with small children.

The consistencies we observed were also contrasted with some areas of greater variability. For example, the preference for words from the Vehicles, Clothing, and Animals categories appeared to be a somewhat coherent dimension in our data, with many (northern) European languages higher on this dimension than non-European languages. Still, substantial caution is necessary in interpreting these results as the sample of non-European languages is small and heterogenous. Finally, we found that acquisition of complex conceptual words reflecting colors, time, and logical constructs was highly variable across languages. For these more complex conceptual domains, order of acquisition may depend on specific cultural practices that govern the use and teaching of these words (e.g., for color) or the linguistic structures of the target language (e.g., for logical words).

13 Morphology, Grammar, and the Lexicon

How does abstract structure emerge during language learning? In some accounts, children's early syntax emerges from direct generalizations from particular lexical items while in others, syntactic structure is acquired independently and follows its own timetable. Communicative Development Inventory (CDI) data can help us decide between these two views. In this chapter, we summarize the state of grammatical development across languages while noting the challenges posed by radically different representations of grammar across CDI forms. We also replicate and generalize analyses linking grammatical generalization to children's vocabulary size (see chapter 15 for links to vocabulary composition). We end by pursuing two relatively more novel directions. First, we investigate the idea that that age modulates the relationship between grammar and the lexicon. Second, we use structural equation modeling to evaluate the directionality of this relationship.

13.1 Introduction

For many children, their first words are spoken in isolation. While these single word utterances sometimes seem to be picking out objects in the world (e.g., *ball!*), others seem to convey more complex ideas or desires (e.g., *up!* for *Mommy, pick me up!*). By two years of age, however, many children have acquired a large repertoire of words and are beginning to use them in two- or three-word combinations (e.g., *Mommy up!* or *kitty sleep here*). These utterances will gradually increase in length and complexity in various ways, forming sentences that increasingly reflect the grammatical structure of their native language (e.g., *Mommy, the kitty is sleeping here*). Children also begin to add more verbs, adjectives, and other predicates to their working vocabularies (see chapter 11), and substantively increase their use of prepositions, articles, and other closed class forms that do grammatical

Authors' Note: An earlier version of some of these analyses was first reported in Braginsky et al. (2015).

work, including the productive use of inflectional morphemes (e.g., the English past tense *-ed* or *-ing*).

Understanding the origins of grammar is critical because children's ability to use morphosyntactically rich language is thought to reflect the uniquely human mental machinery that enables speakers to comprehend and produce novel utterances that have never been heard in the input (Berko, 1958; Pinker, 1991). The questions surrounding the development of grammar are challenging. How do abstract morphosyntactic structures emerge during language learning? What mechanisms underlie the formation of generalizations that support such inferences and allow children to apply them during language production? Does an understanding of the abstract rule structure of language emerge from the interactions of individual words, or is that structure acquired independently and represented separately?

Broadly speaking, theoretical views on grammatical development generally take one of two forms. On nativist theories like Principles and Parameters (Chomsky, 1981; Baker, 2005), grammar emerges independently from lexical knowledge following its own, largely maturational timetable. Moreover, grammatical regularities are mentally represented in a format that is distinct from that used by the lexical system. In contrast, according to lexicalist theories, mental representations of morphosyntactic structure generally emerge from partial probabilistic generalizations that are extracted over exposure to lexical items, and at least early in development, there may be little or no representation of morphosyntactic rules or regularities per se (Tomasello, 2003; Elman et al., 1996). Even when syntactic structures are eventually represented, these representations are directly related to more concrete lexical structures (Bannard et al., 2009).

Historically, the study of individual differences has been critical to this debate. While variation in word learning is generally uncontroversial, individual differences in grammatical development do not illuminate core processes of acquisition under a universalist, nativist perspective. In contrast, lexicalist theories predict that variation in grammatical development is significant, in that such variation should be tightly yoked to variation in lexical development (Bates and Goodman, 1999). Research has shown that, as with lexical development, there is sizable variation in exactly when and how children move into using more grammatically complex utterances in their everyday speech. While some children use primarily multi-word phrases and many closed class forms by 24 months, other children are still primarily producing nouns in single word utterances at that same age (e.g., Bates et al., 1988; Bates and Goodman, 1999). Moreover, there is also variation in the kinds of multi-word utterances that children produce. For example, some children build up sentences from individual words (e.g., *want dat!*), whereas other children seem to produce utterances that reflect "unanalyzed" chunks of more complex speech (e.g., *iwantdodat!*).

13.1.1 Correlations between Grammar and the Lexicon

Associations between individual differences in lexical and grammatical development have been robustly substantiated in the literature. In the original norming data from the English CDI Words and Sentences (WS), children with more sophisticated grammatical productions were also those children with the largest vocabularies (Bates et al., 1994). Using that same dataset, Marchman and Bates (1994) found that size of verb vocabulary was concurrently related to children's overregularization of past tense inflections (e.g., *daddy goed*), productions that are viewed as a major milestone in the development of grammatical rule-based knowledge.

Links between lexical development and grammar have also been reported longitudinally (Bates et al., 1988; Bates and Goodman, 1997); in late talkers (e.g., Paul, 1996; Rescorla et al., 1997, 2000; Thal et al., 1997); in early talkers (Thal et al., 1996, 1997); in Spanish-English bilinguals (Marchman et al., 2004); and in children with neurodevelopmental disorders, such as Williams syndrome (e.g., Singer Harris et al., 1997). Similar relationships have also been demonstrated in many other languages, including Slovenian (Marjanovič-Umek et al., 2013), Hebrew (Maital et al., 2000), Icelandic (Thordardottir et al., 2002), Italian (Caselli et al., 1999; Devescovi et al., 2005), Bulgarian (Andonova, 2015), Finnish (Stolt et al., 2009), Spanish (Mariscal and Gallego, 2012; Thal et al., 2000), and German (Szagun et al., 2006). Thus, these findings should be considered among the most reliable in the literature on CDI analyses.

Finally, and perhaps most intriguingly, in behavior genetic studies of monozygotic and dizygotic twins, the relation between lexical and grammatical level has been found to be strongly heritable (Dale et al., 2000; Dionne et al., 2003). In other words, even though genetic factors contribute relatively modestly to each aspect of language as assessed individually, the genetic factors that influence lexical growth are the same as those that influence grammatical growth and are perhaps operating in a bidirectional manner.

While these studies are consistent with the view that vocabulary and grammar development are strongly associated developmentally, the interpretation of these relations is still under debate. Some researchers have interpreted these links to suggest that domain-general learning mechanisms guide the child's construction of a working linguistic system at many different levels—in this case, learning words and learning grammatical rules (e.g., Elman et al., 1996). As Bates and MacWhinney (1987) proposed many years ago, "the native speaker learns to map phrasal configurations onto propositions, using the same learning principles and representational mechanisms needed to map single words onto their meanings" (p. 163).

In contrast, other proposals suggest that the process of learning words involves learning both their lexical-semantic and their morphosyntactic properties (e.g., in what constructions they can legally appear and what inflectional morphemes are required), and that grammatical knowledge is generally built up on a case-by-case basis (Tomasello, 2003). Early word combinations are often highly routinized and situation specific, suggesting that learning grammar, like word learning, is guided by learning mechanisms that are item specific and frequency dependent. It is only later that grammatical structures become encoded in terms of their abstract syntactic form (e.g., Lieven et al., 1997; Tomasello, 2003). And yet other accounts view the relation as reflecting mechanisms that operate in the opposite direction. On these views, grammatical analysis is a driving force behind word learning such that the process of analyzing sentences into their constituent grammatical parts facilitates the further acquisition of lexical-semantic knowledge (Anisfeld et al., 1998; Naigles, 1990).

Recently, a literature has developed using longitudinal models to assess the directionality of these relations. Such studies use cross-lagged models to investigate whether vocabulary size at earlier time-points predicts grammar at later times and vice versa. These models provide a perspective on growth over time that can help describe the directionality of the relationship, even though they do not allow for causal inferences about that directionality (Rogosa, 1980). Unfortunately, studies using such methods have come to very different conclusions, perhaps owing to the different models and datasets they have used. In one study, Pérez-Leroux et al. (2012) found relations from vocabulary size to grammar in 3- to 5-year-old Spanish speakers. In contrast, Brinchmann et al. (2018) found relations in the opposite direction, with stronger paths from grammar to vocabulary in a sample of 4- to 6-year-old Norwegian children. And using data from Spanish-English bilingual 2 ½- to 4-year-olds, Hoff et al. (2018) found limited evidence for direct relationships between growth in grammar and vocabulary and proposed that the observed relations were actually the result of third-variable explanations, perhaps from common input to both systems. Thus, the evidence at present is complex and unresolved.

13.1.2 The Current Analyses

Here we explore relations between estimates of children's vocabulary size based on the vocabulary checklist and responses on other sections of the WS instruments. Many versions of these instruments provide indices of grammar learning by asking about children's use of inflected forms (e.g., *walked*) and overgeneralizations (e.g., *goed*) as well as the complexity of their multi-word combinations (e.g., *kitty sleeping/kitty is sleeping*). While many studies have examined associations between lexical and grammatical development cross-linguistically, the scope and power of these early studies were limited, with few opportunities for direct comparisons of the nature or extent of these relations across multiple languages at the same time. In contrast, our data allow analyses of lexical-grammar relations with enhanced statistical power and broader cross-linguistic representation.

In addition, we explore a hypothesis that has not been explicitly tested in these earlier studies: that there remains age-related variance in grammatical development unexplained by vocabulary development. While the overall relationship between grammar and the lexicon provides support for lexicalist theories, the identification of age-related variance would suggest the presence of developmental processes that regulate grammar learning, above and beyond those captured by measures of vocabulary size. Such age-related processes could be either maturational or experiential and either domain-general (like working memory) or language-specific (like grammatical knowledge). Importantly, since both nativist and constructivist theories could in principle predict age-linked variance in grammatical development, our goal is not to differentiate these theories but instead to test this novel prediction and explore its implications for future work on understanding the processes of grammatical development.

Further, addressing the question of directionality in the relationship between grammar and the lexicon, we fit structural equation models to longitudinal data from two datasets. We use a variant on the standard autoregressive cross-lag panel model that accounts for stable trait differences among individuals (Hamaker et al., 2015). While such models do not provide conclusive evidence about the nature of developmental links between grammar and the lexicon (cf. Rogosa, 1980), they nevertheless provide suggestive evidence that—at least for young children—vocabulary drives grammatical growth rather than vice versa.

A final contribution of our work is that, due to the size of our dataset, we are able to make more fine-grained distinctions than the initial cut between grammar and the lexicon. In particular, we distinguish morphology from multi-word syntax since morphological generalizations might be more specifically dependent on vocabulary size than those requiring more global, sentence-level syntactic regularities.

13.2 Methods

CDI forms typically contain both vocabulary checklists and other questions relevant to the child's linguistic development. All of the data reported here come from WS type forms that were administered to children ages 12 to 36 months (most in the 16- to 30-month range). In addition to the vocabulary checklist items, these forms typically contain a single item asking whether the child is combining words at all. There is also a Word Forms section, that asks whether the child produces each of around 30 morphologically inflected forms of nouns and verbs (e.g., *feet*, *ran*) and a Complexity section that asks whether the child's speech is most similar to the syntactically simpler or more complex versions of around 40 sentences (e.g., *two foot* versus *two feet*, *there a kitty* versus *there's a kitty*).

Importantly, each instrument for languages other than English is not just a translation of the English form but rather was constructed and normed to reflect the nature and early development of the lexicon and grammar of that language. Thus, there are substantial differences in the content of these items and their coverage of different morphological and grammatical phenomena. The major commonality is that the form developers have attempted to provide a comprehensive and representative survey of important developmental phenomena in their language.

Even more so than many parts of the CDI, responses to individual word form and grammatical complexity items should be interpreted with caution. Such items have not been validated as extensively as other parts of the CDI.[1] In addition, children's early speech is thought to be inconsistent from a grammatical perspective, occasionally containing word orders beyond the standard word order of the language (e.g., Bowerman, 1973), and this variability is not fully reflected in the potential responses to grammatical complexity items. This last point goes double for word forms, as these are often highly variable within individuals (Marcus et al., 1992). Finally, word forms like "went" may be used by the child appropriately to denote motion without understanding the temporal marking that distinguishes it from the less frequent "go."

Keeping these caveats in mind, we derive several measures to analyze lexical and morphosyntactic development. Each child's vocabulary size is computed as the proportion of words on the corresponding CDI form that the child is reported to produce. Similarly, each child's Word Form score is the proportion of word forms they are reported to produce, and their Complexity score the proportion of complexity items for which they are reported to use the more complex form. We compute all of these quantities as proportions to make the scales comparable across languages. Note that different analyses often incorporate different amounts of data due to the presence or absence of specific sections (or data from those sections) in particular language datasets.

13.3 Results

We present four sets of results. First, we show analyses of the "combines" item, a binary item in which parents indicate whether their child is combining words. Second, we analyze the relations between vocabulary size and the Word Form and

1. At the risk of some circularity, we note that given the high degree of correlation with vocabulary (shown), grammatical complexity items should inherit some presupposition of reliability and validity.

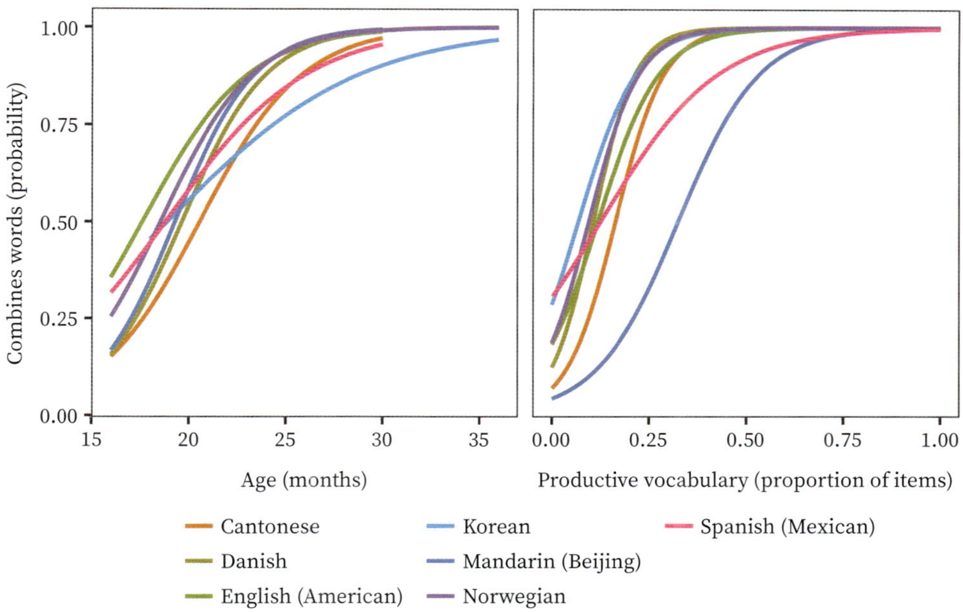

Figure 13.1
Trajectory of the combines item in each language across age (left) and vocabulary size (right).

Complexity items. Third, we follow up on a pattern reflected in the "combining" item, namely age-related modulation of the grammar-lexicon relationship. Finally, we investigate the degree to which the age-related pattern is found in individual items.

13.3.1 Combine Items

Figure 13.1 shows the probability of a parent checking that their child combines words, plotted by the child's chronological age (left) and raw productive vocabulary size (right). As can be seen, across 7 languages, there is some consistency in the developmental trajectories for this item. By 19 months, around 50 percent of children are reported to be combining words; by 25 months, around 90 percent are combining them.

Vocabulary-related trajectories are more variable, however. In general, children who were marked as combining had vocabularies larger than around 100 words. There are several notable exceptions, however. As noted in chapter 5, raw Beijing Mandarin vocabulary in the WS form is unusually high, but the "combines" item does not appear to be comparably accelerated. Thus, Beijing Mandarin–learning children appear to be combining words only after producing substantially more

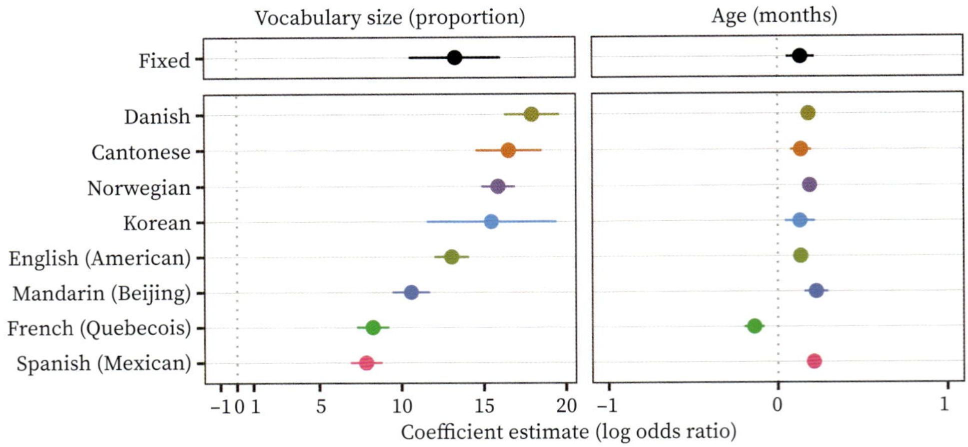

Figure 13.2
Coeffient estimates from the combines model, with fixed effects in the top row and random effects in the bottom row.

words than children learning other languages. On the opposite side, children learning Quebecois French and Korean were reported to be combining with quite small vocabularies.[2]

To investigate the quantitative relationship among word combination (as measured with this item), age, and vocabulary, we fit a logistic mixed effects model predicting whether a child combines as a function of their vocabulary (as a proportion of items), age, and interaction between vocabulary and age. We also included a random effect of language with a random intercept and random slopes for vocabulary and for age. Coefficient estimates from this model are shown in figure 13.2.

This model shows a large effect of vocabulary ($\beta = 13.28$, $t = 9.39$) with a relatively smaller effect of age ($\beta = 0.13$, $t = 3.18$). These effects mean that, for example, a 16-month-old learning American English with a vocabulary size of 50 words has a 31 percent chance of combining words. At the same age of 16 months, a child is more likely to combine by about a factor of 2 if they have a vocabulary size of 120 words (at 63 percent). Conversely, at the same vocabulary size of 50 words, a child is more likely to combine by about a factor of 2 if they are 28 months old (at 63 percent). In addition, there is a substantial negative interaction of vocabulary and age ($\beta = -0.34$, $t = -6.94$), indicating that older children are more likely to

2. It appears possible that the Quebecois French data have some issue for this item given the very flat slope we observed. We speculate that this could perhaps be due to misinterpretation of the way the item is worded. Data from this language are hence not shown in figure 13.1.

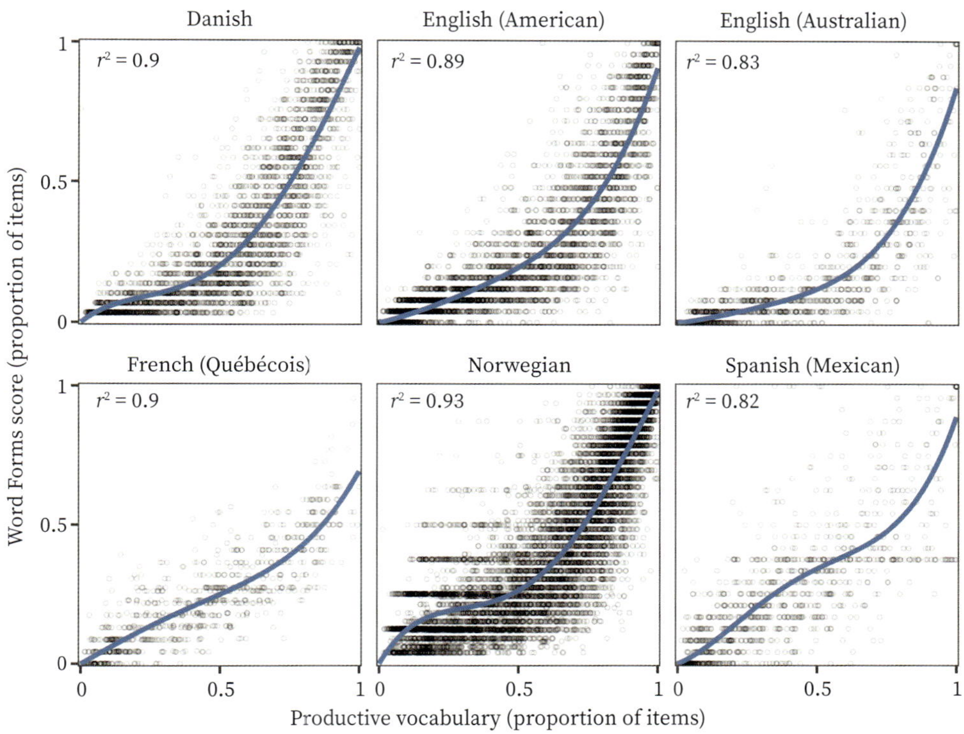

Figure 13.3
Each child's Word Forms score as a function of their vocabulary size in each language (curves show model fits).

be combining words, even with smaller vocabularies. This result parallels others reported below, suggesting that there are age-related components in grammatical performance, at least for production of word combinations, that are unaccounted for by vocabulary alone.

13.3.2 Grammar and Lexicon Relationship

We next examine the correlation between the proportion of Word Form items and Complexity items completed and the proportion of vocabulary items completed. First reported by Bates et al. (1994), these correlations are extremely robust and can be observed in all of our datasets. Figure 13.3 shows this relation for Word Form items. We fit generalized linear regressions predicting Word Form score or Complexity score as a function of linear, quadratic, cubic, and quartic terms of productive vocabulary size (subtracting the intercept to ensure that the function passed through the origin because a vocabulary size of 0 necessarily implies scores of 0). The total r^2 for these relationships ranges from 0.82 to 0.93.

Complexity items show the same relationship (figure 13.4), typically with equal or greater strength (depending on data density and number of items). The values for r^2 vary from 0.75 to 0.94. Some ceiling effects are observed.

Overall, these data add strong cross-linguistic support to the claim of Bates et al. (1994) and others that the emergence of grammatical competence in production is related across individuals to the size of the productive vocabulary.

13.3.3 Age Effects

In our next analysis, we follow up on the relationship between age and grammatical ability found in the "combines" analysis above. In that analysis, we noted that less vocabulary was needed for older children to be marked as combining words, suggesting that other abilities can contribute or supplement vocabulary in achieving this milestone. We investigate this pattern in the full Word Form and Complexity item set by splitting data from each language by age. We plot the same curves as above, except that children who are older and younger than the median age (within each dataset) are separated.

In essentially every language, for both Word Form (figure 13.5) and Complexity items (figure 13.6), we see a higher curve for older children than younger children. This finding is consistent with the idea that grammatical development is faster per unit lexicon for older children (mirroring the negative interaction shown for the "combines" item). That is, an older child with comparable vocabulary to a younger one will likely be better at combining words.

This pattern is further summarized in figure 13.7, where we show the area under the grammar/lexicon curve for younger and older children. The upward slope from younger to older of nearly every line demonstrates the consistency of the age effect, discussed further below. In addition, there is a trend for age effects to be larger in Complexity items compared to Word Forms, suggesting that factors beyond vocabulary size have a larger effect on syntax than on morphology.

13.3.4 Individual Items

In our next analysis, we examine the individual items on the Word Form and Complexity sections. Given the heterogeneous nature of the CDI instruments, particularly in the Complexity sections, we attempted a more fine-grained item-analysis by classifying items as capturing either more morphological or more syntactic phenomena. Items for which the difference between the simple and complex sentences is in the inflection of a noun or verb (such as *kitty go away/kitty went away*) were coded as Morphological. Items were coded as Syntactic if they involved the use of some sentence-level syntactic construction (such as *you fix it/can you fix it*). Items from some languages involved a mix of the two or otherwise didn't fit into this dichotomy, and these were coded as Other.

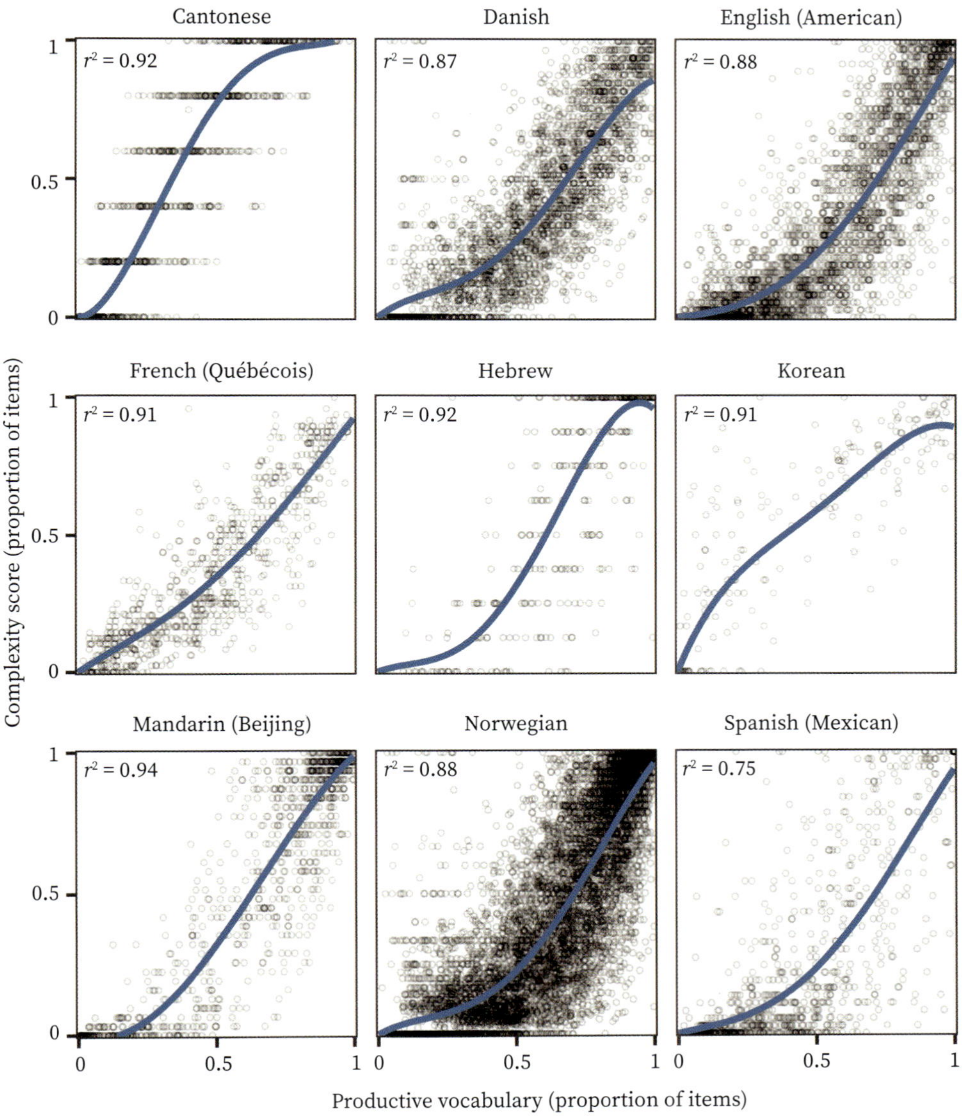

Figure 13.4
Each child's Complexity score as a function of their vocabulary size in each language (curves show model fits).

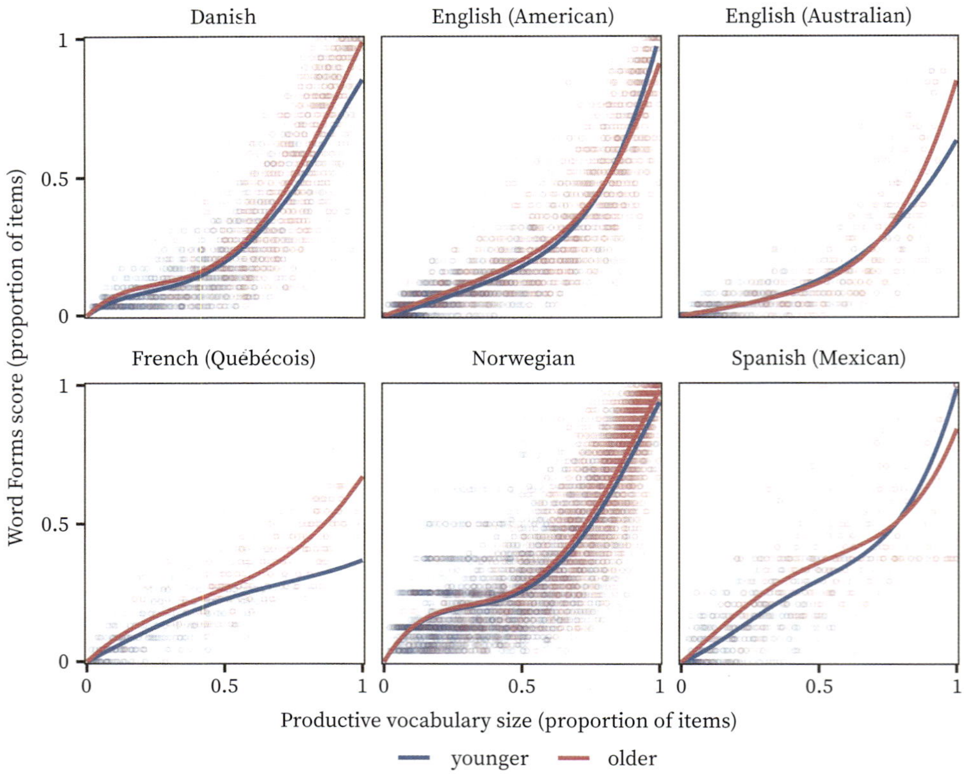

Figure 13.5
Each child's Word Forms score as a function of their vocabulary size in each language for children younger and older than the median (curves show model fits).

We then a fit logistic regression separately for every item predicting whether children produced the item from vocabulary size and age. Figure 13.8 shows the resulting model fits for two example items. For both items, children with larger vocabulary sizes are more likely to produce the item than children with smaller vocabulary sizes (the curves have positive slopes), and older children are more able to produce it than younger children (the older curves are higher than the younger curves). For the item on the left the interaction between vocabulary size and age is very small (all the curves have roughly the same slope) while for the item on the right this interaction is large (the older curve is much steeper than the younger curve, indicating that at larger vocabulary sizes the advantage for older children is larger).

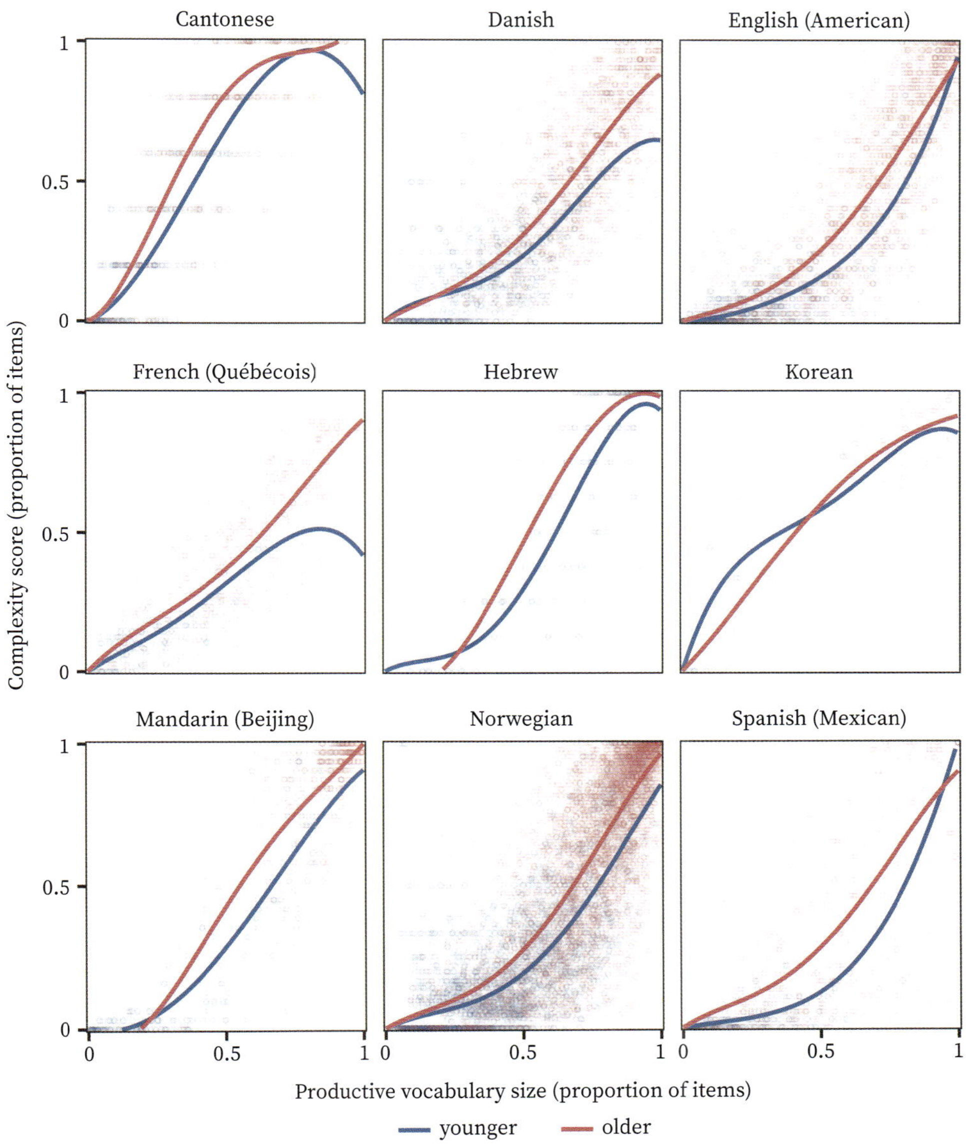

Figure 13.6
Each child's Complexity score as a function of their vocabulary size in each language for children younger and older than the median (curves show model fits).

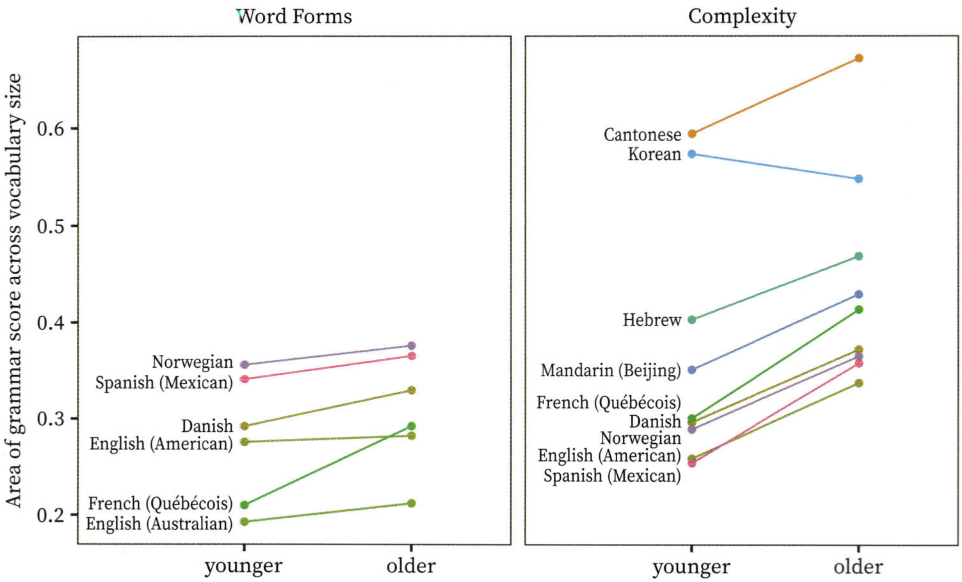

Figure 13.7
Area under model fit curve for Word Forms score and Complexity score as a function of vocabulary size in each language for younger and older children.

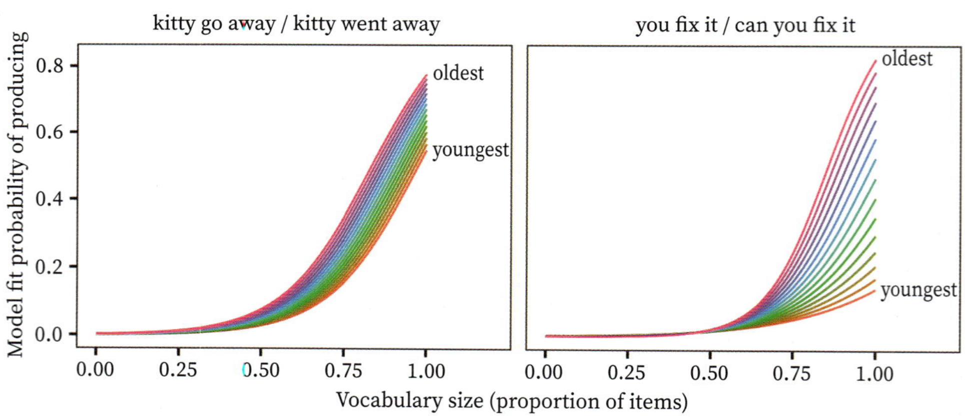

Figure 13.8
Model fits for two example items.

Figure 13.9 shows, for each item, the coefficients of vocabulary size, age, and their interaction. In general, age effects are smallest for Word Form items, larger for Morphological Complexity items, and largest for Syntactic Complexity items, suggesting that more syntactic phenomena likely have greater age contributions.

Building on previous analyses that showed a strong relationship between lexical and grammatical development, we added age into this relationship. Across languages, our measures of syntactic development consistently showed greater age modulation than measures of morphological development. Further distinguishing between items that were more reflective of morphology than syntax, we again found greater age effects for more syntactic items. Thus, this analysis provides evidence for a relationship between syntactic development and age that is *not* captured by lexical development.

13.4 Longitudinal Relationships

In our next set of analyses, we take advantage of the presence of longitudinal data in two languages (Norwegian and American English) to investigate the relationship between vocabulary size and grammatical ability, as operationalized by average Complexity score. As mentioned above, this relationship has been the subject of a variety of previous investigations using longitudinal data (Dionne et al., 2003; Moyle et al., 2007; Pérez-Leroux et al., 2012; Hoff et al., 2018; Brinchmann et al., 2018), but outcomes have been contradictory. Given the heterogeneity of these studies, results could differ for a variety of reasons, including the language background of the participants, the target age range of the study (which ranged from 2–6 years old), and the nature of the specific models being fit to the data.

Further, a specific technical concern underlies one potential source of variance. The standard method for examining reciprocal relationships between longitudinally measured variables is the cross-lagged panel model (CLPM). While this model does not allow for causal inferences (Rogosa, 1980), it is a convenient way to estimate temporal precedence in the relationships among variables measured over multiple time points; many of the previous studies have used variants of this model. However, in cases where the measured constructs show stability within individuals—as is clearly the case for language-related measures—these models are misspecified since they assume that the *only* source of consistency among individuals is temporal autocorrelation (Hamaker et al., 2015). To address this issue, Hamaker et al. (2015) propose the random intercepts CLPM (RI-CLPM) that adds a latent intercept that accounts for individual variation across observation time points (shown in figure 13.10). This is the approach used by Brinchmann et al. (2018), and they find significantly better fits for the RI-CLPM on their vocabulary/grammar data than the standard CLPM. Hence, we use this model in our analyses.

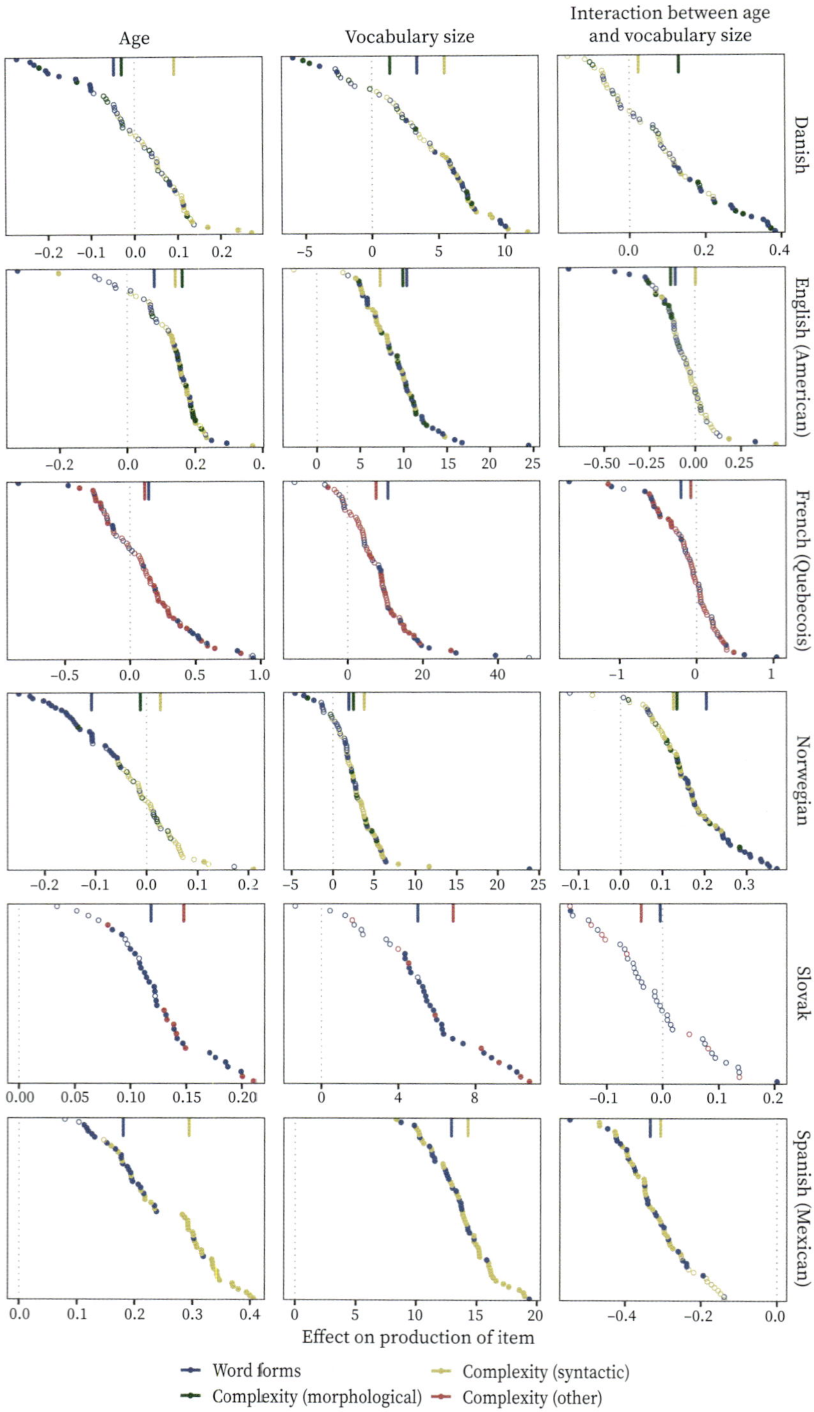

Figure 13.9
Coefficient estimates (as log odds ratio) for each grammar item in each language. Line segments indicate means effect for category.

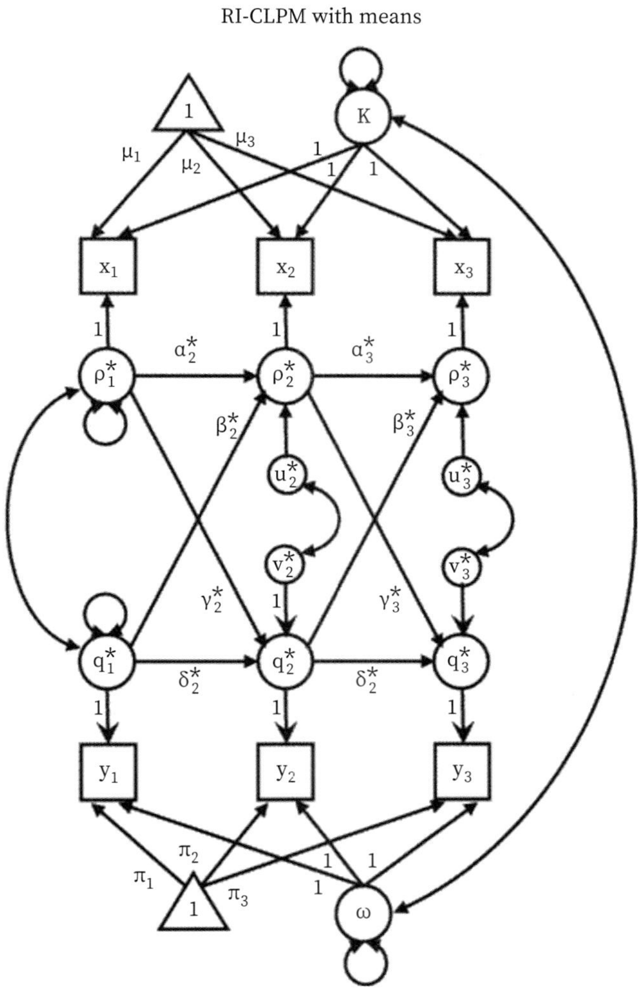

Figure 13.10
The random intercept cross-lagged panel model. Circles represent observed variables, squares represent latent variables, and directed edges represent regression effects. *Source*: Hamaker et al. (2015).

Table 13.1
Number of children in the Norwegian longitudinal dataset with each number of observations.

Number of observations	Number of children
2	794
3	134
4	67
5	72
6	43
7	15
8	13
9	3

We include data from two different longitudinal datasets. The first is a group of children from a study conducted by Virginia Marchman with English-speaking children in Dallas, Texas ($N=247$). This dataset includes children with three waves of data collection centered around 18, 24, and 30 months; we included children if they contributed to two waves ($N=160$) or all three waves ($N=83$) for a total of 243 children. The offsets among waves varied somewhat in these data, but for simplicity, we split all administrations into three time points: 18–21 months, 22–25 months, and 26–30 months.

The second dataset is a large group of longitudinal administrations in the Norwegian normative data and includes a total of 1,565 children. These children's parents were invited to complete multiple CDIs online every month or every few months, with many parents completing two and a small number completing 10 or more. For purposes of discrete analysis, we binned these continuous data into two-month waves from 18 to 35 months (the small number of 36-month-olds were included in the oldest wave). We then included children with two or more administrations, each falling in a different wave ($N=1141$). Table 13.1 shows the distribution of how many observations are available for children across different waves.

Using these two datasets, we fit RI-CLPM models.[3] These models measure the magnitude of the reciprocal relationships between grammar and vocabulary over time. Coefficient weights on the paths between latent grammar and vocabulary are then interpretable as temporal influences. Following the RI-CLPM specification in Hamaker et al. (2015), we posited latent variables κ and ω that describe stability in vocabulary and grammar, respectively.

Figure 13.11 shows the fitted model for the English data. Coefficient estimates from vocabulary to grammar are large and significant ($\beta_{p1-c2} = 0.95$, $p<0.001$,

3. We gratefully acknowledge code from Flourney (2017).

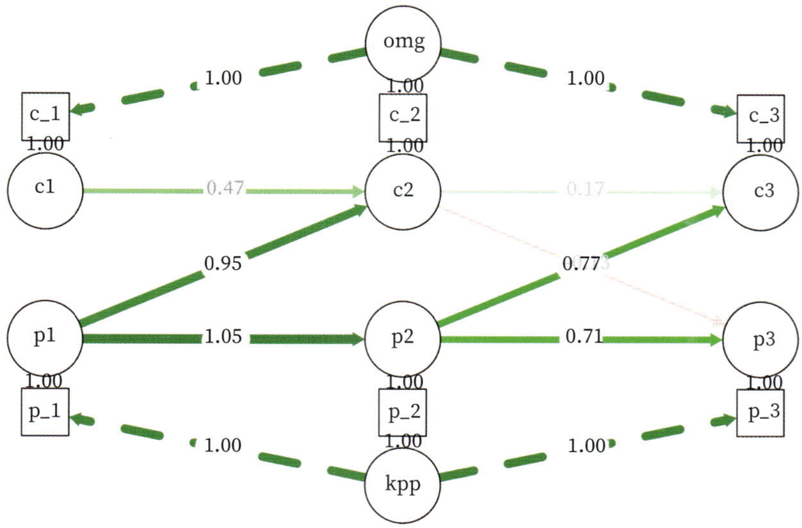

Figure 13.11
RI-CLPM model fit to English (American) data. Path diagram for latent (circles) and observed (squares) variables. Edges show estimated regression coefficients, scaled and colored by estimate magnitudes. Lowercase p variables indicate production coefficients, while lowercase c variables indicate grammatical complexity coefficients. Lowercase square variables are observed data. The variables kpp and omg denote global intercepts.

$\beta_{p2-c3} = 0.77$, $p < 0.001$). These are consistently higher than estimates from grammar to vocabulary (which are both small and nonsignificant).

Figure 13.12 shows the analogous fitted model for the Norwegian data.[4] Coefficient estimates from vocabulary to grammar are large; as in the English data, these are consistently higher than estimates from grammar to vocabulary (which are both small and nonsignificant). Coefficient estimates are shown in table 13.2.

In both of these cases, we see strong evidence for cross-lagged influences of vocabulary on grammar, but *not* for grammar on vocabulary. These results are consistent with the idea that, early in acquisition, learning vocabulary items provides the materials for generalization of grammatical constructions and rules—following the general line of speculation in Bates et al. (1994). In contrast, our findings are prima facie less consistent with results that show smaller relationships (Hoff et al., 2018) or grammar-to-vocabulary linkages (Brinchmann et al., 2018), both in older children. But in fact, we might expect a reversal in directionality for these relationships over time. Early in development, some vocabulary is necessary to get

4. Because of the high degree of missingness in this dataset, fititng these models was difficult, and so we interpret standard errors with some caution.

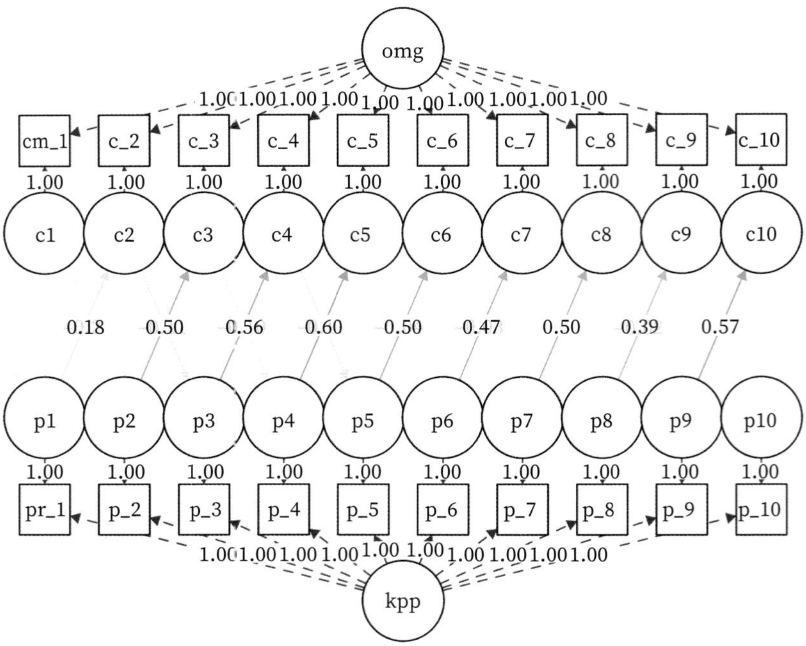

Figure 13.12
RI-CLPM model fit to Norwegian data. Plotting conventions are as above; p indicates production while c indicates complexity.

grammar learning off the ground. But later on, as grammatical rules stabilize (probably after age three), grammar then serves to help elucidate the meanings of complex sentences, allowing syntactic bootstrapping of new word meanings (Gleitman, 1990).

13.5 Discussion

In this chapter, we revisited classic findings on the relationship between grammar and the lexicon and explored novel questions regarding the role of age in this relation. Our results provide general support for a constructivist view, in that, in these 12 languages, variance in vocabulary production strongly aligns with variance in grammar. However, we also estimated additional age-related contributions, specifically contrasting the links to morphological forms versus syntactic constructions. In general, we find that measures of grammar that are more closely aligned with syntax are modulated by age to a greater extent than those reflecting inflectional morphology.

Table 13.2
Coefficient estimates for cross-lag parameters in the Norwegian regression model.

Predicted	Predictor	β	Standard error	p
p10	c9	0.004	0.057	0.943
p9	c8	−0.010	0.037	0.795
p8	c7	−0.028	0.041	0.489
p7	c6	−0.035	0.040	0.388
p6	c5	−0.044	0.041	0.283
p5	c4	−0.113	0.058	0.050
p4	c3	−0.095	0.059	0.107
p3	c2	−0.107	0.093	0.248
p2	c1	−0.022	0.178	0.900
c10	p9	0.568	0.140	<0.001
c9	p8	0.393	0.099	<0.001
c8	p7	0.502	0.085	<0.001
c7	p6	0.469	0.078	<0.001
c6	p5	0.495	0.060	<0.001
c5	p4	0.600	0.067	<0.001
c4	p3	0.563	0.060	<0.001
c3	p2	0.496	0.090	<0.001
c2	p1	0.181	0.205	0.377

As with the correlations reported in chapter 7, parent bias is a potential confound for these correlational analyses. If some parents tend to overreport on their child's language more than others do—for reasons such as sensitivity, optimism, greater time spent with the child—then this overreporting would likely extend across linguistic domains. Thus, in principle, an observed correlation between two sections of a single parent report form could be driven by parent bias acting independently on each section without any connection.

Two studies provide evidence against this deflationary hypothesis. First, Moyle et al. (2007) used a combination of parent report and standardized assessments (e.g., the Preschool Language Scale) to provide evidence for the same relation in both typically developing and late-talking children from 2 to 5 years of age. Second, Brinchmann et al. (2018) gives a similar analysis using cross-lagged structural equation models. Critically, their work relied only on direct testing of the child (not parent report) using standardized instruments. In their model, they found a strong correlation between the time-invariant (trait-like) components of vocabulary and grammar ($r = 0.72$). While this correlation is somewhat smaller than the correlations we report here, it is still quite large—and it appears in a model that also controls for a number of other relationships.

Intriguingly, however, the Brinchmann et al. (2018) study suggests that as grammar-lexicon correlations are cross-lagged, grammar to vocabulary links are stronger than vocabulary to grammar links, although neither could be characterized as strong. We followed up on this finding by fitting similar models to our own data, showing the opposite result (stronger vocabulary to grammar links across two languages). One potential reconciliation of these two findings is that early in development, more vocabulary allows for more generalization, but then later, syntactic bootstrapping effects are more important. Regardless, combined with the Moyle et al. (2007) study, these findings suggest that reporting bias is very likely not the sole cause of the correlations we observed.

Our analyses go beyond earlier work by also investigating the relationship of age to vocabulary and grammar. One possibility is that age-related developments are dependent on maturational factors that operate on grammatical development in a domain-specific way, independent of lexical-semantic processes. Another possibility is that age-related effects are a reflection of domain-general learning mechanisms, such as attention or working memory, that provide differential support for sentence-level processes than word-internal ones (Gathercole et al., 2013). Yet another possibility is that by an older age, children have received more linguistic input from which to generalize grammatical structure. Future studies should also explore the extent to which lexical and age-related processes are shaped, either independently or in tandem, by features of the learning environments that children experience (e.g., Weisleder and Fernald, 2013; Hoff et al., 2018; Brinchmann et al., 2018).

In sum, questions about the nature of morphosyntactic representations in early language have often seemed deadlocked. By mapping out developmental change across large samples and multiple languages, our findings challenge theories from across the full range of perspectives to more fully describe the mechanistic factors underlying the interaction of vocabulary, grammar, and development.

14 Morphological Overgeneralization

Although chapter 13 examined broad patterns of morphosyntactic development in relation to vocabulary size, we did not conduct specific analyses of the development of morphology per se. In this brief chapter, we rectify that omission by examining patterns of morphology in two cases: plural noun morphology and past tense verb morphology. These two morphological systems are well studied in the literature because plural (e.g., *cats*) and past tense forms (e.g., *walked*) are some of the earliest words to be produced by young children (Brown, 1973). Moreover, plural and past tense forms are viewed as a window into the mechanisms underlying productive language use because children will sometimes make overgeneralization errors, such as *tooths* or *goed*. The developmental time course of children's overgeneralization errors has also been an influential case study for language learning—and mental representation more broadly (Rumelhart and McClelland, 1986; Pinker and Prince, 1988; Pinker, 1991; Elman et al., 1996). We begin by describing cross-sectional patterns of overgeneralization across languages and then turn to characterizing longitudinal change in overgeneralization in two languages which have different morphological systems—English and Norwegian.

14.1 Introduction and Methods

When a child gleefully says, "Mommy, I brushed my tooths!," a proud parent might rejoice in their child's accomplishment. At the same time, they might worry because the child has *overgeneralized* the regular plural -s inflection to an irregular form (*teeth*). Overgeneralization errors are not uncommon in child speech, although like all aspects of early language development, there is considerable variation across children in how many errors children make (Marcus et al., 1992; Maratsos, 1993). Importantly, overgeneralization errors have been viewed as a positive sign that the child has abstracted the regularities of their language (e.g., add -s to indicate there is more than one thing) and applied that regularity in a productive way (i.e., the child has not likely heard the adult use that form).

The developmental timeline of generalization has been investigated extensively for the English plural and past tense. One classic study in child language (Berko, 1958) explored children's ability to be productive with plural and past tense forms in an elicitation task using novel forms (e.g., *Here is a wug! Now there are two of them—there are two wugs!*). Other classic studies using naturalistic designs (Cazden, 1968; Brown, 1973) noted that correct irregular forms often appear early in development alongside correctly inflected forms. Only later in development are overgeneralizations produced, *after* children have been successful in producing the correct irregular form, apparently unlearning what they already knew. With further development, overgeneralizations are less likely to occur—although they are still more difficult for children with language disorders (Marchman et al., 1999) and sometimes produced by adults, especially under stressed conditions (McDonald and Roussel, 2010).

The particular mechanisms underlying this developmental time line have been subject to considerable debate in the literature (Marcus et al., 1992; Elman et al., 1996). A particular focus of interest has been the so-called "U-shaped" developmental pattern (correct irregular—overgeneralization—correct irregular). On one view, the onset of overgeneralizations after correct productions may signify a grammatical rule coming "online" after a period in which correctly inflected forms were learned by rote (Pinker, 1998). Alternatively, the onset of overgeneralization errors may be the consequence of an accumulation of lexical exemplars from which children extract the general patterns (Marchman and Bates, 1994; Plunkett and Marchman, 1989). These views make different predictions about the universal nature of the U-shaped pattern, both across children and across individual noun or verb forms, as well as providing different accounts of the sources of variability that might influence the developmental time course of this pattern.

This debate has become a case study for understanding the representational formats that form the foundation for human learning and generalization (Pinker, 1998). Yet few studies have had the opportunity to explore patterns of the onset of overgeneralization errors in large datasets like those available in Wordbank. Moreover, most of the published work in this debate came from English. For a broader view, it is critical to begin to evaluate patterns of overgeneralization errors cross-linguistically, given that morphological systems vary widely across languages (Marcus, 1995; Clahsen et al., 1992).

Here we explore the consistency and variability of the time course of children's use of plural and past tense overgeneralizations using Communicative Development Inventory (CDI) data from the English and Norwegian samples. Our approach here depends on the fact that the original developers of the CDI were interested in this debate and included overregularizations as options in the morphological sections of the form. For example, an item like *foot* includes *foots* and *feet* as possible options to be checked. To take advantage of this, we coded overgeneralizations on each of the 5 Words and Sentences–WS forms for which we had access to data from these items (two from English dialects). We then used this coding to count the total proportion overregularizations by child, both overall and within noun and verb categories. We first show analyses of these overregularizations for cross-sectional data and then present more in-depth analyses of longitudinal data in English and Norwegian.

14.2 Cross-Sectional Data

We begin by describing overregularization within the cross-sectional data. For any dataset with longitudinal administrations, we include only the earliest administration for each child. We then compute the proportion of nouns and verbs that each child is reported to overregularize. The mean proportion of overregularizations (table 14.1) reported is relatively low overall, both for nouns and verbs, and ranges from 2 percent to 12 percent. For children who overregularize at least one item, the overall overregularization proportion is of course higher, ranging from 9 percent to 30 percent, but even these rates are low, suggesting that those children who overregularize are not overregularizing all or even most forms at the same time.

Individual items vary in how often they are overregularized (tables 14.2 and 14.3): in English, items range from *childs* and *satted*—words that are virtually never overregularized—to *feets* and *blowed*—words that are overregularized by 12 percent and 9 percent of children, respectively.

Figure 14.1 shows each child's proportion of overregularized items plotted by age for children who overregularize at least one item overall. The major generalization from these data is that curves tend to be surprisingly flat—there are not large and consistent developmental increases for any language. The inset r^2 for each panel in this graph shows the variance explained by quadratic models of overregularization by age, quantitatively confirming the visual impression of limited developmental change; the largest of these r^2 values is only 0.05.

Overregularization is more tightly correlated with vocabulary size than with age (figure 14.2), with r^2 values ranging from 0.45 to 0.70. However, the relationship between overregularization and vocabulary tends to be far weaker than the relationship between correct morphological inflection and vocabulary (as described in chapter 13), for which r^2 is around 0.9.

Another perspective on the structure of these data comes from comparing overregularization across lexical categories, as shown in figure 14.3. While a substantial proportion of children (0.03–0.19) do not overregularize any items, those who do are doing so across nouns and verbs quite consistently for all of the

Table 14.1
Proportion of noun and verb overregularization across languages.

		All children		Children who overregularize	
Language	*N*	Nouns	Verbs	Nouns	Verbs
Danish	3714	0.05	0.03	0.12	0.09
English (American)	3721	0.04	0.02	0.20	0.13
English (Australian)	1497	0.04	0.03	0.20	0.14
Norwegian	5131	0.07	0.05	0.19	0.12
Slovak	1058	0.11	0.12	0.30	0.21

Table 14.2
Least overregularized items in each language.

	Nouns		Verbs	
Language	Item	Proportion	Item	Proportion
Danish	mænde	0.01	såddede	0.00
English (American)	childs	0.01	satted	0.00
English (Australian)	childs	0.01	wented	0.01
Norwegian	barner	0.02	lyvet	0.00
Slovak	k jazerovi	0.04	zjednem	0.01

Table 14.3
Most overregularized items in each language.

	Nouns		Verbs	
Language	Item	Proportion	Item	Proportion
Danish	fodder	0.11	drikkede	0.14
English (American)	feets	0.12	blowed	0.09
English (Australian)	teeths	0.08	falled	0.07
Norwegian	boker	0.16	hjelpte	0.14
Slovak	bábätkovi	0.22	česaj	0.29

languages in our sample. The r^2 values for the relationship between noun and verb overregularizations range from 0.36 to 0.59.

In summary, our data suggests that when aggregated across children, overregularization does not appear to have a set relationship with age and does relate to vocabulary size, though not nearly as strongly as does correct morphological inflection. Additionally, overregularization is highly variable across children but relatively stable within children but across lexical categories.

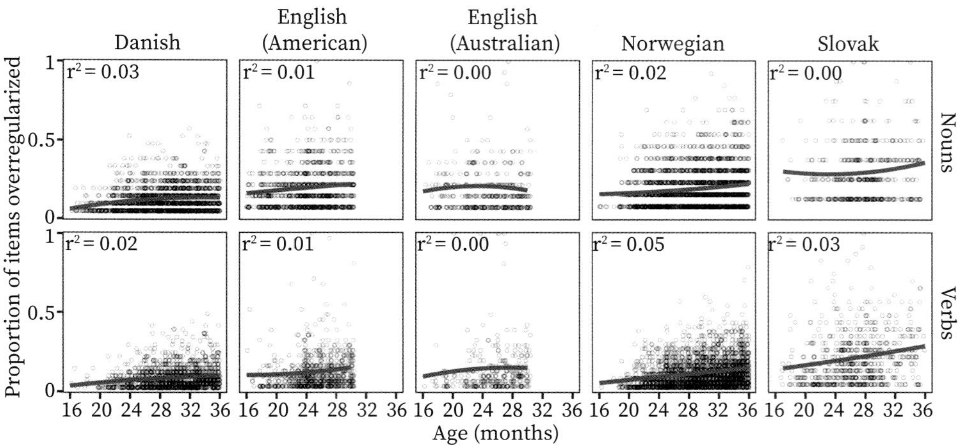

Figure 14.1
Each child's proportion of items overregularized as a function of their age in each language (curves show model fits from models predicting overregularization proportion from quadratic and linear terms for age).

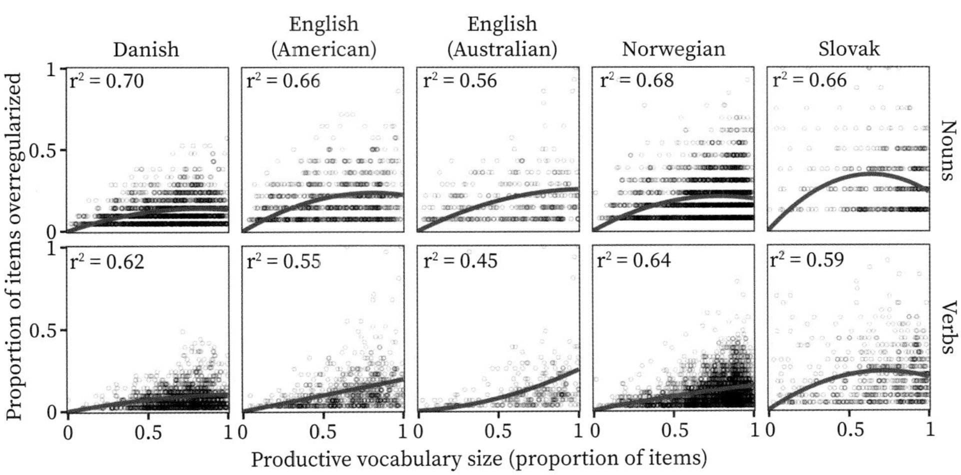

Figure 14.2
Each child's proportion of overregularization as a function of their vocabulary size in each language (curves show model fits from models predicting overregularization proportion from quadratic and linear terms for vocabulary size).

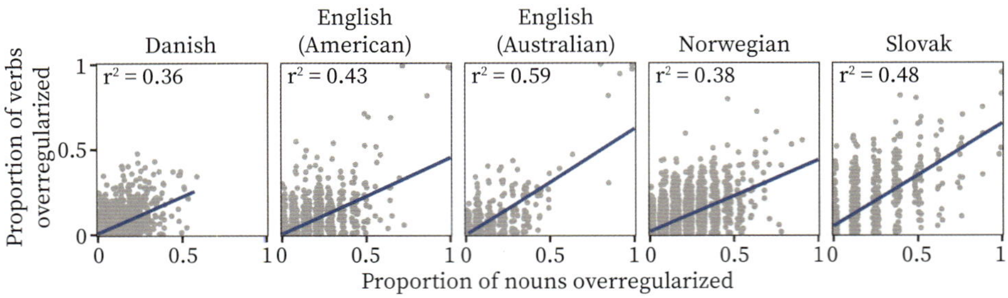

Figure 14.3
Each child's proportion of verbs overregularized against proportion of nouns overregularized (lines show linear model fits from models predicting verb proportion from noun proportion).

14.3 Longitudinal Data

Next, we ask whether individual children's overregularization trajectories can paint a clearer picture of the overall developmental time course of overregularization. As in other chapters, we make use of the longitudinal data from the American English and Norwegian datasets. These data allow us to examine changes in generalization across individuals. Since data are sparser in the English data than the Norwegian, we pursue slightly different approaches in our analysis.

14.3.1 English

In the American English dataset, out of children who overregularize at least one item, there are 85 with three longitudinal administrations, two with four administrations, and none with more than that. For each child, we compare their three overregularization values—youngest, middle, and oldest (with the two middle values averaged for the four administration children)—to categorize their overregularization trajectory as:

- **an increase** (/) if youngest to middle *stays the same* or *increases* and middle to oldest *stays the same* or *increases,*
- **a decrease** (\) if youngest to middle *stays the same* or *decreases* and middle to oldest *stays the same* or *decreases,*
- **a recovery** (Λ) if youngest to middle *increases* and middle to oldest *decreases,* and
- **a retreat** (V) if youngest to middle *decreases* and middle to oldest *increases.*

Figure 14.4 shows the trajectories of each child, grouped by this classification. For both nouns and verbs, the vast majority of children (72 percent and 84 percent, respectively) **increase** in overregularization over this age range, a substantial

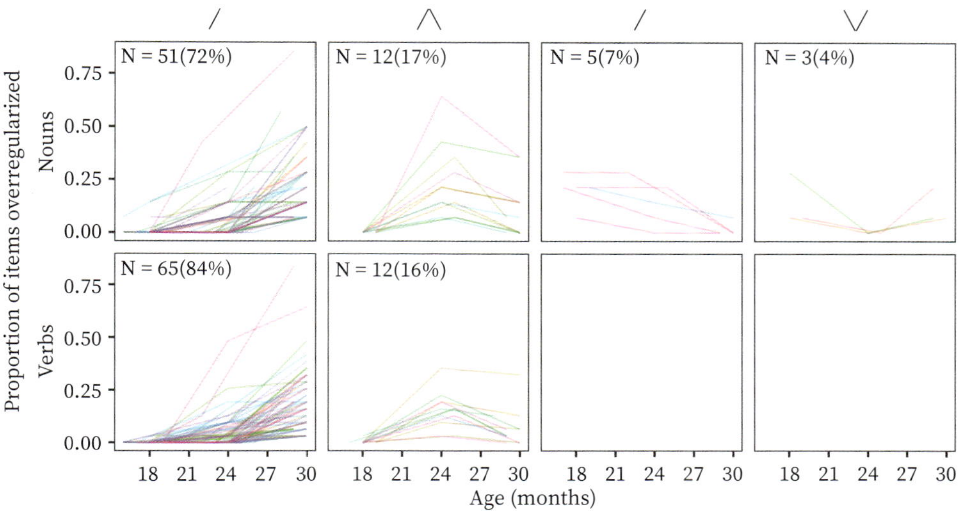

Figure 14.4
Empirical overregularization trajectories for American English children with at least three observations, categorized by overall shape.

minority (17 percent and 16 percent) **recover** from overregularization, and very few **decrease** or **retreat**. So we see that by 30 months, most children who have shown any sign of overregularization to date are continuing to overregularize more and more while some are recovering from an earlier peak rate.

14.3.2 Norwegian

Because longitudinal data are so much more plentiful in the Norwegian dataset, we can quantify children's trajectories more rigorously. For the 449 number of children who overregularize at least one item and have at least four administrations, we fit a logistic regression for each child predicting how many items they overregularize from their age. For each child, we select either a model with a linear effect of age or both linear and quadratic effects of age based on which one fits the data better (i.e., has the lower Akaike information criterion [AIC]). We then classify each child's trajectory as:

- **an increase** (/) if the best fit model is *linear* and the effect of age is *positive*,
- **a decrease** () if the best fit model is *linear* and the effect of age is *negative*,
- **a recovery** (Λ) if the best fit model is *quadratic* and the quadratic effect of age is *negative*, and

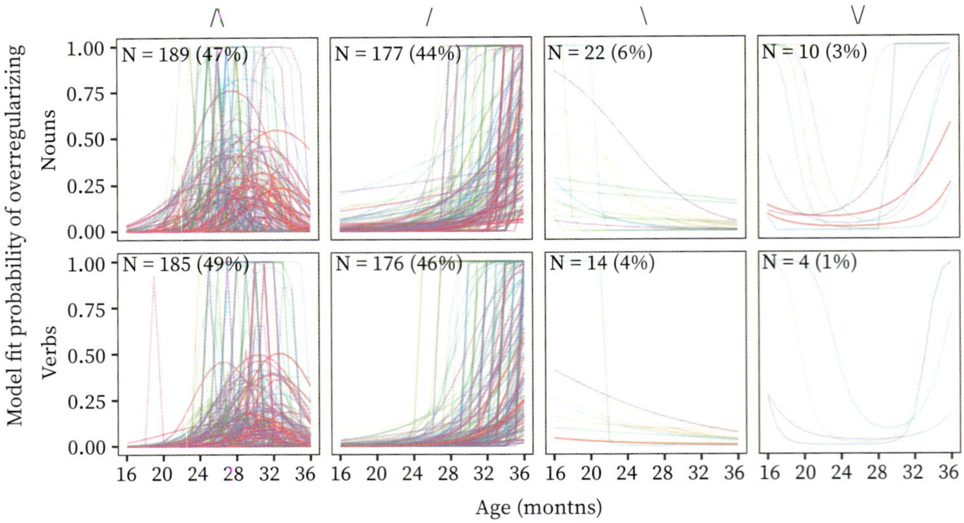

Figure 14.5
Model fit overregularization trajectories for Norweigian children with at least four observations, categorized by overall shape.

- **a retreat** (V) if the best fit model is *quadratic* and the quadratic effect of age is *positive*.

These classifications are analogous to the three datapoint categories from the English data but take advantage of the higher density in Norwegian to have smoothed trajectories and model-based estimates. Figure 14.5 shows the grouped trajectories of each child. In this dataset, which extend to 36 months, many more children exhibit recovery trajectories for both nouns and verbs (47 percent and 49 percent) and fewer exhibit increase trajectories (44 percent and 46 percent), for an approximately even split between the two types. While the English and Norwegian data may differ in a variety of ways, it's plausible that the difference reflects that many children are recovering from overregularizing between 30 and 36 months (an age range that is not represented in the English dataset).

Additionally, we see evidence that trajectory shape is correlated between nouns and verbs: out of the children who have **recovery** noun trajectories, 62 percent also have **recovery** verb trajectories; conversely, out of the children who have **increase** noun trajectories, 59 percent also have **increase** verb trajectories (see figure 14.6). This result echoes the observation from the cross-sectional data that while there is substantial variation among children, there is some degree of consistency within children across lexical categories.

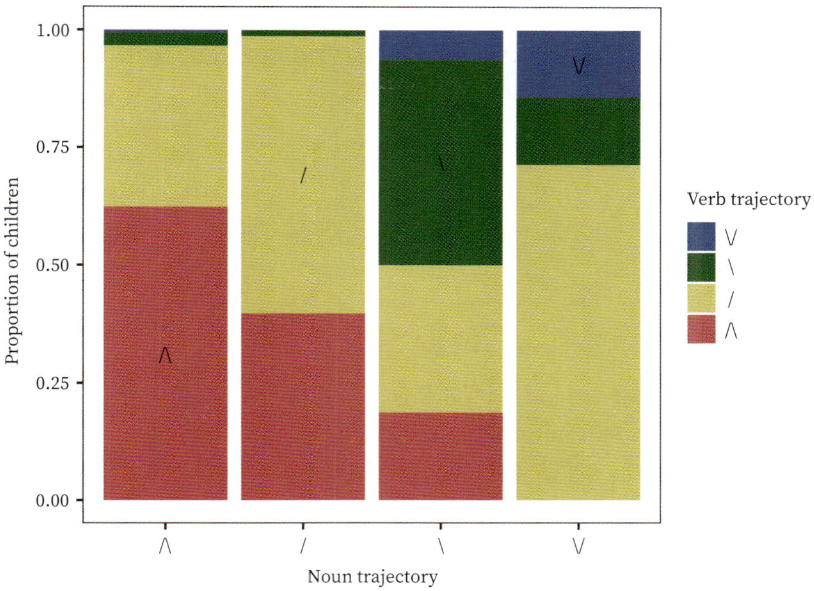

Figure 14.6
Proportion of children exhbiting each verb overrregularization trajectory type as a function of noun overregularization trajectory type.

Lastly, for a more detailed look at recovery from the overregularization, we examine the trajectories of the 30 Norwegian children with the most observations (at least 9) who have recovery-type trajectories for both nouns and verbs (figure 14.7). We observe that for many of these children, the trajectories for nouns and verbs not only have the same overall type but also hang together over the course of development. The median age when these children are most likely to overregularize is 29 months for nouns and 30 months for verbs, consistent with the speculation above that 30 months marks the beginning of more widespread recovery.

14.4 Conclusions

This chapter examined overregularization data from the noun and verb morphology items of the WS form in English and Norwegian. A central takeaway from our analyses is that there is tremendous heterogeneity in these data: some children are never reported to overgeneralize within the age range of the sample while others do substantially more, without a set developmental time course across children.

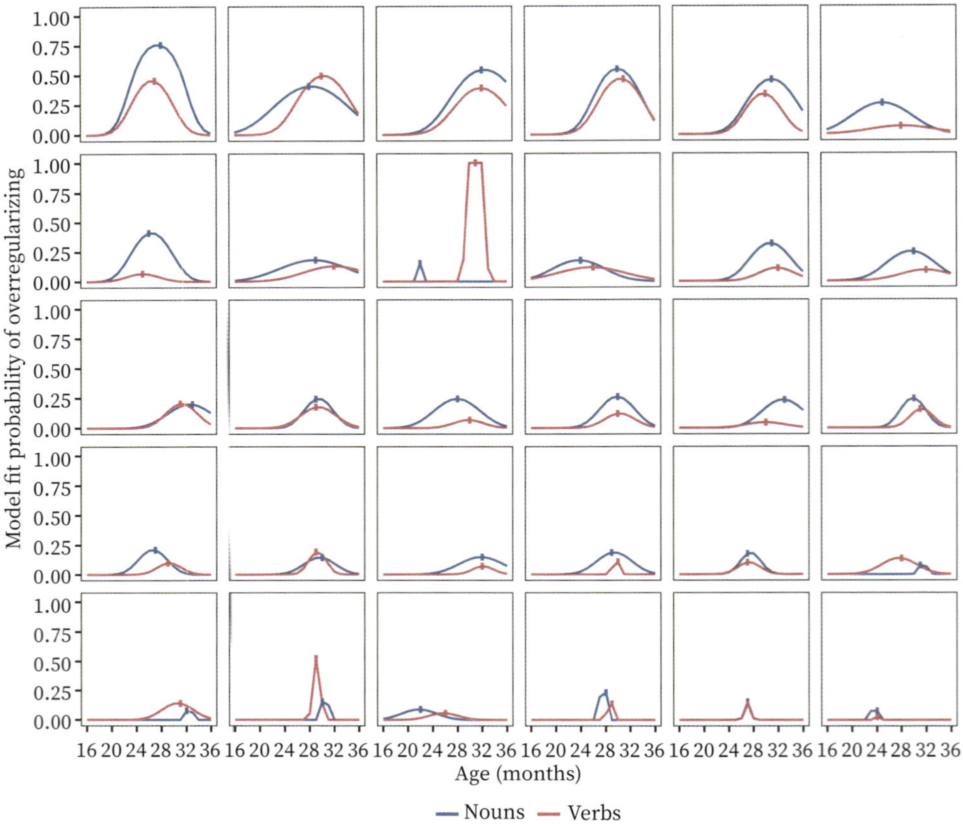

Figure 14.7
Model fit overregularization trajectories for Norwegian children with recovery-type trajectories and at least nine observations. Marks in each line show the inflection point of the recovery.

Children with larger vocabularies tend to overregularize more. Overregularization also tends to proceed in tandem for nouns and verbs in both cross-sectional and longitudinal data. Further, individual children's trajectories can be fruitfully categorized by overall shape, leading to the observation that, for children who overregularize, approximately half begin recovery toward correct inflection at around 30 months and half continue to overregularize more and more through 36 months.

One important caveat to these findings is the possibility that parents are less keen observers of morphological generalization than they are of vocabulary growth more broadly. It is possible that the lack of systematicity we observed is due to inconsistency in which parents report overgeneralization—perhaps only some parents are sensitized to the somewhat metainguistic observation that their child is using a frequent ending incorrectly (e.g., *foots*). This kind of bias would be consistent with the noun-verb overregularization correlation we observed—only systematic validation outside of the CDI would truly dispel this worry. On the other hand, the kind of variability we observed is not unlike the variability observed in naturalistic studies of overgeneralization (e.g., Marcus et al., 1992).

In sum, our work here suggests caution in the received narrative of morphological generalization as just another stage in the predictable progression of language learning. Unlike, say, word combination (cf. chapter 13), we do not see as clear and consistent developmental progression in this more elusive behavior.

15 Individual Variation in Vocabulary

> Of all the individual differences described to date in the literature on early child language, variations in rate present the least interesting challenge to traditional "universalist" models of development. If it can be shown that all children go through the same basic sequence, activating a common set of structures and processes, then small variations in the onset time for specific language milestones might represent little more than a minor perturbation to a maturational theory (like variations in the onset of puberty). Putative variations in style of development are more problematic, because they raise questions about the order in which structures are acquired, and the mechanisms used to acquire those structures. (Bates et al., 1994, p. 86)

Preceding chapters have dealt with the degree of variability in vocabulary size among individuals in chapter 5 and the stability of individuals' percentile rank in chapter 4. Then chapters 6 and 9 explored demographic factors as (partial) explanations of this variability. Throughout, our treatment of variability has focused on issues of learning rate in the sense discussed by Bates et al. in the quote above. With the exception of Chapter 9, which examined the rate of learning individual words, we have not dealt yet with issues of *style*. This chapter rectifies that omission, investigating three major topics: (1) whether there is stable individual variation in the category composition of vocabulary, (2) whether there are "spurts" in vocabulary growth, and (3) whether there is stable variation in the relationship between production and comprehension. To summarize, our answers are "yes," "no," and "difficult to determine," respectively.

15.1 Introduction

What does it mean for two children to show differences in language learning style? Intuitively, children who vary in style of learning should differ in some aspect(s) of the learning process. Unfortunately, from the data that we use here, we cannot observe the process—we can only observe the outcome of learning: knowledge. Thus, children must exhibit some differences in the way their vocabulary grows.

This difference could be distributional and inferred indirectly from cross-sectional data, or it could be shown directly through longitudinal data (though this strategy limits the number of datasets we can use from Wordbank).

One prominent candidate for a stylistic difference in language acquisition is the distinction between "referential" and "expressive" children. In an early report on individual differences in vocabulary acquisition, Nelson (1973) noticed that there was substantial variation in how many nouns children had in their vocabularies. Children who had more than half of their vocabulary devoted to nouns were named "referential" children. They tended to have speech that was less syntactically complex than other children and showed faster vocabulary growth. In contrast, "expressive" children had speech that was more syntactically complex and had fewer nouns. Dore (1974) proposed a related version of this distinction, focusing on speech acts from two children in the middle of the second year and labeling them as "code oriented" (focused on labeling, similar to "referential") versus "message oriented" (instrumental, social requests, similar to "expressive").

Since this seminal work, the referential/expressive distinction has become enshrined in the literature as a canonical aspect of variation in children's style. Yet further debate about the consequences of this stylistic distinction has continued in the literature. For example, Bates et al. (1988) observed a correlation between the proportion of nouns in the children's vocabulary and their vocabulary size and suggested the possibility that referential style might simply reflect a more effective strategy for learning. Reacting to this claim, Pine and Lieven (1990) argued that the direction of causality might be reversed, however: perhaps having a bigger vocabulary (at least at a certain point) would lead to a greater representation for nouns.[1]

While exploring the referential/expressive phenomenon more rigorously, Bates et al. (1994) presented analyses largely supporting the Pine and Lieven view: much systematic variation in the "referentiality" (proportion nouns) in children's vocabularies was due to the size of their vocabulary. As children's vocabularies grow, they tend to increase in their overrepresentation of nouns (as shown also in chapter 11). Thus, two children of the same age who have different proportions of nouns may also have different overall vocabulary sizes. After controlling for overall vocabulary size, Bates et al. (1994) found only limited evidence for stylistic variation. In

1. Caught up in this issue is the question of whether there is a route into language via the memorization of "frozen phrases." This is an independent theoretical question that is difficult to address with Communicative Development Inventory (CDI) data as it is a question about the repetitive use of chunks of language in production. One observation is that, due to evidence of early verb generalization (e.g., Gertner et al., 2006), the original discussion about limited generalization in children's early syntax has been somewhat subsumed into a discussion about differences between general comprehension and conservative production.

the first section of this chapter, we pick up these analyses and apply them broadly to our dataset.

A second area of stylistic variation that has been much discussed is whether some children go through a vocabulary "spurt," defined as a change in the rate of vocabulary growth. The idea of a "spurt" or "explosion" occurs in a wide variety of discussions of early vocabulary (e.g., Nelson, 1973; Kamhi, 1986; Bates et al., 1988; see, e.g., Dapretto and Bjork, 2000, for review). Clearly, from the average growth rates observed in chapter 5, children's vocabulary growth accelerates dramatically during the period following their first birthday and the emergence of first word production. Investigations of this acceleration have focused on two distinct issues: its explanation and its variation across children.

Although there has been a tremendous amount of discussion of this issue in the literature, we will not investigate the possible *explanations* of accelerations in vocabulary growth—that is, the vocabulary spurt. Our data do not allow us to investigate issues of mechanistic process directly or of the cognitive or social changes which are contemporaneous with the vocabulary spurt. In addition, a short, but convincing, analysis by McMurray (2007) suggests that such acceleration is overdetermined in the sense that it will likely emerge from almost any plausible acquisition mechanism(s). Assuming that words vary in difficulty as a normal distribution, vocabulary growth will naturally accelerate with increases in a child's ability. Similar arguments have been made by Plunkett et al. (1992); see also chapters 3 and 4 in Elman et al. (1996). Thus, making reverse inferences from the presence of acceleration to a particular mechanism is unwarranted.

The second issue—variation in acceleration across children—is related to our aims in this chapter, however. Does every child's vocabulary acceleration follow the same general pattern, or is there substantial variation in the type of growth that is followed—for example, in the point at which acceleration begins, or the specific shape of the growth curve? Ganger and Brent (2004) report a systematic study of a subpart of this issue: whether there is a discontinuity in growth rate for individual children. Defining a spurt specifically as a change in the rate of acceleration, they conducted a quantitative analysis of whether such a change occurs for all children using original data and data from several classic studies (Goldfield and Reznick, 1990; Dromi, 1987). Their results indicated that a change in rate of acceleration best captured the growth trajectories for only a small subset (about one in five) of the children analyzed. Our second set of analyses follows up on this general issue.

In the third section of this chapter, we address another stylistic issue—the dissociation between comprehension and production. It has been long acknowledged that children tend to understand more than they can say (e.g., Clark and Hecht, 1983). But to what extent does this generalization vary across children? That is, are there some children who not only understand many words but also are fluent producers versus others who have a large vocabulary in comprehension but have more limited production? At the clinical extreme, significant variation in this respect must be present because there are some children with various apraxias of speech that have strong comprehension but serious production difficulties. At the other extreme, delays in early comprehension have clinical relevance and are often thought to be a more reliable indicator of language delays compared with production alone (e.g., Rescorla, 2009; Thal et al., 1991). While both of these extremes are interesting and important, the question for us is whether, in a typically developing population, we can detect and possibly identify correlates of systematic variation on this dimension over the full range of language abilities.

15.2 Variation in Vocabulary Composition

In this section we examine the question of variation across children in the degree to which their vocabularies reflect a "referential" versus "expressive" style. Following Bates et al. (1994), we operationalize the notion of "referential" children as those with a relatively higher than average proportion of nouns compared with other children (the definition of "noun" is the same here as in chapter 11). While there might be other more nuanced measures that we could construct, this one has the advantage of being directly related to the analysis framework used in chapter 11; thus, we use that same framework to investigate vocabulary composition in individuals here.

15.2.1 Measuring Vocabulary Composition in Individuals

The proportion of children's vocabulary that is made up of nouns is shown in figure 15.1. There is a general trend for an overrepresentation of nouns as shown by the blue line (representing the smoothed mean proportion nouns) being above the red dashed line (total proportion nouns on the form). The size of this overrepresentation across languages is the topic of chapter 11. Here, we examine its variability across children.

Is a referential style associated with having a larger vocabulary? If so, then the proportion of nouns in a child's vocabulary should be a predictor of vocabulary size, over and above age.

A simple model of this hypothesis is a generalized linear model predicting the number of Communicative Development Inventory (CDI) words a child produces as a function of age and proportion of nouns.[2] The coefficients of such a model, fit

2. We omit interactions from most of the models for interpretability; including interactions leads to unstable coefficient estimates.

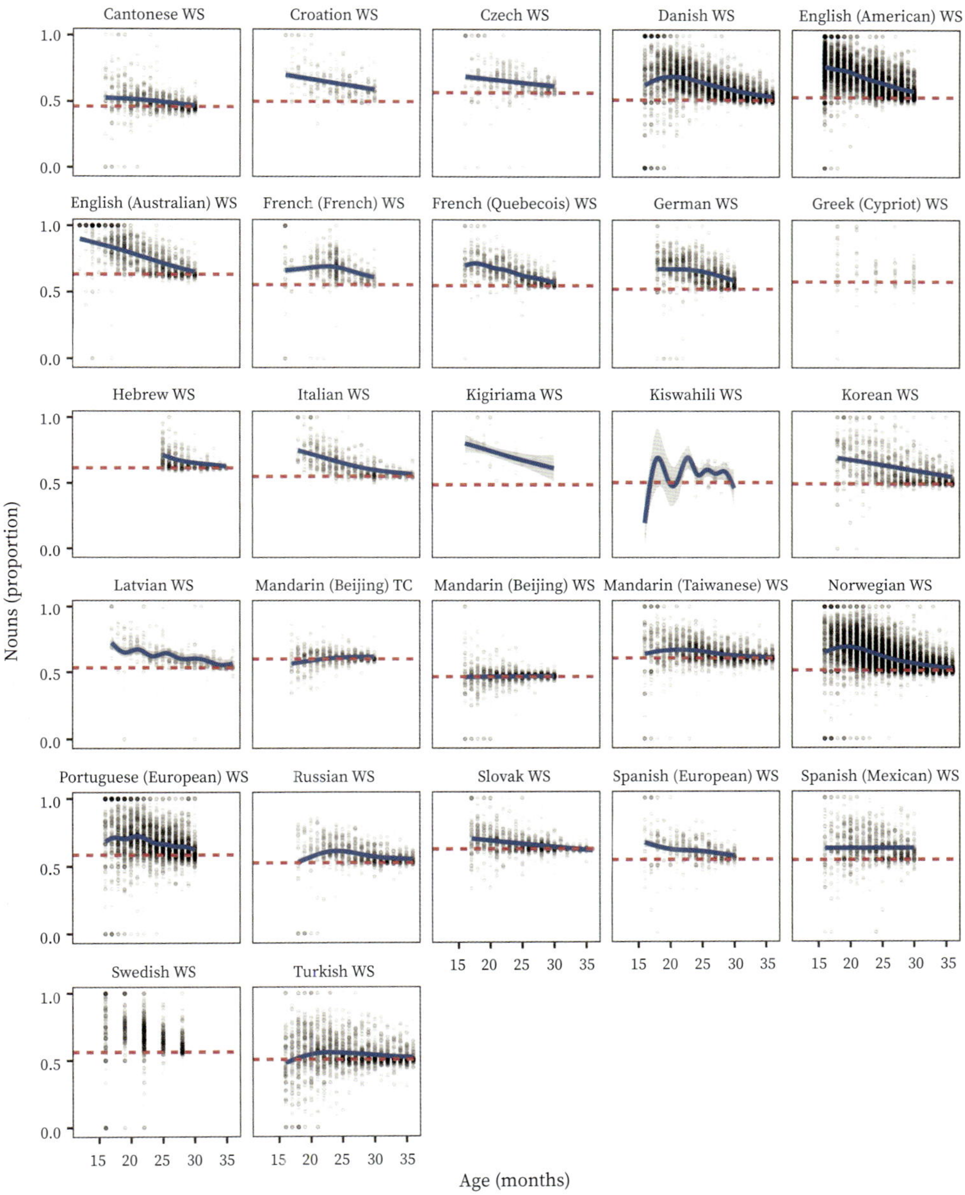

Figure 15.1
Proportion of nouns that each child produces as a function of age, with the blue curve showing a smoother fit and the red line indicating the proportion of nouns on the form.

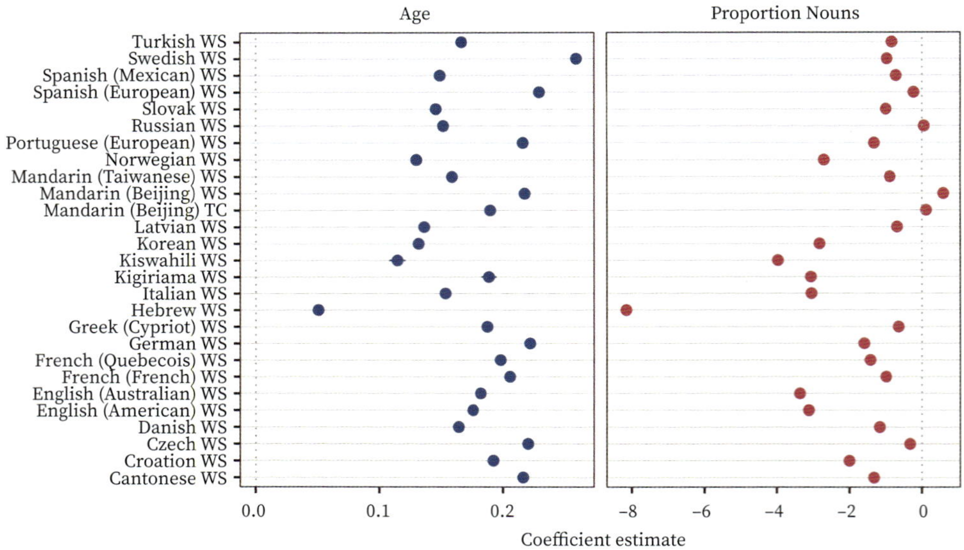

Figure 15.2
Effects of age (left) and proportion of nouns produced (right) on vocabulary size in each language.

to the data from each language, are shown in figure 15.2. Age coefficients are positive, indicating more words with age. Controlling for age, the proportion of nouns is also negative, indicating that having more nouns is related to a smaller vocabulary. (Confidence intervals are plotted but are typically tiny and hence invisible.) This result appears to provide support for the *opposite* of the claimed relationship between the referential/expressive distinction and vocabulary size. Controlling for age, those children with more referential vocabularies have *smaller* vocabularies.

The trouble is that these variables—age, noun bias, and total vocabulary size—have a complex relationship to one another. For a young child, having a bigger noun bias is correlated with having a bigger vocabulary because they are on the early part of the "noun overrepresentation" curve (as shown in chapter 11). In contrast, for an older child, having a bigger noun bias is correlated with having a smaller vocabulary because they are on the later part of the curve. To put it simply, the directionality of the relationship that you discover is largely determined by what part of your sample is densest.

Put another way, proportion of nouns could be predictive of vocabulary size not because children with a particular style have bigger vocabularies but because having more nouns in their vocabulary simply indicates that the child is further along a standard progression. Perhaps all children follow the same trajectory through the noun bias. Even in this scenario, knowing the size of a child's noun bias will tell you something about vocabulary size without that implying that the child is following a different trajectory. This point is made in different ways by both Bates et al. (1994) and Lieven et al. (1992).

15.2.2 Growth-Corrected Vocabulary Composition

One way to circumvent this critique is to measure whether a particular child has a greater-than-average, vocabulary-adjusted noun bias. In other words, if we remove the average correlation with noun bias and vocabulary, can we still find a relation with an individual's degree of noun bias and their vocabulary size?

Figure 15.3 shows both the English noun-proportion data (plotted now by vocabulary size) and the residuals of that distribution when it is predicted via a cubic model. The next question we can ask is how this "residualized style" relates to other variables like age, grammatical ability, and (in longitudinal data) further vocabulary growth. Note that we cannot directly compare this variable with other aspects of concurrent vocabulary size because features like the number of closed class items in the vocabulary are nonindependent—if more of the vocabulary consists of nouns, necessarily less of it consists of closed class items.

Our first analysis looks at the correlation between vocabulary-residualized noun bias and age. Figure 15.4 shows coefficient estimates on this analysis. Now, most age coefficients are reliably negative, suggesting that a greater residual noun bias is associated with being *younger*. Two of the main outliers here are from Mandarin, which, as discussed in chapter 11, has the smallest noun bias of any of the languages in our dataset.

15.2.3 Vocabulary Composition and Grammatical Ability

Lieven et al. (1992) were interested in the possibility that an alternative route into combinatorial language from a referential style would be the use of construction-based generalizations. To test this hypothesis using our bias-corrected measure, we examine correlations with grammatical ability.

For grammatical complexity scores (figure 15.5), we see that vocabulary-residualized noun bias is reliably related to grammatical complexity as indicated by estimates that are to the left of the vertical dashed line. (Again, Mandarin is an exception.) Because of the residualization procedure, this relation reflects effects that are over and above the correlations between grammar and lexicon (as reported in chapter 13). Thus, those children with more nouns than expected for their vocabulary size produce *less* complex language. As seen above, they are also *younger* than expected.

We next repeat the analysis of complexity while controlling for age and total vocabulary size. The coefficients are shown in figure 15.6. Vocabulary size, of course, has a positive relation to grammatical complexity as does age (see chapter 13).

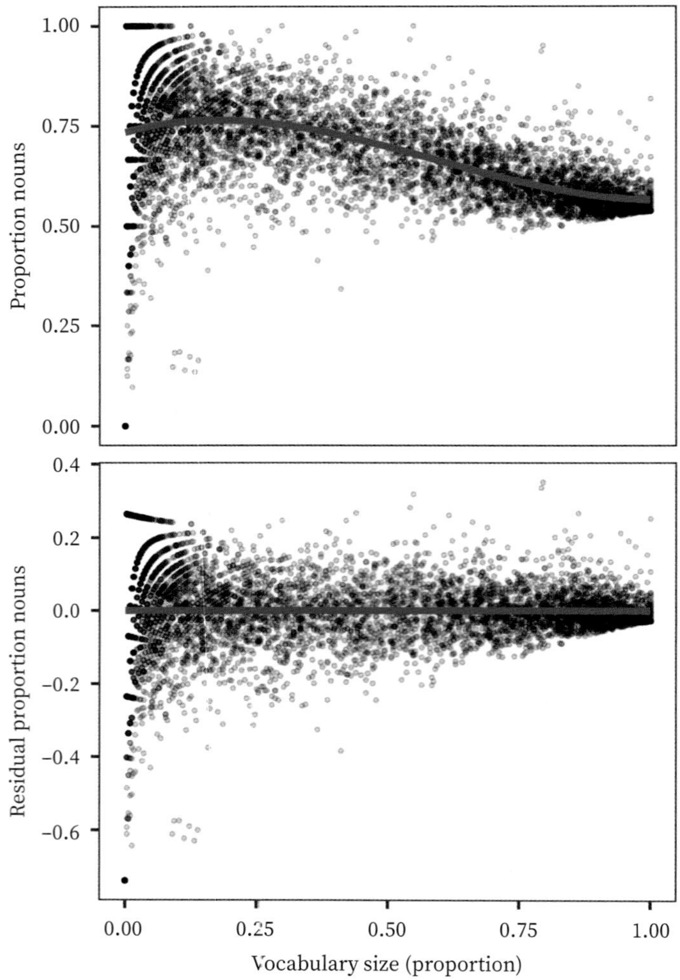

Figure 15.3
For American English data, proportion of nouns produced by each child as a function of their vocabulary size (top) and the residual proportion of nouns regressing out vocabulary size (bottom).

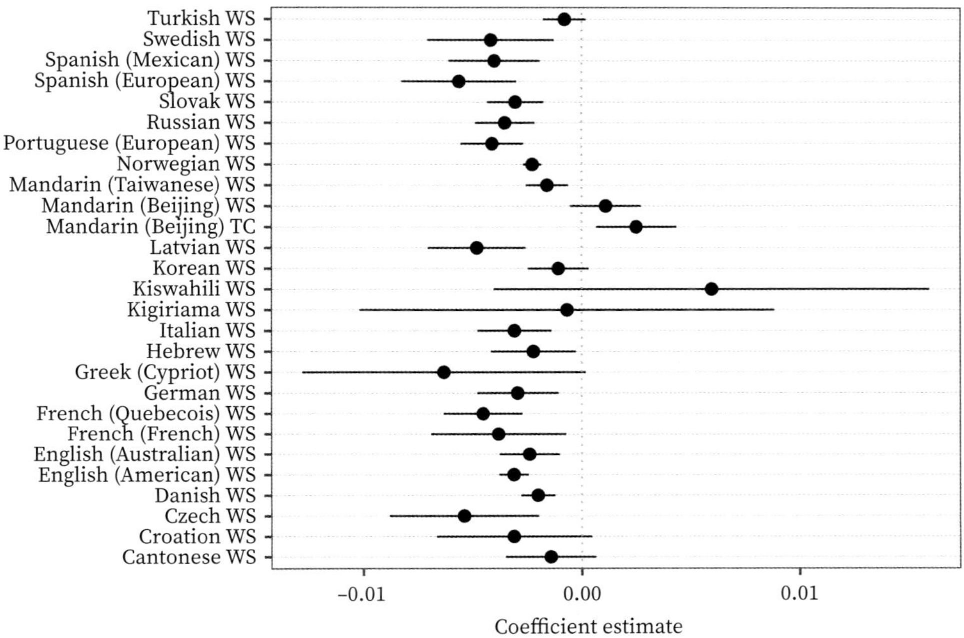

Figure 15.4
Effect of age on vocabulary-residualized noun bias in each language.

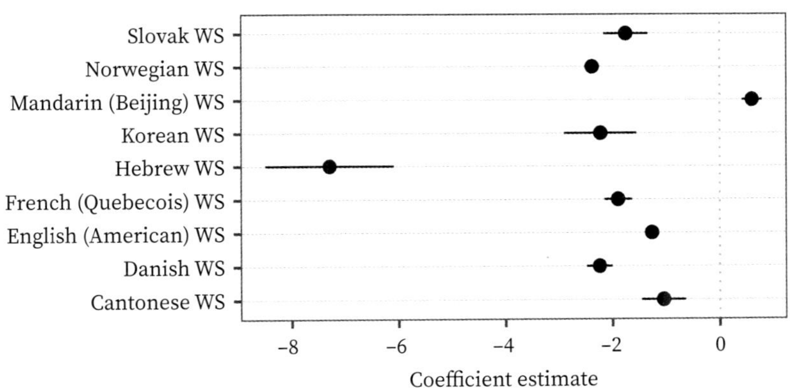

Figure 15.5
Effect of vocabulary-residualized noun bias on complexity score in each language.

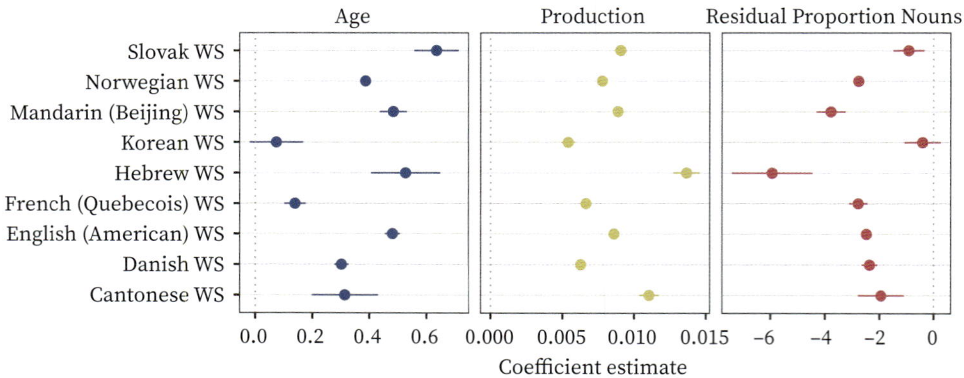

Figure 15.6
Effects of age (left), vocabulary size (middle), and vocabulary-residualized noun bias (right) on complexity score in each language.

Even controlling for these two factors, however, we still observe a consistent negative relation between residual noun bias and complexity.

15.2.4 Vocabulary Composition Differences across Siblings

Nelson (1973) noted that later-born children had larger expressive vocabularies than firstborn children. This observation is consistent with the confounding of vocabulary size and composition since later-born children tend to have smaller vocabularies (see chapter 6). But there are other reasons to suspect that vocabulary composition could differ for children who are firstborn versus later-born. Tomasello and Mannle (1985) observe that older siblings tend to be more directive with their younger siblings. Since more directive speech has been argued to lead to an "expressive" style, perhaps later-born children have a different vocabulary composition due to the input they receive (Tomasello and Todd, 1983).[3]

We first examined uncorrected vocabulary composition by birth order. Figure 15.7 shows this analysis. There is a trend for young children (15–20 months) to show slightly "nounier" vocabularies—consistent with Nelson (1973)'s observation—but this trend is very small and not observed in all languages.

Further, when we examine the trend in the residualized noun proportion, it appears even weaker and is observed in even fewer languages. This pattern is shown in figure 15.8. Thus, we conclude that differences in vocabulary composition

3. Technically, the observation that directive speech leads to expressive style could also be due to confounding of amount of speech; more directive parents also tend to be those who talk to children less (e.g., Hart and Risley, 1995).

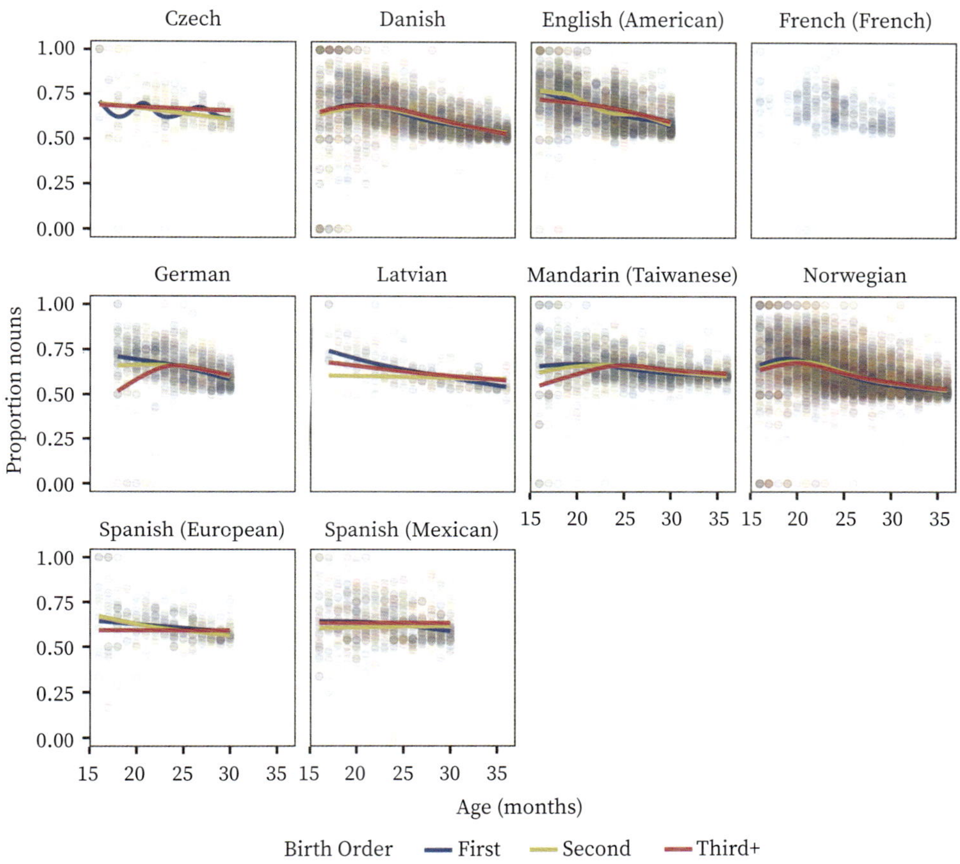

Figure 15.7
Proportion nouns in vocabulary by age, with color showing birth order.

across birth order are likely to be small and may be, at least in part, due to faster learning by earlier-born siblings.

15.2.5 Conclusions

To summarize, in this section we were interested in whether we found cross-linguistic evidence for different styles of language learning—in particular, individual differences that mapped onto the referential versus expressive distinction. Operationalizing this distinction, we asked whether children with a larger noun bias were different in their language-learning trajectory than children with a smaller noun bias. The relations between overall noun bias and vocabulary are complex to interpret due to the nonlinear relationships between these variables

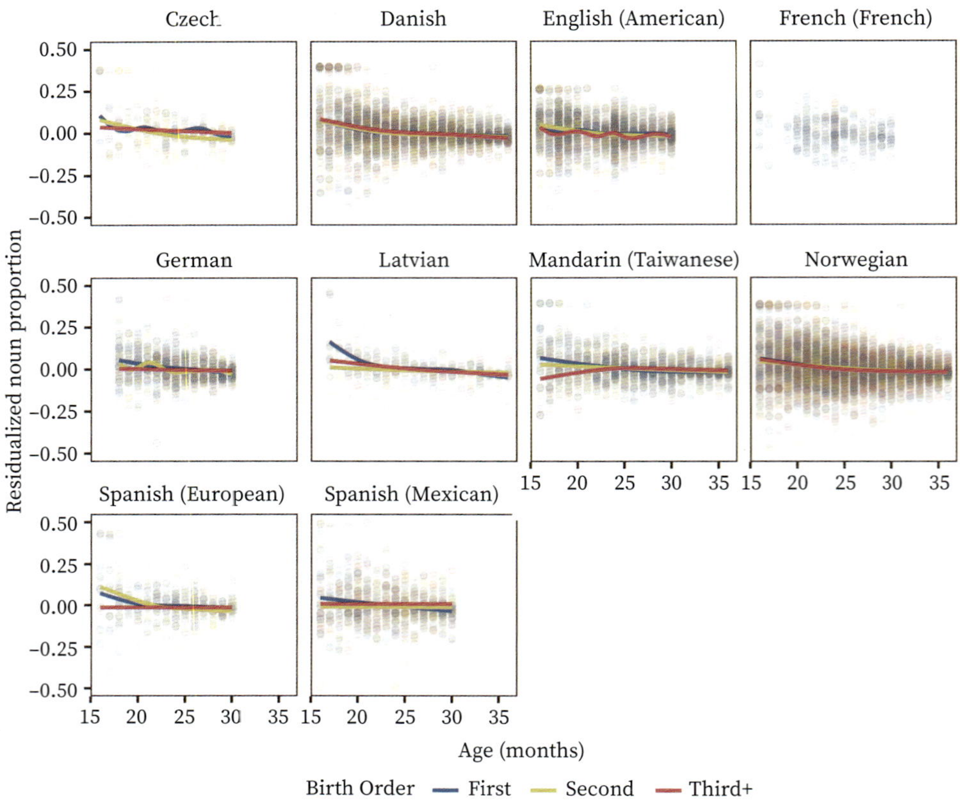

Figure 15.8
Vocabulary-residualized noun proportion by age, with color showing birth order.

and age. To circumvent this, we examined a residualized noun bias measure that controls for total vocabulary size. Across languages, this measure was related to age and to grammatical complexity: children with relatively more nouns in their vocabulary tended to be younger and to be producing less complex speech. On the other hand, there was only limited evidence that these differences were manifest across birth orders when controlling appropriately for overall vocabulary size (contra early speculations by Nelson, 1973).

Taken together, these data provide some cross-linguistic support for the idea that children show variability beyond differences in rate of vocabulary acquisition. One dimension of variability is that, for a particular level of vocabulary size, there appear to be children who are younger, know more nouns, and combine words less (perhaps the "referential" children referred to in previous literature); other children will be older, have a more diverse vocabulary (including more predicates), and will tend to use more complex word combinations.

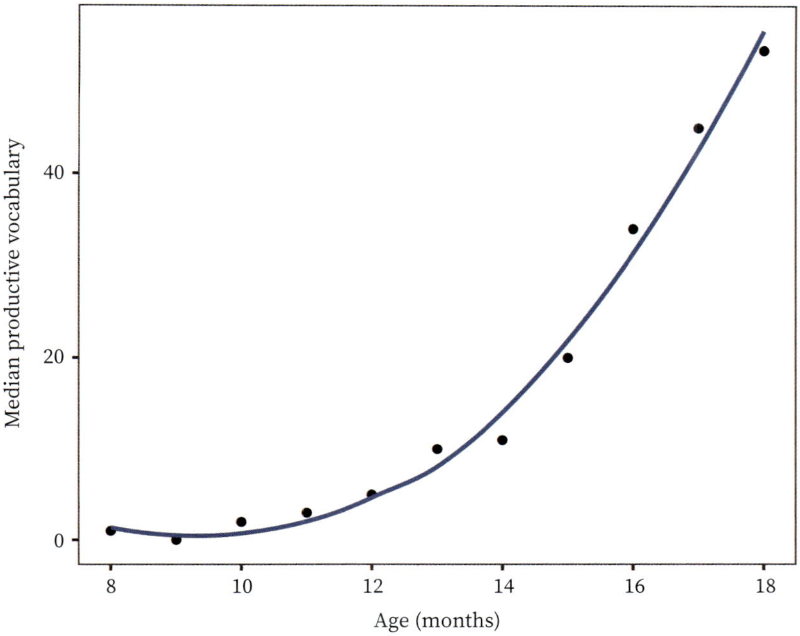

Figure 15.9
For American English data, median productive vocabulary as a function of age (curve shows smoothed fit).

15.3 Spurts in Vocabulary

Perhaps the most obvious aspect of early vocabulary growth is that it picks up speed over the second year after birth. First words are hard-won, but soon children's vocabulary acquisition appears to "explode." This acceleration is easily visible in figure 15.9, which shows the median productive vocabulary from 8 to 18 months (for American English Words and Gestures [WG] data). (We could measure the rate of learning by taking the derivative of this curve, but this rate calculation is slightly misleading for cross-sectional data, and so we postpone this analysis for later in this chapter). This increase in the rate of vocabulary learning has been much remarked on in the literature, as discussed previously.

Our starting point here is a study by Ganger and Brent (2004) who provide a detailed, curve-fitting analysis of the question of vocabulary "spurts." They evaluate whether individual children's longitudinal growth patterns are better fit by a model with constant acceleration in growth rate or whether some children have a discontinuous "step" in terms of their growth rate.

In our view, this is a productive approach but requires access to sufficient appropriate data that capture children's growth rate generally. For example, using longitudinal data, we can simply examine features of rate and acceleration and how they change with time. For these analyses we focus on longitudinal production data from Norwegian and English Words and Sentences (WS) and WG forms. Because we are interested in computing (potentially nonlinear) changes in rate, we need four datapoints from each child as a minimum (for intercept, linear, and quadratic components). In addition, because we are interested in early changes, we set the restriction that the first time point reported should have fewer than 50 words reported (50 words is often used as a semi-arbitrary cutoff for the onset of the vocabulary spurt; Dapretto and Bjork, 2000; Ganger and Brent, 2004).

The decision to exclude children with larger vocabularies at their first recorded measurement means that there is likely to be an overrepresentation of children with slower rather than faster rates of vocabulary growth. In particular, from the WS data, we exclude a substantial proportion of children even from the youngest groups (e.g., 22 percent of Norwegian 16-month-olds). So as not to bias the analysis further by including a large proportion of older, slower learners, we only include children younger than 21 months in this sample.

We now have a population of children for whom we can evaluate the rate of vocabulary growth and how it varies as a function of age. This winnowing leaves us with data from 290 children, whose growth curves are shown in figure 15.10.

We exclude datapoints associated with large decreases in vocabulary (negative rates). Although some small negatives would be expected based on measurement

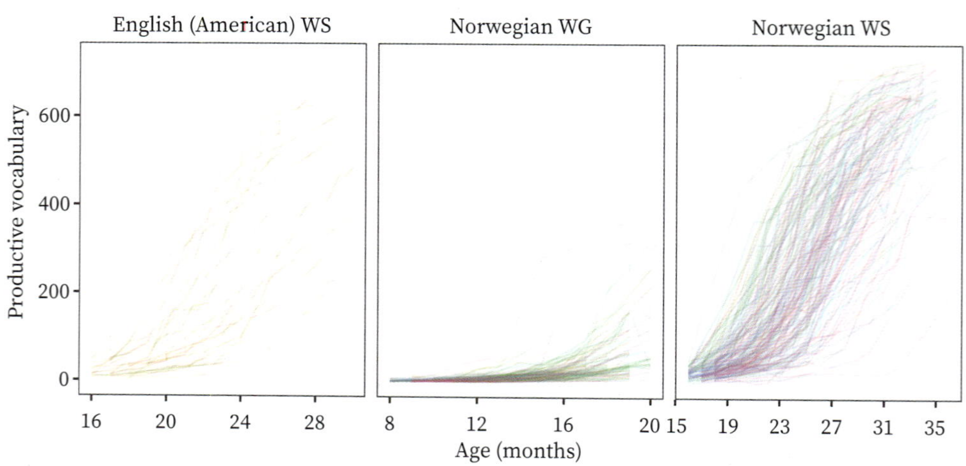

Figure 15.10
Vocabulary size as a function of age for included longitudinal administrations.

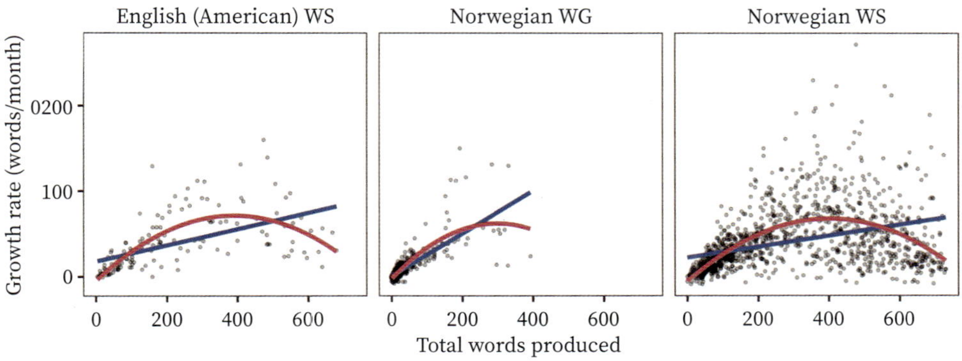

Figure 15.11
Vocabulary growth rate as a function of vocabulary size for all included longitudinal administrations (red line shows quadratic fit, blue line shows linear fit).

error or forgetting, large negative spikes are rare and likely due to errors in the data (e.g., a partially filled form). We exclude rates of –10 words per month and below (1.3 percent of data).

Figure 15.11 shows children's estimated growth rates in the period leading up to that month (number of words learned since the last longitudinal measurement divided by number of months since that measurement) plotted by total words produced in that month. There is a clear quadratic shape to this pattern, almost certainly caused by ceiling effects for children who are "running out of words" on the form.

The definition of a vocabulary spurt, as posed by Ganger and Brent (2004), concerns the *way* that vocabulary growth rates may or may not increase for individual children. Clearly vocabulary growth rates increase, because vocabulary growth picks up speed generally. The question is, instead, do growth rates change smoothly, indicating constant acceleration? Or do growth rates move from one equilibrium to another (indicating a "spurt")? Ganger and Brent (2004) proposed analyzing children's growth rates not as a function of age but of total vocabulary. To examine this question, we need to focus on the initial 250 words when the average rate appears to be increasing linearly (before ceiling effects are found; see red curve above) and identify children with more than 4 CDIs before this time to ensure sufficient density. Further, we need to examine the rate trajectories of individual children. Figure 15.12 shows this analysis for a randomly sampled subset of children in our available datasets.

Ganger and Brent (2004) analyzed the question of developmental spurts by fitting different curve types to the rate function in their data. They compared the

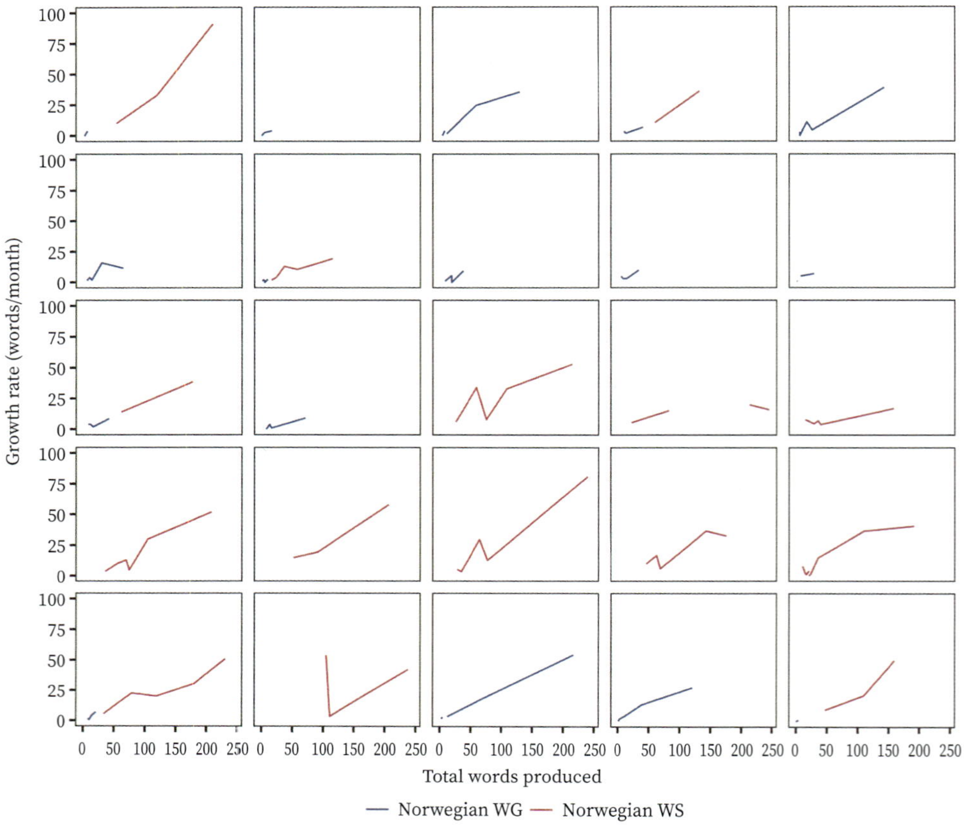

Figure 15.12
Vocabulary growth rate as a function of vocabulary size for 25 randomly sampled children.

likelihood ratio of quadratic and logistic rate curves for each child's data. The quadratic curve represented the hypothesis of smooth growth in rate (smooth acceleration). In contrast, the logistic curve was of the form

$$R \sim \frac{\alpha}{1 - e^{-\beta(W-\gamma)}}$$

where R is the rate of acceleration and it is assumed to be distributed as a function of α, the asymptotic rate, β, the slope of the change between the initial and final rates, W (the total vocabulary), and γ, the point at which the spurt occurs. This curve captures a discrete "spurt"—a movement from one equilibrium to another.

We fit these functions (as well as a simple linear function) to our data. Fitted curves for children with more than 7 datapoints are shown in figure 15.13. The basic

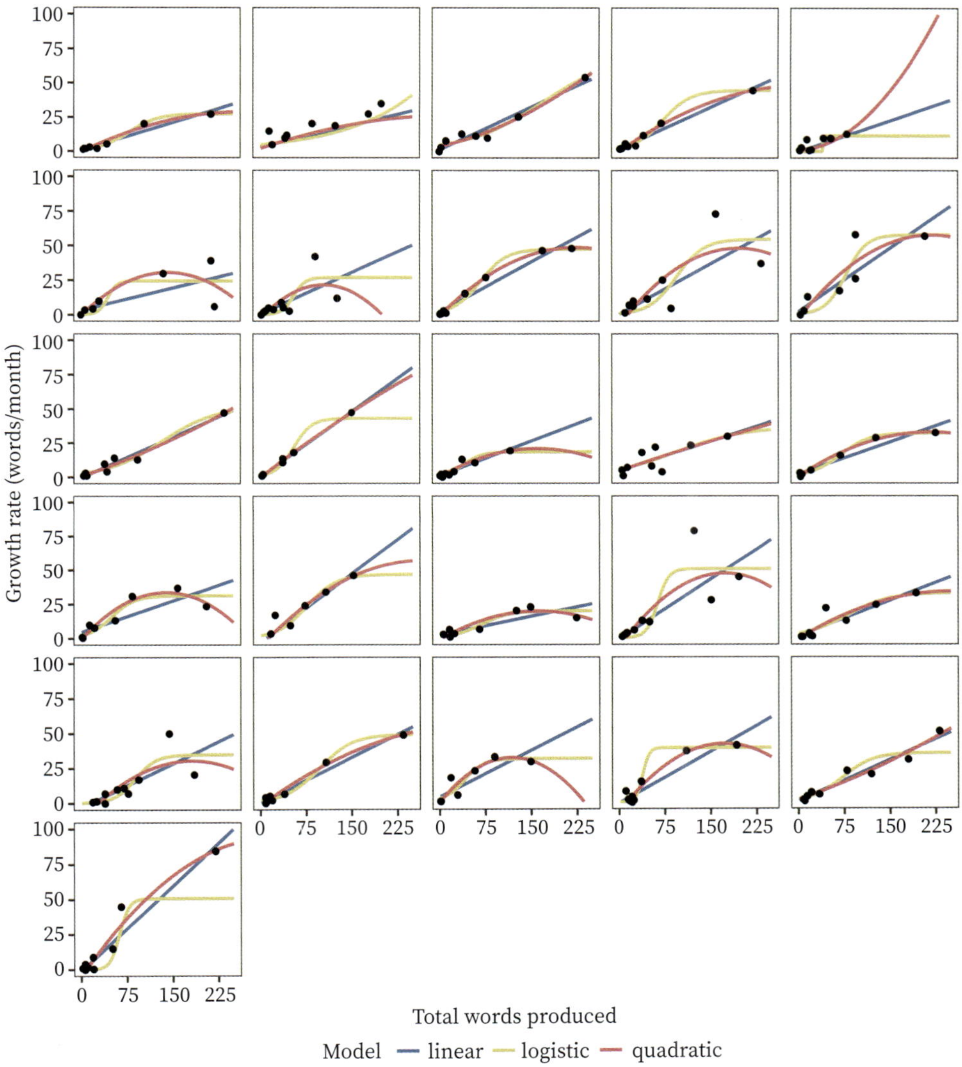

Figure 15.13
Vocabulary growth rate as a function of vocabulary size for children with more than 7 datapoints, with curves showing various model fits.

Table 15.1
Number of children for whom the best fitting model falls into each model type.

Instrument	Linear (no intercept)	Linear	Quadratic	Logistic
English (American) WS	2	1	0	0
Norwegian WG	35	2	17	20
Norwegian WS	11	3	4	6

visual impression from these data is that, even with the longitudinal depth we have for individual children, there is substantial uncertainty in the best-fitting curve. Nevertheless, it does not appear that many children show something that looks clearly like a spurt. A few children rise quickly and then show one datapoint that levels off.

But, there is a puzzle. We can only confirm that the rate has leveled off if we have data at higher levels of production. Yet, for vocabulary sizes over 300 words, *every* child will level off in their rate because of the potential for ceiling effects—that is, they will have "run out of words" on the form. This example illustrates some of the difficulties in making strong inferences from data of this type.

We next conducted the full model comparison analysis that Ganger and Brent (2004) conducted over the 101 separate longitudinal records included.[4] In Ganger and Brent's analysis, they used only two models (quadratic and logistic), which had the same number of parameters. Accordingly, they compared only the likelihoods of the data under these models. In contrast, we compared a linear function with intercept at 0 (one parameter), standard linear (two parameters), quadratic (three parameter), and logistic (three parameter) curves. To make up for the difference in parameters, we computed Akaike's Information Criterion (the AIC, a measure of model goodness of fit, where smaller is better) for each model for each participant.

Table 15.1 shows the proportion of children in each dataset for which different model types fit best. Overall, children were split between models, with some children best fit by the logistic. The linear functions, which were simpler, however, fit more participants better, with the one parameter linear model with no intercept providing the best fit for more children than any other.

The 21 children (out of 101 total) with a best-fitting quadratic model are shown in figure 15.14. Some of these children do appear to have data that are well fit by the quadratic model. But for many, this fit appears to be the product of a single datapoint; assuming some error, a more parsimonious model (e.g., simple linear) might do just as well. Thus, with more children but less density, our conclusion

4. This number consolidates data between WG and WS forms when a child has both to maximize our use of the available longitudinal data.

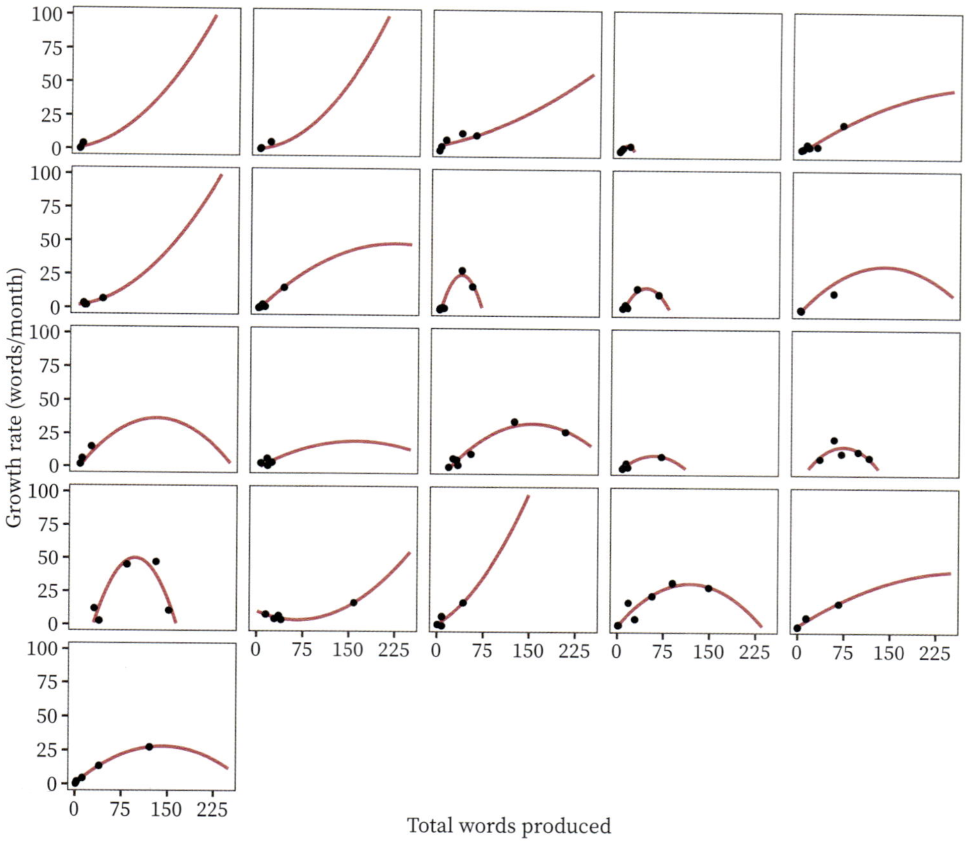

Figure 15.14
Vocabulary growth rate as a function of vocabulary size for children for whom the best fitting model is quadratic, with curves showing quadratic model fits.

is strikingly similar to Ganger and Brent (2004): there is limited evidence for a vocabulary spurt in most children.

To further examine this issue in a denser dataset, we used data from Roy et al.'s (2015) in-depth study of a single child. This is an ultra-dense dataset with millions of words of transcribed speech and hand-checked age-of-acquisition data for over six hundred words up to the child's second birthday. The comparable curves for this dataset are shown in figure 15.15. Using the same AIC method, the quadratic model fits best, but the linear model is clearly close as well.

Stepping back, we examined the growth rate of children's vocabulary. To a first group-wise approximation, children's vocabulary growth accelerates linearly with vocabulary size during the initial period (up to around 250 words). After this point, we run into substantial measurement issues because the CDI does not contain

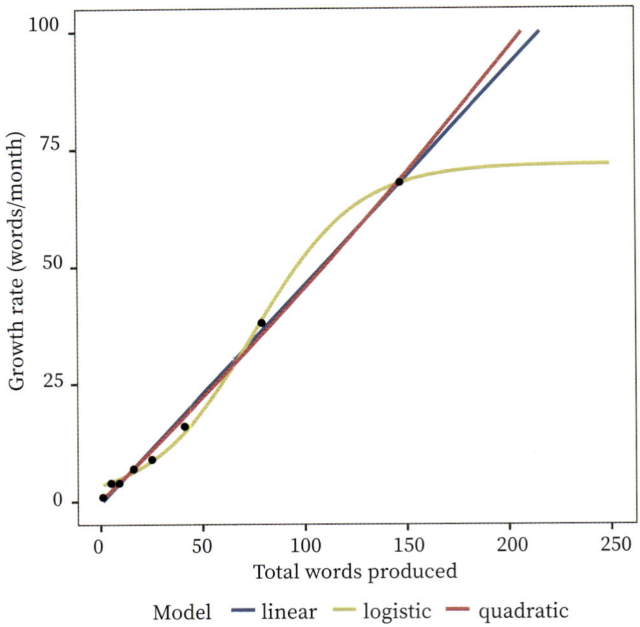

Figure 15.15
Vocabulary growth rate as a function of vocabulary size for the Roy et al. (2015) data.

enough words to be certain of the pattern of growth. Further, when we examined individuals' growth, it also often appeared to be linear or quadratic; only in a minority of individuals was there any evidence for a "spurt" (a discrete change). This conclusion was tempered, however, by the difficulty of drawing conclusions without even denser longitudinal data concerning the very beginnings of language.

15.4 Variation in Comprehension Versus Production

Our next investigation concerns the question of how tightly comprehension and production are yoked within CDI data. Our assumption is that there is variability among children on this dimension—while some children can say a large portion of the words that they understand, others have low production scores despite appearing to understand substantial amounts. How does the ratio of production to comprehension vary across ages and across languages?

It is important to be clear that some of the pattern in the relation between comprehension and production could be due to variation between parents in under- or overreporting comprehension (or for that matter, production, but we assume—and chapter 4 confirms—that production reports likely carries more

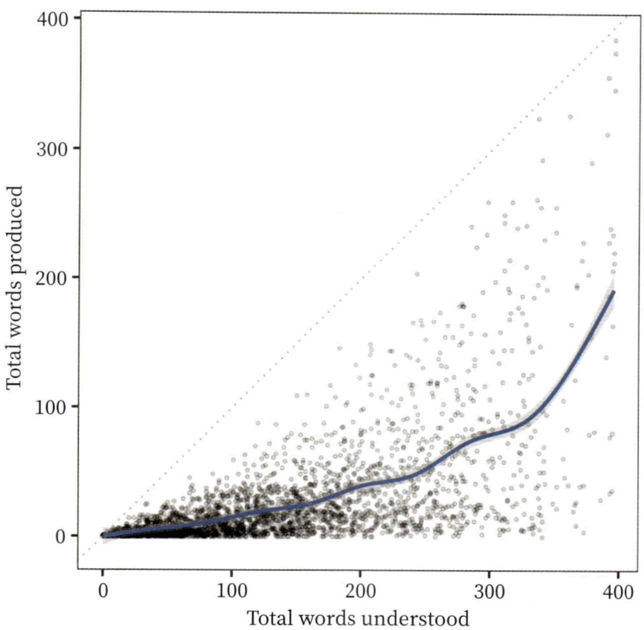

Figure 15.16
For American English data, each child's productive versus receptive vocabulary size.

signal). For example, compared with the more observable facts of word production, the variation might reflect that comprehension is a more difficult concept for parents to grasp, so there will be differences in across parents in what "comprehends" means. Some parents might be very liberal and recall a generally understood story that included a particular word while others might be searching for a specific anecdote that clearly illustrates comprehension of that word. We might also be detecting variation in the threshold at which parents assume that a response indicates comprehension. Nevertheless, there is evidence that children's level of early comprehension is a useful metric for identifying possible language delays beyond production data alone. Specifically, children with only a few words who also have low comprehension scores are at increased risk for language delays compared with children with a few words who nonetheless appear to be understanding language well (e.g., Rescorla, 2009).

Of course, this type of analysis can only be conducted on WG-type forms that include both comprehension and production information. We begin by investigating the American English WG data as an example.

Figure 15.16 shows individual children's comprehension and production plotted against one another. The diagonal indicates a child who comprehends and

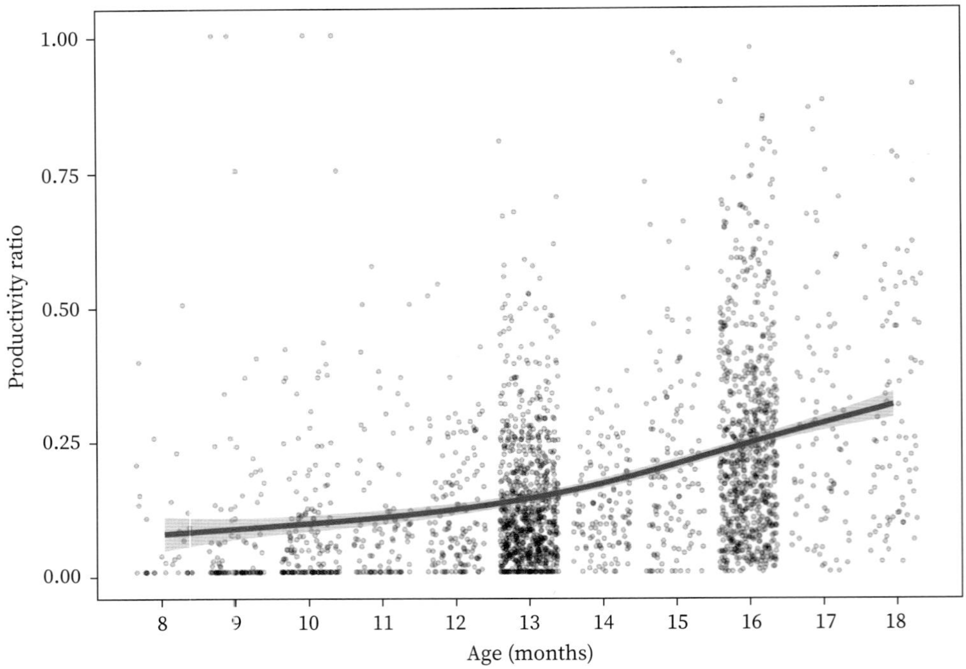

Figure 15.17
For American English data, each child's productivity ratio as a function of age.

produces exactly the same number of words. In practice, this measure is always below the diagonal because, by the design of the form, a child cannot "say but not understand" a particular word; they can only "understand" or "understand and say."

We can convert these data into a productivity ratio:

$$\text{productivity} = \frac{\#\text{produced}}{\#\text{understood}}.$$

Figure 15.17 shows this ratio for all children.

The resulting scatterplot is quite interpretable. It contains a few outliers at the very top of the range for very young children (whose parents report them producing and comprehending the same number of words). But, for most others, the ratio is low, increasing from about 10 percent to 30 percent by the top of the form.

Figure 15.18 plots these productivity ratios by language for an age-restricted subset between 8 and 18 months. Plots are sorted by the mean productivity ratio. While the majority of languages show the same pattern as English (an increase from around 10 percent to 30 percent), there are some outliers (Kigiriama, European French) that show a flatter slope.

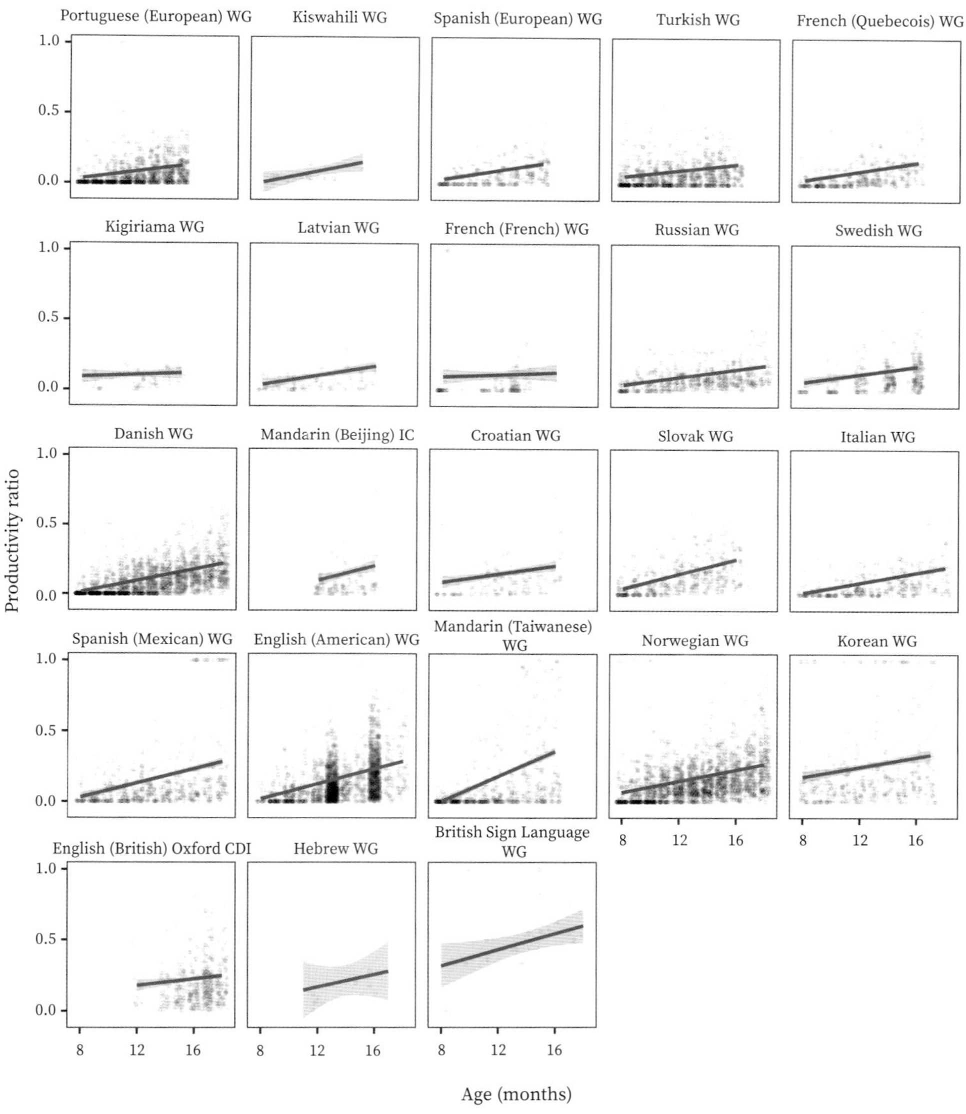

Figure 15.18
Productivity ratio as a function of age for each child in each language (lines show linear model fits).

We can see this pattern even better by plotting the best-fit lines across languages (figures 15.19 and 15.20). Nearly all of these go up with age and have similar slopes.

However, in nearly every language, to one degree or another, we see some children with productivity ratios > 0.95. Figure 15.21 shows the proportion of children showing more than 95 percent productivity. A number of samples have substantial proportions of parents reporting comprehension in this way. While it is possible that these numbers represent actual children whose production is synchronized with their comprehension, a more parsimonious explanation is that there are local variations in administration, leading to some fraction of parents who are not completing the form properly. In particular, it does not appear that these "no comprehension without production" children are the tail of a shifted distribution of productivity ratios; instead, they appear to be due to a separate small population. Yet, despite that, they also appear to have an outsized effect on our estimates of the development of productivity ratios across languages.

In sum, although the relationship between production and comprehension is a fascinating locus for individual differences, we may not be able to measure this relationship effectively using cross-linguistic comprehension data. Further, these analyses underscore the importance of developing instructions for parents that more effectively convey the concept of what it means for a child to "understand" a word. Outside of these concerns, while there very well may be reliable variation in the comprehension-production ratio across children, we unfortunately do not have access to independent signals that could validate this quantity.

15.5 Discussion

In this chapter, we have explored three classic indices of individual variation in children's "style" of learning—that is, variation in how children approach the task of language learning. Compared with differences in rate/timing of learning, stylistic differences have been proposed to be a stronger test of the proposal that language acquisition is a constructed process. Only a few studies to date have had the power and scope to explore the extent of these differences cross-linguistically.

We can summarize the conclusions from our various subanalyses. First, we were not able to substantiate the disparities among individuals in production/comprehension differences. It is tricky to use comprehension data to estimate variability among individuals in how much they produce versus comprehend, likely due to differences across sites (and thus, across languages) in the uptake of instructions regarding what "comprehension" means. In contrast, we did see support for the traditional referential versus expressive distinction from cross-linguistic analysis. We found that, at a given level of vocabulary size, some children tend to be younger, use more nouns at a given vocabulary size, and combine less;

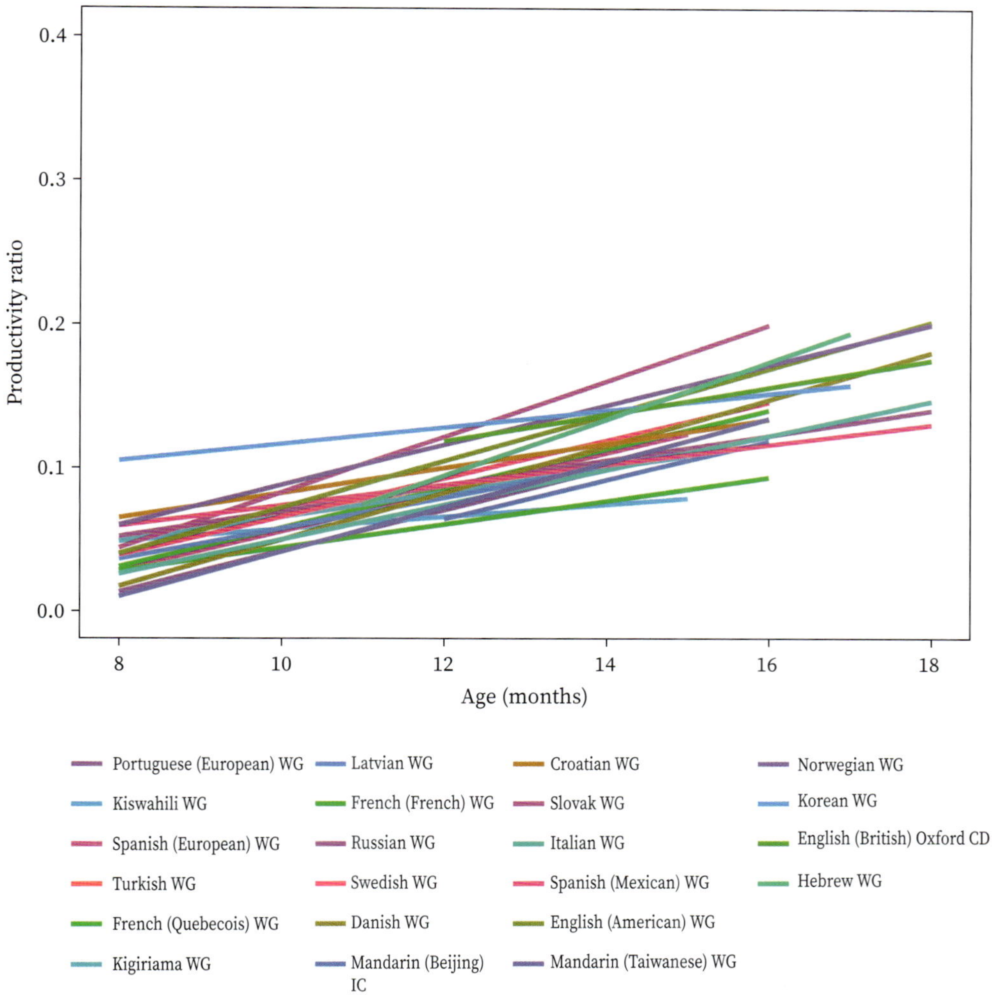

Figure 15.19
Linear model fits of productivity ratio as a function of age for each language.

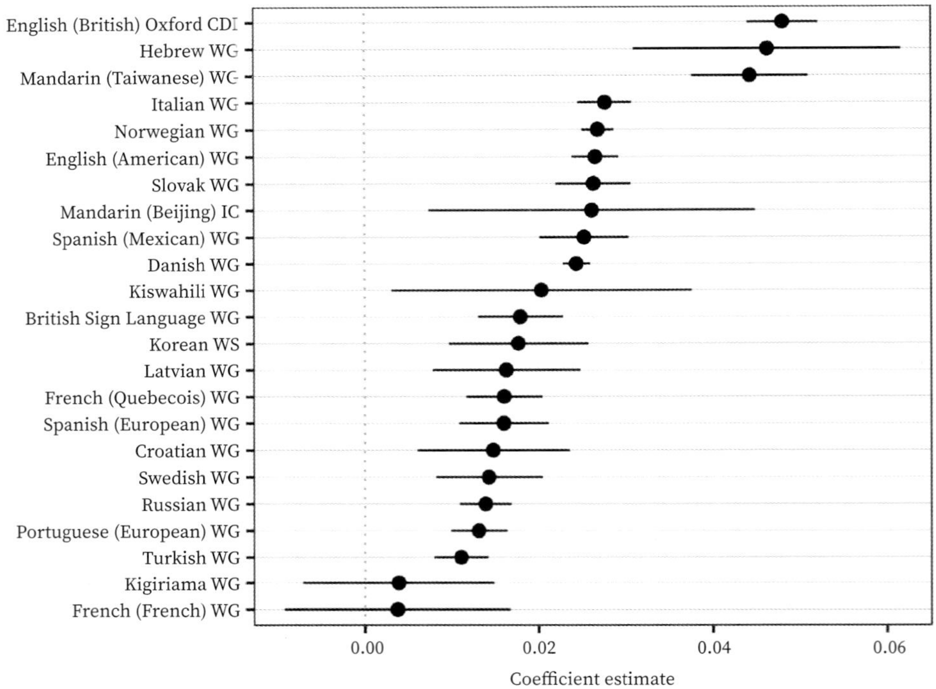

Figure 15.20
Effect of age on productivity ratio in each language (ranges indicate 95 percent confidence intervals).

others tend to be older, know more predicates, and combine words more. Finally, confirming previous work by Ganger and Brent (2004), we found no evidence supporting the notion that a "vocabulary spurt" occurs universally for all children. Instead, we found that, while vocabulary growth accelerates, that acceleration is approximately linear for the first 250 words, and hence a "vocabulary spurt" is at most observed in only a small portion of children.

Together, these analyses do confirm the existence of stable individual variation in language learning "style." Our work also highlights the methodological difficulty of this enterprise, however; despite the tremendous amount of data we have access to, our analyses still suffer from the limitations of the parent-report format of the CDI. It is often difficult to tell whether we are measuring children's style of learning or parents' style of reporting. Overcoming this limitation would require non-parent report measures with equivalent domain breadth, psychometric stability, and sample diversity to the CDI. At present to our knowledge, no publicly available datasets of this type exist.

Further, we want to highlight one potential source of some of the variation we observed: differences in children's speech-motor ability. Children's ability to

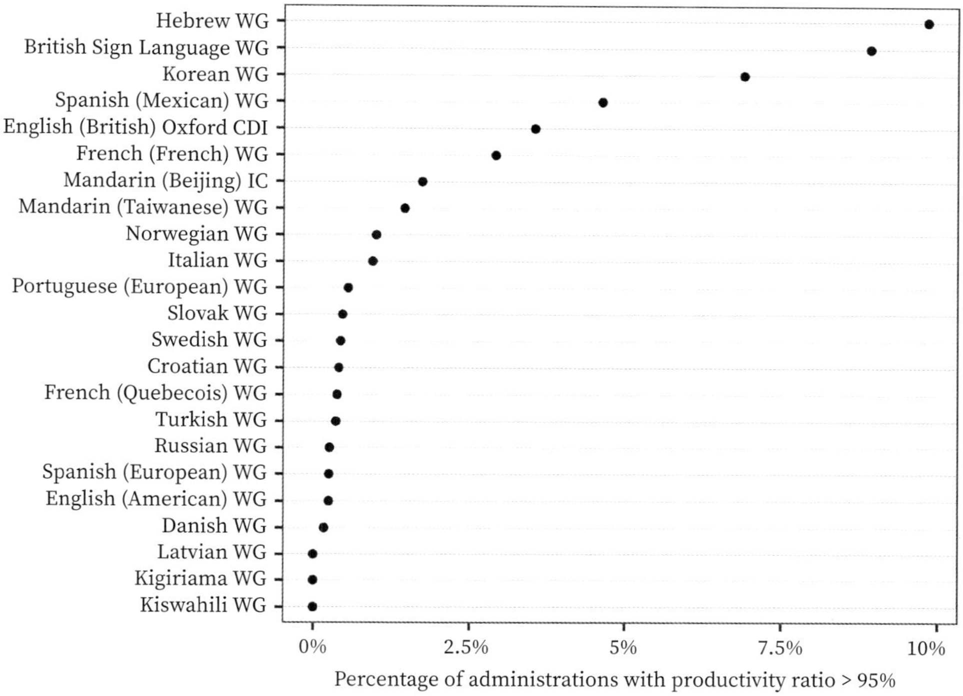

Figure 15.21
Percentage of administrations in each language for which the productivity ratio is greater than 95 percent.

produce complex phonological forms (e.g., Dromi, 1987; Clark, 2016) and complex phrases (e.g., Demuth and Tremblay, 2008) change substantially during the developmental period we are examining. Although we have not had much to say here about these changes (due to the emphasis of the CDI on content rather than on form), variability in this developmental pattern could be an important driver of the changes we observed here. Of course, speech-motor variability could be one source of variability in production/comprehension differences: children who have a harder time producing distinct word forms might nonetheless have a large vocabulary. In addition, speech-motor variability might also play a role in children's position on the referential-expresssive continuum. Children with slower speech-motor development could be those children who are relatively more expressive in style. They would be older for their relative vocabulary size due to their delays, and due to their greater language experience they might also be better at combining words. Testing this sort of speculation would be a worthy goal for a data-collection effort aimed at characterizing variability in children's naturalistic speech across a large, consistent sample.

16 Variability and Consistency within and across Languages

In each of the analytic chapters of this book (roughly speaking from chapter 5 to 15), we have presented analyses of specific phenomena of theoretical interest. Wherever possible, we have generalized these analyses across languages so that their relative consistency can be examined and discussed. This chapter brings together a selection of the analyses from these different chapters into a single analytic framework, implementing the general idea discussed in chapter 1.

To outline, we identify "signatures" for each of the preceeding chapters: measurements that we believe are theoretically interesting or central. We then quantify variability in these signatures across languages using a single measure. These estimates of cross-linguistic variability quantify the degree to which aspects of language development are more or less similar across different languages and cultural contexts. While there are some limitations to the method we employ, this method nevertheless allows us to synthesize results across very disparate analytic frameworks and methods.

16.1 Analytic Method

We begin by identifying a small number of measures computed in each chapter to serve as the "signatures" to be promoted into this analysis. For each measure, we compute its cross-linguistic variability using a standardized measure of variance, the coefficient of variation (CV), where μ is the mean and σ the standard deviation:

$$CV = \frac{\sigma}{|\mu|}.$$

This measure can range from zero (indicating a phenomenon that is completely invariant across languages) to infinity (with higher numbers indicating greater variation). The CV provides a single common measure to allow comparability of otherwise very different quantities, allowing inferences across analyses and datasets.

Each measure for which we compute the CV will have both a different base unit and a different number of languages contributing. For example, when considering the correlation between grammar and the lexicon (chapter 13), we will compute the CV of a set of r^2 values over languages. In contrast, when we look at the size of the noun bias (chapter 11), we will be looking at a bias estimate that ranges from –.5 to .5 and is typically close to 0, computed over 27 languages. To assist in interpretation, we include only those measures that can be computed in 6 languages or more; we provide the N contributing languages for all analyses and we compute an estimate of the standard error (SE) of the CV ($SE \approx CV/\sqrt{2N}$).

The set of signatures we include in this analysis are necessarily a subjectively determined subset of the possible measures we have examined in this book. And of course, those in turn are a subset of the measures we could have computed. Wherever possible we have attempted to make reasonable decisions, but some of these are, by necessity, somewhat arbitrary. An example of such a decision comes from the summary of chapter 5. In that chapter we noted that population variability appears quite consistent across languages. We summarized population variability in production via a statistic—mean absolute deviation from the median (MADM)—but what is the appropriate range of ages to include in a single estimate? We also observed that there appears to be a ceiling effect in the later ages. Thus, we decided to adopt as our measure the variability in production from 12 to 24 months. But this decision is data dependent, and so of course, there is a risk of circularity. We point the issue out not to undermine this particular analysis; we believe that the ceiling effect is quite clear and that other aspects of the age choice do not lead to much change in the CV estimate. Rather, we intend to highlight that the summary we give is not a theory-neutral estimate but rather a "best guess"—an attempt to navigate the myriad choices involved in our analysis in a reasonable way.

One further example of such a choice is that we have made the decision throughout to omit estimates of early production from WG-type forms. Our judgment was motivated by the fact that such estimates are routinely quite noisy and difficult to interpret, likely due to the small size of early productive vocabularies. In chapter after chapter, we found unreliable or uninterpretable results that are plausibly due to data sparsity; thus, we choose to omit these patterns from our broader synthesis effort.

Another interpretive caution comes from the nature of the CV measure. The goal of the CV measure is to provide a unit-less way to compare variability across many measures. But because it is a ratio, when μ is small, CVs can be arbitrarily large. Figure 16.1 shows this relation. Concretely, this property will cause difficulties for us when we have phenomena within a particular grouping that have both small means and small standard deviations relative to other phenomena. For example, in chapter 12, the bias for household words is negligible—as is the variation across

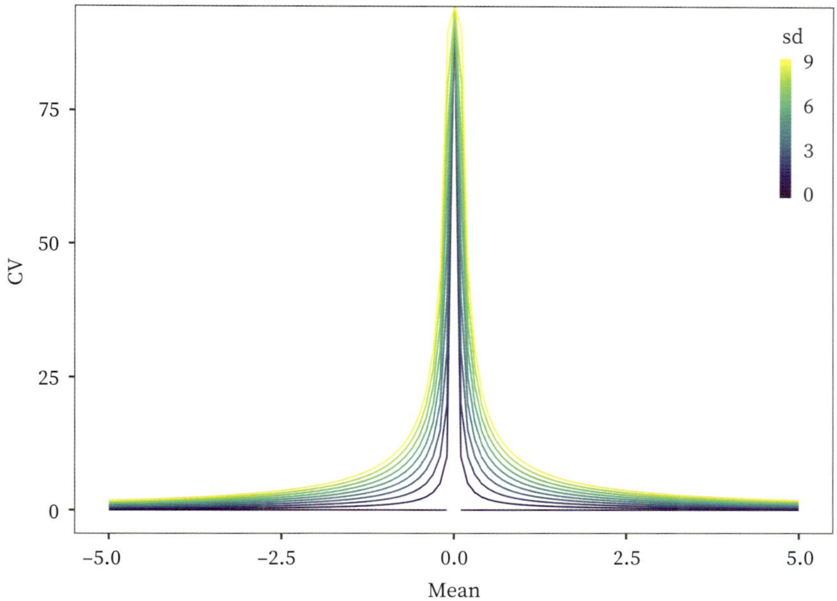

Figure 16.1
A theoretical analysis of the properties of the CV method. CV is plotted by its two components: mean and standard deviation. Mean values are shown on the horizontal axis, with colors indicating different standard deviations.

languages. But, because the mean bias is very close to zero, the coefficient of variation appears gigantic. To assist interpretability, we have simply omitted negligable effects like this one from the analysis below, recognizing that this move limits the generality of the approach.

16.2 Results and Discussion

Figure 16.2 shows the coefficient of variation across languages for selected measures. For the sake of our analysis, we have divided measures from the preceding chapters into four categories.

1. *Measures of the composition of vocabulary* (chapters 11 and 12). These measures describe the over- and underrepresentation of various word categories in vocabulary. The units over which the CV is computed are bias scores; these are bounded from –0.5 to 0.5 (deviation from unbiased acquisition of a particular category).

2. *Predictors of word difficulty* (chapter 10). The consistency of different regression predictors of age of acquisition are represented here by their cross-linguistic consistency. This analysis is distinct from the analysis presented in that chapter

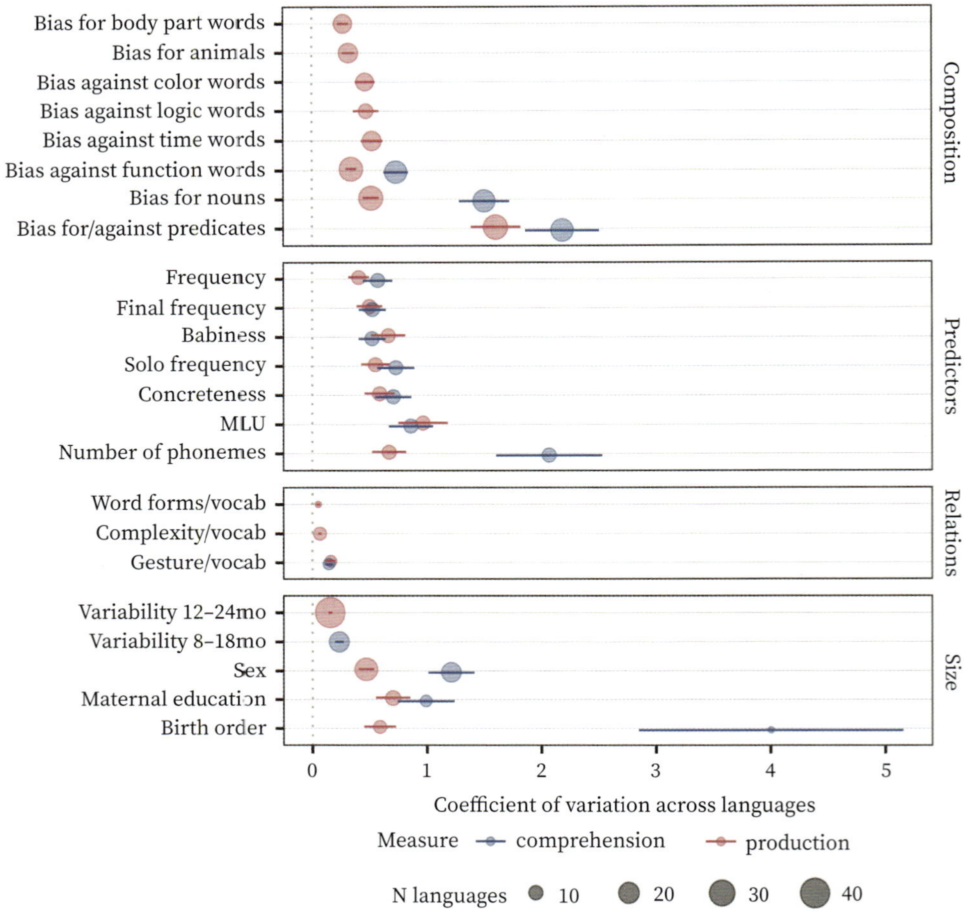

Figure 16.2
Coefficient of variation across languages for signatures of language development corresponding to four different categories (panels). Each point gives an estimate, with the point size corresponding to number of languages. The color indicates whether comprehension or production is measured. Error bars give the standard error of the coefficient of variation.

(which focused mostly on the *magnitude* rather than *variability* of the coefficients themselves). Despite that, we include it here for comparison with other signatures. The units over which CV is computed are standardized regression coefficients.[1]

3. *Relational measures* (chapters 7 and 13). These measures are originally correlations between vocabulary size and other aspects of early language.

4. *Vocabulary signatures* (chapters 5 and 6). These measures document patterns in the overall size of vocabulary across individuals and demographic groups. The original units are themselves variability-based (MADM scores).

A number of local patterns are immediately apparent. First, comprehension is almost always more variable than production, even when a comparable number of languages are included. Why would comprehension be more variable across languages than production, especially given evidence that comprehension vocabulary tends to be *less* idiosyncratic than production (Mayor and Plunkett, 2014)? One strong possibility is the psychometric properties of comprehension versus production reports. As described in chapter 4, while comprehension scores are likely still a reliable and valid index of children's abilities in the aggregate, individual comprehension questions tend to carry less information. Thus, there may simply be more noise in these measurements, leading to less cross-linguistic stability. This regularity illustrates a point we made in chapter 1: the inferences from consistency and variability are asymmetric. In the case of consistency, we can make relatively strong inferences about some kind of shared process or mechanism. In contrast, in the case of variability, there are many sources of variance (including measurement error) that can account for a specific pattern of performance.

Second, relational measures are highly invariant across languages. These relations include correlations between the size of children's lexicon (in production or comprehension) and measures of gesture, morphology, and grammatical complexity. These findings can only be measured in a relatively small set of languages (due to limited data availability for the gesture and complexity items on the form). Nevertheless, the high level of consistency is striking. In chapter 17, we summarize this pattern as the finding that "the language system is tightly woven"—different parts of early language correlate highly with one another, and these correlations themselves are quite consistent across languages (the current finding).

Third, the role of demographic predictors—birth order, maternal education, and sex—is somewhat variable, likely reflecting at least some cultural differences in mechanisms by which demographic variation relates to language development. The most variable demographic category is the relation of maternal education with vocabulary. Maternal education is plausibly a proxy for socioeconomic status

1. We have excluded Valence and Arousal coefficients here as their CVs are quite high due to the very small mean effects for each.

(SES); in turn, the relation of SES to vocabulary is likely mediated by many local- and national-level policies, including access to childcare, parent leave, prenatal and postpartum education, and services. Thus, we view variation on this dimension as highly plausible a priori.

Fourth, and in contrast to demographic variability, the consistency of children's variability is quite striking: the variability of children across instruments is almost completely constant, especially for production. Around the world, toddlers appear to be have similar levels of variability in their level of production. As we explored in chapter 5, one suggestive explanation of this finding is that much of this variability is endogenous to the child or to the possible ways that child-caregiver dyads interact rather than being a product of variation in the environment that can be measured with global indices, such as maternal education. We discuss this finding further in chapter 17.

Fifth, generalizations about vocabulary composition span the range from extremely consistent (early overrepresentation of body parts) to extremely variable (bias for—or against—predicates). Overall, however, it is interesting to see that there are some consistencies in the things that children talk about early in their lives: in particular, the bias for body parts and animal words seems consistent with some of the results on the predictive power of "babiness" (things babies like to talk about) as noted in chapter 10. In contrast, vocabulary for abstract conceptual domains was typically more variable—for example, for color, time, and logic words (as well as function words more generally).

16.3 Conclusions

Our analysis in this chapter attempted to provide a single metric for the cross-linguistic consistency of the phenomena discussed in the remainder of the book. While developing a scale-free unit for this analysis is a challenge, we still see a number of robust generalizations emerge. We make use of these generalizations in the next chapter, which discusses broad conclusions from the study of language development at scale.

17 Language Development at Scale

This chapter synthesizes knowledge gained from the broader enterprise, attempting to identify generalizations about early language development that hold true across the different datasets in our sample. We begin by briefly reviewing the content of our findings across the preceding substantive chapters. Next, we discuss three synthetic generalizations from these findings. We then turn to the question of particular process universals that might underly this empirical picture.

17.1 Summary

We began with the question of what parent report can tell us about children's early language. We are now in a better position to answer this question and can summarize our findings across the content chapters of the manuscript.

In chapter 4, we reviewed evidence for the reliability and validity of parent report instruments for assessing children's early language, with generally positive conclusions. Further, we contributed two sets of novel analyses. First, examining large-scale longitudinal datasets, we estimated the falloff in test-retest correlations over developmental time, suggesting a relatively high level of reliability overall. Second, we used item response theory (IRT) models to examine the measurement properties of individual words on the Communicative Development Inventory (CDI). Overall, most items had strong psychometric performance, and even comprehension reports appeared to have consistent informational value about a child's overall abilities. The broad picture from this chapter was positive with respect to the psychometric properties of the CDI, motivating further use of these data.

Chapter 5 then examined trends in the growth and distribution of individual vocabulary estimates. Across languages, and for production and comprehension, early vocabulary development starts slow and grows rapidly in the second year after birth. There are both absolute and relative differences among languages in the estimated rate of vocabulary growth, however. While some of these may reflect true differences (for example, Danish being famously difficult to acquire, at least at the early stages; Bleses et al., 2008), others likely stem from differences in instrument, sample, and administration conditions. Cross-language variability was dwarfed by cross-individual variability, however. Further, we observed a striking consistency in this variability across languages. On average, language emerges quickly, but individual toddlers around the world are highly variable in their own learning rate.

Chapter 6 considered demographic predictors of vocabulary size. We found a general, though small, female advantage in early vocabulary that was more pronounced for production than comprehension. This advantage did not appear to stem from reporting bias and was relatively consistent in size across languages. We also observed a firstborn advantage and an advantage for children with more educated mothers (treating maternal education as a proxy for general socioeconomic status). Chapter 9 followed up these analyses by identifying whether specific words were the locus of some of these differences. The conclusion of these analyses were that there were *both* general demographic differences in vocabulary *and* word-specific differences. For example, girls on average have a higher probability of knowing any word, and they also have a much higher probability of knowing words like *dress* and *doll*.

Turning to gestural communication, chapter 7 showed that children's uses of early gestures are both quite consistent with one another and quite consistent across languages. Gestures appear to be acquired together as part of a suite of communicative abilities rather than being learned piecemeal as isolated items. A small set of similar gestures are often the earliest emerging across languages, likely arising from relatively universal dyadic needs (e.g., a signal for "pick me up" or "give me"). Finally, within individuals, variation in early gestures is highly correlated with variation in early vocabulary.

Like early gestures, early words are often surprisingly similar across languages. In chapter 8, we quantified this similarity and showed that it was related to typological relatedness. However, a set of core concepts were encoded by words in children's earliest vocabulary, even across quite dissimilar languages. Like early gestures, these early vocabulary items likely reflect relatively universal communicative needs in infant-caregiver dyads. (We return to this theme below in section 17.2).

In chapter 10, we created predictive models of the ease or difficulty of individual words. These models take into account the average environmental input for a learner of a particular language (estimated from corpus resources). Although a number of factors, including input, conceptual factors, and phonological complexity, interact to predict when words are first understood or produced, the profile of predictors across languages was quite similar. This finding points to a relatively consistent set of processes that operate for children learning different languages in different contexts.

Following the thread of cross-linguistic differences, chapter 11 examined the question of the syntactic composition of vocabulary across early development. Consistent with previous reports, we found that most languages exhibit a bias toward nouns, although the strength of that bias varied across languages. Further, essentially all languages showed a developmental bias against function words. On the other hand, the bias for predicates (verbs and adjectives) was more variable across languages, with most languages showing a negative bias but a few (Mandarin and Cantonese, in particular) showing a neutral or positive bias. This pattern of variability points to some language-specific factors—perhaps syntactic structure, perhaps other linguistic or interactional factors—that influence learning of words from particular syntactic categories.

We next applied this same approach to semantic categories (chapter 12). Computing the relative bias for or against particular semantic categories, we found some cross-linguistic consistency in early biases: biases for body parts, games, and onomatopoeia were quite consistent; biases against words for places and time were as well. These biases suggest that there are likely some attentional and conceptual biases that facilitate or inhibit word learning in similar ways across languages.

Turning back to syntactic development, chapter 13 showed that children's morphological and grammatical ability appears tightly coupled with their vocabulary size. This generalization holds very consistently across languages, providing support for Bates' speculation that there is a general core of language development that is shared across vocabulary and grammar learning. This chapter extended earlier analyses of these relations by identifying a moderating effect of age such that older children appear to gain more "units grammar" per "unit vocabulary"—that is, grammar emerges slightly faster for older children as compared to younger children. Further, analyses of longitudinal data suggested that, for children in the CDI age range, vocabulary learning is temporally prior (and perhaps, causally prior as well) to grammatical complexity, supporting a view in which grammatical abilities emerge from generalizations based on learned vocabulary.

Chapter 14 examined children's morphological overgeneralizations specifically. Here, the data supported the idea of a phase of over-regularization and recovery, with some evidence that over-regularization was correlated within children across distinct morphological generalizations in nouns and in verbs.

The last empirical chapter, chapter 15, examined individual variability in the ways children approach the task of learning or "style." This chapter presented evidence for the "referential" versus "expressive" distinction in children's learning trajectories across languages. Even controlling for overall developmental level, some children appear to have larger, "noun-ier" vocabularies while others have smaller vocabularies but are more likely to be reported to combine words. Similarly, some children certainly produce relatively more of the words they understand as compared with other children, though these measurements were difficult to validate independent of reporting factors. However, we did not observe strong evidence for the idea that children's vocabulary growth is particularly "spurt-y"—rather, once statistical artifacts are accounted for, vocabulary growth appears to accelerate relatively smoothly in all children, albeit with variation in acceleration rates across children.

17.2 Generalizations

What is the picture of language development that emerges from these individual findings? Making use of the patterns of cross-linguistic variability and consistency that emerged in chapter 16, we now return to the theoretical project of chapter 1—to use these patterns to provide constraints on theories of early language learning. We begin by considering three major generalizations from our data and then turn back to the question of possible learning processes that could support these generalizations.

17.2.1 The Language System Is Tightly Woven

It could easily be the case that children are "pointy" with respect to language—that is, highly proficient in some parts and less proficient in others. A visitor to a toddler classroom might plausibly observe the following pattern: some children primarily gesturing to communicate, some others knowing the names of many things but not yet combining words, and still others fluently combining words into multi-word utterances. Further, these children could be distributed across a range of ages, with language ability only somewhat predicted by relative age. Seeing these differences, this visitor might conclude that the variation they observed was supportive of a multifactorial viewpoint where some children are better at words, others are better at grammar, and still others are better at gestures.

While there is some true variation of this sort (see point three below), the kind of observations described above will tend to overstate it. Instead, our analyses suggest that the language system is tightly woven together, such that on the whole, children who are good at gestures tend to have larger vocabularies and children who have larger vocabularies also tend to inflect and combine words more proficiently. These correlations within individuals are large, and they are almost completely consistent across the languages in our sample. Further, they are also largely borne out in a variety of observational and behavioral datasets that do not rely on parent report (e.g., Brinchmann et al., 2018; Rowe and Goldin-Meadow,

2009). How can the consistency and coherence of these distinct aspects of early language be reconciled with the type of variation a preschool observer might see?

Consider a model in which learning follows a single, coherent path through a set of discrete stages. Initially, gestures and early nouns and social routines are learned. Nouns follow, leading into verbs. Verbs in turn promote the use of word combinations and morphology. In such a model, children's rate of learning those particular features of language could be nonlinear. For example, their noun bias could increase and then decrease just by virtue of whether they had begun learning verbs (Bates et al., 1994). And, the relation between lexicon and grammar could be flatter at one period than another (Dixon and Marchman, 2007).

Under this kind of model, the children in the preschool classroom could simply be at different points along the same trajectory. A younger child who knows a surprising number of names might be further along this trajectory relative to their age and hence firmly in the "noun bias" part of the general path. In contrast, an older child who primarily gestures might be a child with a slightly slower growth rate. In other words, observed variation need not indicate that the subparts of language do not "hang together" (to return to the Batesian phrasing).

Alongside its support from the cross-linguistic consistency in cross-domain correlations that we observed, this unitary view of early language is supported by several other bodies of work. The first is a set of longitudinal analyses by Bornstein and colleagues. Bornstein and Haynes (1998) and Bornstein et al. (2004b) collected data from a sample of American English–learning children at two and four years of age and measured language using a variety of instruments, including parent report, transcripts, and behavioral tasks. In a reanalysis of these data, Bornstein and Putnick (2012) found that the core construct of early language that emerged from the submeasures was both highly correlated with each of the submeasures and also highly stable over time.

Research on international adoption provides additional support for this unitary view. International adoptees often have to let go of their native language completely and begin learning a second language, sometimes years after they have begun acquiring their first native language. Work by Snedeker et al. (2007; 2012) suggests that these children pass through the same general stages of acquisition as learners with a single native language but that they do so much more quickly on average than native learners.

In sum, the language system is much more tightly woven than casual observation might lead an observer to expect. Although our data provided multiple opportunities for aspects of language development (communication, vocabulary, morphology, grammar) to dissociate into modular subsystems across individuals, measures of ability instead hung together very tightly. Further, the strength of these connections varied little across the languages we examined.

17.2.2 Children's Similar Interests Drive Their Communication

A second broad generalization from our work here is the content similarities in children's early communication—both vocabulary and gestures—across languages. Put simply, regardless of the language they are learning, children in the early stages of language learning appear to talk about similar things, especially regarding people, social routines, small objects, and body parts. This conclusion builds on analyses by Tardif et al. (2008), who compared the distribution of children's first ten words across three languages, Mandarin, Cantonese, and English. Many of the content generalizations that held for those three languages in fact hold here as well in a much more broad array of languages. Further, children's early gestures are also strikingly similar across languages and communicate showing, giving, and the desire to be picked up before their first birthday with high consistency across languages.

A naive observer might suppose that children in different cultures begin communicating about quite different things. They could imagine a culture in which children primarily begin by describing the properties of objects ("soft!" "small!") or another in which children first describe places or things around the house. Or, perhaps it is just American children who are obsessed with animals while children in other cultures are less so. But it turns out that children in different cultures are much more similar in the content of their early communication than we might have expected.[1]

We also observed a strong correlation in the order of acquisition for the earliest words and gestures. Those words and gestures that were typically learned very early in one language tended to be very early learned in others. Correlations in the ordering of words were attenuated for later-learned words, however. Looking at the vocabulary as a whole, we also saw that some semantic categories were overrepresented across languages, including words for body parts, games and social routines, and sounds. (In contrast, predicates and function words as well as words for time and places were under represented.) When we looked for systematic predictors of the ease of learning words across language, the predictor "babiness"—which captures association with infants—was a strong predictor, especially for early learned words.

Where do these similarities come from? They likely *do not* emerge from simple linguistic or environmental frequency. Not only is frequency statistically

1. The caveats expressed in chapter 1 still hold, however. Most of our languages are Western, educated, rich, industrialized, and democratic (WEIRD; Henrich et al. 2010), and the majority are Indo-European. A more diverse set of languages will be necessary for assessing the generality of our claims about the content of children's early communication.

controlled in some of our analyses (e.g., those in chapter 10), but also there are clear dissociations between linguistic frequency and children's comprehension and production. The words for mother and father are relatively infrequent in language directed to children, and yet these words are learned very early. Many function words are highly frequent yet are produced much later than expected. And many omnipresent environmental stimuli, from couches and carpets to diapers, are referred to only much later than the people, animals, and small objects on and in them.

Instead, similarities in children's early communications across languages likely emerge from a combination of factors: (a) children's particular attention and interests, (b) the communicative priorities of child-caregiver dyads, and (c) the informational structure of language. These factors work together to render certain aspects of language easier to learn than others.

What children are interested in and what they pay attention to are powerful determinants of children's behavior, as anyone who has spent time with toddlers will likely attest. The red puppet Elmo, who has a squeaky (some might say annoying) voice, is optimized to draw children's attention and interest, despite some parents' best efforts to minimize his frequency in the environment. More generally, toddlers tend to make a beeline for animals and toys that they perceive as interesting. These preferences may stem from preferences for animates, perceptual preferences (e.g., for bright red and high voices), or a mixture of these, but their impact on children's attention—and from there, their vocabulary across languages—is undeniable.

Although toddlers have their own interests, their language exposure takes place in an environment that is largely constructed by their caregivers. Much of the language they hear is "functional talk" that takes place in the context of routines like dressing, diapering, mealtime, and bathtime. These dyadic priorities mean that children's vocabulary often contains words that help them name and navigate these routines. Roy et al. (2015) even argued that the contextual distinctiveness of words that appear in these routines—signaled by their distribution in space, time, and linguistic context—may lead to earlier acquisition. Across cultures, if parents are going through the same functional routines, this similarity would lead to cross-linguistic consistency in what words their children learn.

Finally, some words may be learned earlier than others simply because of the informational structure of language (Snedeker et al., 2007). Verbs may be learned later than nouns simply because one needs to know some nouns to figure out the meanings of the verbs (Gleitman, 1990; Gillette et al., 1999). To the extent that these regularities extend across languages, they would impose a variety of ordering constraints on acquisition that could be reflected in our analyses.

17.2.3 Children Take Different Routes into Language

In the original CDI norming study monograph, Fenson et al. (1994) noted that variability is perhaps the primary and most striking fact about children's vocabulary learning. Our observations confirm this conclusion. Despite the aggregate similarities across children and across cultures, the speed with which children acquire language is highly variable in the first years of life. From a biological perspective, this variability is quite unprecedented. As a comparison, variation in heights for toddlers is tiny compared with variation in vocabulary: the mean height for a 24-month-old is around 33 inches, with a standard deviation of a little more than an inch, leading to a coefficient of variation around 0.03. This measurement is almost two orders of magnitude smaller than the coefficient of variation on vocabulary.

Where does this variability come from? We can only speculate. Some must come from variation in input across households, which has been amply documented to relate to children's early vocabulary (e.g., Hart and Risley, 1995; Hoff, 2003; Weisleder and Fernald, 2013). Further, some component of this input-correlated variation is likely to be genetic (e.g., Hayiou-Thomas et al., 2012), such that some children inherit a tendency toward slower vocabulary growth from parents who themselves talk relatively less and use relatively less diverse vocabulary (Dale et al., 2015).

The degree of variability we observed is itself a constant across cultures, however. Examining the variability in English-learning children's vocabulary documented by Fenson et al. (1994), it was easy to think that the spread of children's outcomes was due to the demographic and parenting variability found in the United States, which—even in the restricted set sampled in that initial study—was large. But a look at the variability estimates found in our broader sample quickly falsifies this hypothesis.

Further, children vary stably in the nature of their acquisition path, with some naming more objects and others combining words relatively more. Controlling for the nonlinear trajectory of acquisition described above, we found stable differences in "referential style" among children. Although the language system is tightly woven and moves through a relatively consistent learning trajectory across individuals, there is nevertheless an interesting, second-order component. Some children, especially the faster learning ones, appear to learn more nouns. Others, often the slower learning ones, tend to combine words more frequently. As we saw in chapter 13, it is these older children who appear to gain more "grammar per unit lexicon." Maybe these early word combinations are actually "unanalyzed wholes" that are syntactically less complex than they might appear. Perhaps those children who are learning more slowly are then able to bring more mature working memory to the task of grammatical induction. Or perhaps these children have heard more language, and hence the natural statistics of language are more apparent to them. Regardless, these differences appear as a stable aspect of the relatively consistent developmental course we observed.

Does variation in rate or style of learning persist beyond the range of our study? One of the most intriguing aspects of our rate analysis was the suggestion that, accounting for ceiling effects, variation in ability does not compress in the range we measured. An important direction for future work would be to ask about the range of variation observed within and across cultures on other standardized measures of language in school-age children. One possibility is that less variability will be observed for older children on most standard vocabulary and grammar measures simply because functional communication is possible for nearly all speakers—ceiling effects, in essence. Instead, variation will be measurable in more "leading edge" domains like literacy or discourse comprehension.

17.3 Learning Processes

In chapter 1, we introduced the idea of "process universals." These cannot be universals of content, since all of the content being reported by parents filling out CDI forms is language-specific. Instead, similar to Slobin (1973)'s "operating principles," we are interested in processes that operate in different language contexts to produce the observed pattern of phenomena (cf. Clark, 1977). What more can we say about potential process universals, building on the generalizations above? We highlight three process-level connections that appear consistent with our data.

17.3.1 Language Grows through Interactional Input

Without input, there can be no uptake. The importance of language input is a fundamental tenet of all models of word learning, from the simplest accumulator model (McMurray, 2007) through to more complex models (e.g., McMurray et al., 2012; Frank et al., 2009; Fazly et al., 2010). In all of these models, what is learned by the model is a function of the frequency and statistical distinctiveness of the learner's input. This conclusion is amply supported by a large body of correlational research linking observed speech from children's caregivers to their vocabulary size (e.g., Hart and Risley, 1995; Hoff, 2003; Huttenlocher et al., 1991; Hurtado et al., 2008).

But quantity is not enough. Beyond the first order correlation of input quantity to language outcomes, a body of research now provides nuanced qualification. In a number of studies across cultures, child-directed speech is a better predictor than overall speech, even in cultures where this kind of speech is relatively rare (Weisleder and Fernald, 2013; Shneidman and Goldin-Meadow, 2012). One hypothesis is that child-directed speech provides more grounded moments in which word meanings can be inferred from context (Cartmill et al., 2013). Indeed, in one study, a variety of measures of input quality—including joint engagement as well as the presence of rituals and routines—were better predictors of vocabulary size than pure quantity (Hirsh-Pasek et al., 2015). Of course, high-quality input must be developmentally appropriate, and for older children, language that is more syntactically complex supports complex syntax acquisition (Huttenlocher et al., 2002). Presumably the evidence for the importance of grounded, engaged communication is at least in part due to its specific importance for younger learners who are engaged in discovering word meanings (e.g., Clark, 2007; Clark and Estigarribia, 2011); other aspects of language gain importance for other learners (Hoff, 2006; Hoff and Naigles, 2002; Rowe, 2012).

We see many of our conclusions as fitting well with this broader picture. While we do not have predictors of individual children's input, the research presented in chapter 10 suggests that even aggregated, average measures of input are relatively powerful predictors of word-by-word uptake. The most obvious example of this is the repeated finding that word frequency is a useful predictor of age of acquisition, especially for nouns. Even if high quality, grounded instances are the appropriate learning input, greater frequency overall will—all else being equal—lead to greater frequency in the appropriate learning context. We further observed effects of solo frequency (being used in a one-word sentence) and mean length of utterance. These results are both consistent with the idea that it is easier to learn meanings for words in shorter sentences, which pose both fewer word segmentation challenges and fewer word-meaning mapping ambiguities. In addition, the relative cross-linguistic consistency in these predictors suggests that the input-uptake connections that have largely been documented for English learners are likely to be robust for learners of other languages as well (though the strength this conclusion is moderated by the relative lack of typological diversity in the sample of languages we have available).

The demographic differences we observed in vocabulary are also consistent with interactional-input theories of vocabulary development, in which the more higher quality the input the child receives, the faster vocabulary grows. Under this hypothesis, children who are firstborn and who have mothers with more education are likely to receive more and more higher quality input. Firstborn children receive more input through their greater allocation of parent attention (though as we note in chapter 6, firstborn children's parents may also be more aware of their vocabulary). Children with mothers who have more education receive more and more higher quality talk through the availability of more parent time, different values around talk, greater awareness of the role of interaction for young children, and perhaps differing parental practices (Evans, 2004; Farkas and Beron, 2004; Rowe et al., 2012).

17.3.2 Individual Word Meanings Must Be Inferred Based on (Cross-Situational) Evidence

Cross-situational learning is the proposal that children use the statistical properties of how words are used across contexts to help them infer meaning (Gleitman, 1990; Siskind, 1996). This specific proposal has been instantiated in a variety of associative learning experiments with both adults (Yu and Smith, 2007; Yurovsky and Frank, 2015) and children (Smith and Yu, 2008; Vlach and Johnson, 2013). The specific mechanisms underlying learners' performance in these tasks are still controversial (e.g., Medina et al., 2011; Trueswell et al., 2013; Yurovsky et al., 2013; Yurovsky and Frank, 2015). Despite this controversy, the general model of learning as proposed in these studies has rapidly become the default instantiation of how interactional input translates into learning (e.g., the models of Frank et al., 2009; Fazly et al., 2010; McMurray et al., 2012). The basic idea is that children bring multiple information sources—including statistical, social, and grammatical information—to bear on resolving the meaning of each utterance. Then, this information is accumulated and brought to bear on resolving reference in subsequent utterances (Frank et al., 2009; Bohn and Frank, 2019).

Several findings from our investigations are consistent with these general ideas. As noted above, the predictors of word difficulty we found in chapter 10 are consistent with the general input-uptake viewpoint. In fact, it is very easy to see some of these effects being generated by cross-situational learning models. Frequency effects are ubiquitous in cross-situational learning models (though their specifics vary somewhat from model to model; e.g., Kachergis et al., 2012). Further, on the cross-situational perspective, shorter sentences and especially single-word utterances are less confounded in terms of the information they provide about the relations between words and their meanings, simply because they include fewer extra words that would create spurious co-occurrences with the referent of the sentence.

Above, we also speculated about the effects of children's idiosyncratic interests and attention on their early communication. Such effects have been accommodated into models of cross-situational word learning where attention and salience are posited to shape the mappings that are learned (Kachergis et al., 2012; Roy and Pentland, 2002; Yu and Ballard, 2007). More generally, the viewpoint on cross-situational communicative mapping that we have advocated provides a convenient theoretical tool for integrating external, situational constraints on input with the children's internal priorities. While external circumstances (e.g., caregivers' priorities) provide the distribution of language that is heard, internal factors like motivation and attention can nevertheless shape what is learned from these distributions.

Further, the noun and verb bias findings reported in chapter 11 can be accommodated in the cross-situational viewpoint as well. On this view, nouns are easy to learn because the more they are heard, the more opportunities children get to build consistent mappings between the words and their contextual referents. And this regularity should be especially true for concrete nouns that are more likely to be found in grounded contexts (Gentner and Boroditsky, 2001). In contrast, verbs and other predicates cannot be acquired as easily from basic co-occurrence. Syntactically "light" verbs like *make* or *do* require some nominal information to constrain their meaning in context. And "heavier" verbs may still be systematically ambiguous without syntactic information (e.g., *chase/flee*; Gleitman, 1990). Thus, verb learning—and likely the learning of other predicates and many function words as well—relies on a base of nouns and a basic comprehension of the syntactic structures in which they appear to infer meaning in context (Gleitman, 1990; Gillette et al., 1999). These effects appear in a probabilistic model of cross-situational noun/verb learning (Abend et al., 2017).

Finally, the information-sequencing viewpoint predicts that verbs should be acquired relatively later, with relatively more support from shorter, easier-to-parse utterances. Our finding that mean utterance length is a stronger predictor of acquisition ordering for predicates, rather than nouns, is consistent with this idea. This account explains cross-linguistic exceptions like Mandarin as cases where many early learned verbs are semantically transparent enough to be learned cross-situationally without syntactic information (following Tardif, 1996). Though there is much more work to do here, the cross-situational viewpoint is a useful tool for exploring noun and verb biases in vocabulary development.

17.3.3 Generalizations Appear Gradually

Syntactic and morphological structures on the CDI are likely to be "item-based" in the sense of Tomasello (2003). Since all syntactic information is instantiated in particular items, the knowledge that the CDI assesses is not as abstract as that posited by high-level syntactic theories. Although there may well be broader, more abstract generalizations that underlie the growth of word order, we simply do not have the signal to address the presence or absence of such abstractions. That said, our evidence is broadly consistent with the view of language as growing through the gradual induction of syntactic regularities through a reciprocal interaction with the acquisition of individual content words.

Versions of these viewpoints exist throughout the broad theoretical space of language acquisition (e.g., Yang, 2016; Tomasello, 2003; Meylan et al., 2017), but all proposals rely on a learning mechanism in which generalizations about structure are graded and rely on the amount of evidence available. All such mechanisms would presumably predict some relation between individuals' grammatical and lexical abilities. An important target for future theoretical work in this area is to explore how tight these correlations are predicted to be on different viewpoints. Our suspicion is that only the most construction-based views will predict the level of coupling we observed, however (Bates et al., 1994).

The relation between grammar and the lexicon is reciprocal. As content vocabulary grows, it provides both the groundwork for generalization of constructions (Tomasello, 2003; Goldberg, 2006) and also the specific content words necessary for the induction of predicate meanings (as described above in the case of verbs). In the other direction, as grammar grows, it also enables better disambiguation of predicate meanings (Gleitman, 1990) and creates myriad other opportunities for learning. Both of these directions of longitudinal influence can be observed, but in our study we found differential support for vocabulary driving later grammatical ability. In contrast, Brinchmann et al. (2018) found support for grammar enabling vocabulary learning, although that study focused on older children. One plausible synthesis is that early grammar is based on generalizations from the growing vocabulary while later vocabulary is acquired based on contextual inferences supported by grammar.

As we discussed in chapter 1, we do not see our work here as resolving longstanding debates about the nature of abstract syntactic representations. Instead, we hope our contribution is somewhat different—we sought to refocus the debate away from phenomena that make only occasional contact with the gross regularities of children's early language use in context. Instead, we focus on the pattern of observable changes in complexity and diversity of early language. We hope that this focus leads to theorizing—ideally accompanied by quantitative modeling—that takes these observables as its primary predictive target.

17.4 Conclusions

Developmental psychology often appears to be divided between two groups that do not communicate with one another. On the one hand, there are researchers interested in the ontogenetic and phylogenetic origins of knowledge—the epistemological project of understanding how we reason about objects, communicate using language, or learn from other people (Carey, 2009; Spelke and Kinzler, 2007; Tomasello, 2010). On the other hand, there are researchers interested in growth, change, and variation across individuals in constructs like executive function, working memory, and personality (Diamond, 2013; Gathercole et al., 1994; Ainsworth et al., 2015). The first type of research has led to a productive union of philosophical and psychological ideas, addressing exciting questions in the history of philosophy using empirical methods (e.g., Gopnik et al., 2004; Carey, 2009). In contrast, the second type has had its impact in connections to clinical practice and to educational and public policy (e.g., Nelson, 2007; Diamond and Lee, 2011).

"Origins" research and "variation" research traditions have often appeared to be at odds with one another. While there are, of course, exceptions to this split, in many cases, researchers in these two traditions read different journals, record different measures, use different research designs and statistical models, and often appear to be pursuing different goals. Yet language learning is an area where these traditions come together. The knowledge being acquired—the stock of words in the child's lexicon—is both the epistemic construct whose origins we are studying *and* the psychometric construct whose reliability across individuals we wish to assess. By studying the variability and consistency in the early lexicon, we can study both the origins of knowledge and the processes and factors by which human beings differ from one another across different cultural contexts, family environments, and genetic endowments.

In our work here, we sought generalizations in the universals and variations that hold across cultures and languages and took steps to tie these generalizations to processes of language learning. These generalizations are based in the "variations" approach via the Batesian project of using variability to constrain theory. Yet they are also about the "origins" of a very specific type of knowledge: knowledge of language. The ultimate aim of theorizing in this area must be broader than either "origins" or "variations" alone.

We hope that our work in compiling measurements of consistency and variability across languages is a beginning rather than an ending. Although the number of children represented in our analyses is large, the number of languages and their diversity across families are in the end quite small. And our models of acquisition—especially those linking input to uptake—are only a rough sketch of what is possible. Our hope, however, is that by showing what is possible in the quantitative study of language development, we illustrate a broader set of possibilities for a data-driven science of developmental change.

18 Beyond the CDI

Throughout this book, we have attempted to gain the broadest possible understanding of children's language learning by engaging deeply with data from the Communicative Development Inventory (CDI) family of instruments. We hope that readers agree that this exercise has been very fruitful in uncovering a variety of patterns that can inform our understanding of language learning. But it has also uncovered a wide variety of limitations to the CDI, in turn restricting the issues on which we can comment. In this chapter, we begin by discussing some methodological morals from this work for psychology more broadly. We then turn again to the limitations of the CDI to address the question of how language acquisition research can move beyond the CDI.

18.1 Methodological Morals

As we noted in chapter 1, psychology has recently been plagued by concerns about reproducibility (e.g., Hardwicke et al., 2018) and replicability (e.g., Open Science Collaboration, 2015). Our work here was in part inspired by our considering these issues and their impact on the field of language development. The ultimate goal of research in the area of language learning is to create a quantitative theory that allows for precise predictions and principled explanations of developmental phenomena (Dupoux, 2018). Such a theory cannot be built on a series of nonreproducible findings and binary conclusions (Frank et al., 2017).

Wordbank is one reply to this situation; by compiling the extant CDI datasets into a single open database, researchers can reproduce previous and new research conclusions that use these data. The analyses we report here are computationally reproducible through the availability of the code necessary to build the book and all its figures and analyses. In addition, by seeking a level of scale beyond previous efforts, we have attempted to avoid the variability inherent in "small-N" studies.

Further, our work is built on the notion of replication. Nearly every one of the preceding chapters is in some sense a "replication" of previous work—an analysis from previous research where one CDI dataset was applied (sometimes with modifications) to other datasets (and languages). The result is not a judgment or referendum on the original; we do not declare a binary success or failure of the replication attempt—but if we did, our success rate would be very high! Instead, we are interested in the *degree* to which a particular quantitative estimate varies across languages and cultures.

This sort of analysis is superficially similar to the idea of "hidden moderators" that has plagued the replication debate (Van Bavel et al., 2016; cf. Inbar, 2016). That line of explanation has been an attempt to contextualize failures to replicate particular experimental effects by invoking unknown sources of variability across contexts. In contrast, our efforts here allow us to quantify variation across "replications" of the same effect and use these estimates as the signal rather than as noise to be discarded or averaged out.

One notable feature of our analytic strategy is that we rely very little on binary decision-theoretic inferences using null hypothesis significance testing. There are a handful of p-values throughout the book, but few of these license any prominent inferential conclusion; mostly they exist to provide a quick check that a particular slope is likely to be nonzero. Instead, our goal has been to measure quantities of interest with high precision, looking for statistical measures that relate to our theoretical goals. For example, the existence of a noun bias is a fascinating observation, but the observation alone gives limited leverage to differentiate theories. In contrast, the precise magnitude of a noun bias for a particular sample provides more leverage for quantitative theorizing. And the distribution of magnitudes across many of the world's languages gives greater leverage still.

Yet our analytic strategy is only as useful as the data it relies on, and these have substantial limitations. Some of these limitations are imposed by the specifics of our data, and others come from fundamental limitations of the CDI.

18.2 Limitations of Wordbank and the CDI

Reprising our discussion in chapter 1, despite the large number of children represented in the Wordbank dataset, there are still a number of major omissions. First, because the data are typically from normative studies, data on atypical development are not represented, even though the CDI has been used profitably with several developmental disorder populations (Heilmann et al., 2005; Luyster et al., 2007). Second, for similar reasons, bilingual data are almost entirely absent, though exciting work with multilingual children is beginning to emerge (Bilson et al., 2015; Floccia et al., 2018). Third, the number of children with consistent longitudinal observations is still relatively small. Although we used longitudinal data in chapters 4, 13, 14, and 15, the extant data provide at best a limited picture of change over time within individuals and across languages since our conclusions were drawn almost exclusively from English and Norwegian data.

Beyond these data availability issues, the CDI as an intrument is simply not the appropriate tool for asking every kind of question about child language development. Following the metaphor we introduced in chapter 1, the CDI is a "macroeconomic" indicator. It tells us about the global profile of a child's linguistic abilities rather than revealing the local "microeconomic" dynamics of learning at a particular point in developmental time. The local dynamics of children's learning, language use in communication, and comprehension in the moment have all been important targets for empirical investigation (e.g., Clark, 1988; Fernald et al., 1998; Smith and Yu, 2008). At best, a CDI can provide some emergent average of these processes over time, much the same way the gross domestic product of a nation describes a summary of the impact of all the contributing markets.

The use of parent report to provide a global picture of the child's entire language system—from gestures to vocabulary and grammar—is also a weakness when it comes to addressing detailed questions about the representation of specific words. Because parents are not linguists, they cannot be profitably asked the kinds of targeted questions that might shed light on a variety of theoretical issues. For example, a tremendous amount of research has investigated the development of children's phonological systems (e.g., Vihman, 1996). Research with the CDI must remain silent on this topic; we instruct parents to check "says" if the child produces *any* appropriate phonological form.

Similarly, an important target for research on vocabulary learning is the type of semantic generalizations that children make, including whether words are initially over- or undergeneralized (e.g., Clark, 1973) and how their appropriate extension is found (e.g., Markman, 1990; Xu and Tenenbaum, 2007). The CDI depends on parents to, on average, detect appropriate production and comprehension of words in specific contexts. But these averages of individual uses necessarily reveal relatively little about the nature of the semantic representations underlying the uses of the word—even for concrete objects, but especially for descriptive or closed-class words. A clear example of this phenomenon appears in chapter 11. We see that time words are underrepresented in children's vocabulary, but they are still present. Yet, according to Tillman and Barner (Tillman and Barner, 2015; Tillman et al., 2017), 2 1/2-year-olds probably have incorrect or incomplete semantics for essentially all of these words, even though they still utter the words in appropriate contexts. If the semantics were probed more carefully, however, gaps with adult-like representations would become readily apparent. Experimental methods are likely to be more effective than parent report in these sorts of cases.

In sum, Wordbank and the CDI itself are the right tools for certain kinds of questions. But there are many other questions—some of which arise naturally in and from our work here—that cannot be addressed with these tools. How do we move forward beyond the CDI?

18.3 What Comes Next?

While parent report is no substitute for laboratory observations and experiments, new technical developments suggest that there may be ways to get traction on the questions discussed above by developing successors to the CDI. These approaches are typically inspired by the CDI and the potential of parent report, but they are also not limited by the design features of the specific assessment. We briefly discuss three promising directions for this type of work: web-based assessment, adaptive testing, and app-based assessment.

18.3.1 Web-Based Assessment

CDI forms have traditionally been administered to parents on paper; in general, the CDI community has been relatively slow to transition to online administration despite the prevalence of online methods in experimental and survey research during the last ten years (Buhrmester et al., 2011). One major exception has been two northern European normative datasets included in Wordbank, namely the Norwegian and Danish datasets. As discussed in Kristoffersen et al. (2013), with the creation of an appropriate administration system, these groups were highly successful in recruiting large samples of parents to complete the CDI, and their experiences suggest that it is possible to get high quality data—even deep longitudinal data—through online administration.

Adoption of web-based methods throughout the CDI community has thus far been spotty, however. One obstacle is the varying copyright status of different CDI instruments, which creates legal roadblocks to a simple adaptation of the instrument to a uniform online format. Our personal experience is that some research groups ignore these restrictions and create their own ad hoc surveys using platforms like Google Surveys, Survey Monkey, or Qualtrics. These ad hoc versions often do not have consistent administration instructions or formatting and sometimes contain errors or omissions as well due to the challenges of porting form content to a new format.

To address these challenges, we have created `WebCDI` (http://webcdi.stanford.edu), an online platform that allows researchers to administer a growing range of CDI forms to participants by generating shareable hyperlinks. The system contains a study-management interface so that researchers can create batches of administrations that are specific to a particular participant group (and can reuse demographic data for multiple administrations to the same participant). At the time of writing, we are working to navigate the legal restrictions on a number of instruments so that WebCDI can be used by researchers studying a wide variety of languages. (Although we have not yet implemented adaptive testing in the WebCDI system [see also below], in principle, this method would be relatively straightforward to include as an "instrument" in the broader system.)

A flexible and general web-based administration system like WebCDI has the potential to facilitate efforts to broaden the Wordbank dataset. First, more representative segments of particular national populations can be reached by advertising through email and social media. Such efforts could substantially improve the generalizability of the demographic analyses reported in chapter 6. Our group is currently experimenting with using social media advertising to recruit lower income and racial/ethnic minority groups that are typically underrepresented in research on language development in the United States. Further, as shown by Kristoffersen et al. (2013), online administration substantially simplifies the creation of longitudinal datasets. In the years to come, we hope that other groups (including our own) can leverage WebCDI to create longitudinal datasets that span the full time period of the emergence of language. If they came from a typologically diverse sample of languages, such datasets would be especially valuable in generalizing our analyses of grammatical and morphological development in chapters 13 and 14.

Finally, though there has been some important and impactful bilingual CDI research (e.g., Bilson et al., 2015; Floccia et al., 2018), substantial challenges remain in this domain. For example, in many areas, the population of bilinguals is very diverse, and so the provision of the appropriate CDI forms with language-specific instructions is a nontrivial challenge. Further, the scoring of CDI forms across languages requires conceptual mappings across words (such as those used in chapter 10), so that the child's total conceptual vocabulary—the number of concepts the child has names for across all languages, regardless of the language used—can be established (Core et al., 2013). Both of these challenges can be ameliorated by the design of an appropriate platform. Although the technical challenges are not insignificant, in principle, WebCDI can make it easy for researchers to share a wide variety of forms with parents (with customized, language-specific instructions), gathering data remotely from a more diverse sample of bilinguals. Further, the platform can leverage cross-instrument mappings from our work here to provide measures of both total vocabulary and total conceptual vocabulary.

18.3.2 Adaptive Approaches

Although we have made much use in our analyses of the extensive item list for the CDI, the long forms of the CDIs are simply overkill for some applications. For researchers and clinicians who wish to recover a single percentile score for a child's overall vocabulary size—for example, for purposes of language screening—the CDI is far too long. To ask parents to make hundreds of responses typically takes from ten minutes to upward of a half an hour, and the reliability coefficients across words are high enough that the forms can be shortened substantially without losing assessment fidelity for determining a child's overall vocabulary size.

To address this issue, researchers have developed CDI short forms (e.g., Fenson et al., 2000b; Jackson-Maldonado et al., 2013). These forms typically contain around 100 words that are chosen for their ability to distinguish children across a range of abilities at both younger and older ages. Like the long forms, there is substantial evidence for their validity (e.g., Can et al., 2013). Indeed, some studies even have used so-called "short-short" forms with as few as 25 to 50 words that are chosen to get a rough measure of variation in vocabulary size using much shorter vocabulary lists (Andreassen and Fletcher, 2007). We have not focused on including data from short forms in Wordbank in part because they provide relatively little traction on issues of vocabulary composition and relations between parts of the language system (issues that are critical to our syntheses in chapters 16 and 17). But, for many users, they are more efficient and appropriate instruments than the long-form CDIs for obtaining an overall score that reflects a child's standing relative to same-aged peers.

One exciting possibility is to achieve further gains in efficiency in the context of web-based or app-based administration through the use of adaptive methods. In these sorts of methods, which are common in educational assessment settings, words are chosen to provide maximal information about the child's place in the ability distribution. For example, Makransky et al. (2016) used IRT simulations to demonstrate that an adaptively chosen set of 50 words could in principle recover percentile ranks with a correlation of 0.95 with the full sample. A correlation of 0.85 could be achieved with as few as 10 words, provided that they were chosen appropriately. And using a computational approach based on the use of Wordbank data for the creation of empirical norms, Mayor and Mani (2018) showed good performance in simulation and an empirical validation using the German FRAKIS CDI (Szagun et al., 2006).[1] Thus, adaptive methods have substantial promise for quick global assessments. Further, adaptive methods that aim for global vocabulary estimates could be paired with in-depth questions about specific semantic domains or word classes for an efficient approach to exploring particular theoretical issues.

1. Strictly speaking, results from Mayor and Mani (2018) are not comparable with those in Makransky et al. (2016) since the Mayor and Mani system makes use of age and gender information in the classification process, allowing it to achieve higher accuracy with less data.

18.3.3 App-Based Approaches

A final promising direction for parent reports about children's language is the use of mobile apps. Mobile technology has already been used productively for eliciting emotional information through experience sampling methods (Pejovic et al., 2016) and for engaging large groups of adults in cognitive measurement tasks (Steyvers and Benjamin, 2019). Researchers on children's sleep have even leveraged data about infants' developing sleep cycles (Mindell et al., 2016). The next frontier is the use of app-based methods to allow reseachers (and parents) to collect data about children's language.

We are currently at work on an app called Wordful that would allow parents to keep a running diary of their child's vocabulary. In our current prototype, the Wordful interface allows parents to transcribe specific words that their child has uttered. But it also provides a novel interface that allows parents to enter words more quickly by swiping "cards" showing particular words, indicating whether their child produces the word on the card with a single manual response. The content of these cards can then be chosen adaptively to maximize the information content of a given amount of swipes. Unlike other adaptive testing methods, however, in the case of Wordful, parents can decide how many words they wish to enter and can come back again and again over weeks or months to update their estimates.

We have conducted a preliminary trial of Wordful in which we recruited parents through social media and invited them to fill out an online CDI using WebCDI (Meylan et al., 2019). We then encouraged them to use the Wordful app for three to four weeks, providing notifications periodically to draw them back into the app. At the end of the study period we gathered a further WebCDI administation. Our data ($N = 97$) suggest reasonable correlations between WebCDI and Wordful data ($r = 0.49$ and 0.54 for the initial and final CDI administrations respectively), comparable with correlations between the two CDI administrations (0.59) though lower than the longitudinal correlations reported in chapter 4. Further, by virtue of its flexible data collection interface, Wordful is not limited to the inclusion of a single word list—instead, we were able to broaden the sample of words that parents were asked about and recover age of acquisition information for new words.

Mobile app-based methods like Wordful provide a platform for beginning to address some of the fundamental weakenesses of CDI data that we enumerated above. For example, while parents will never be trained phoneticians, they can be asked follow-up questions about the nature of their child's utterance, and this could shed light on the phonological trajectory of specific word forms. They could even be encouraged to record and upload audio of specific word forms of interest.

A mobile platform could be used to investigate semantic generalization issues as well. Imagine that parents might log a concept like "cat": immediately afterward, parents could be asked to show their child pictures on the mobile device and ask "is this a cat?" By varying the type of picture (e.g., canonical cat, noncanonical cat, dog, tiger, etc.), such an experiment could probe the child's generalizations at relevant times during the process of acquisition. While many empirical challenges remain to be solved to make this method practical, the combination of mobile experimentation and parent observation could be very powerful for probing word learning in the moment that it is happening.

Overall, we are hopeful about the potential role of technology in broadening the applicability of parent report to fundamental questions in language development. While no single approach is right for every question, we hope that efforts to broaden and deepen the CDI approach will yield insights into yet a wider range of facets of early language.

18.4 Conclusions

In addition to the specific methodological limitations of the instrument, our work here represents a potential upper bound on what can be done with data of this type. Even if we continue to accumulate contributions to Wordbank, we are unlikely to gain access in the near future to a deeper set of longitudinal data or a far broader sample of languages. Future work on early vocabulary is likely to require new datasets, perhaps gathered by using the tools described above.

Yet we feel hopeful that there are broader morals to our approach here that can be applied to other datasets. In particular, we hope the fundamental theme of our analyses—consistency and variability—can be applied more broadly. For any dataset like ours that includes samples of children distributed across groups, phenomena can be assessed in terms of their relative consistency across these groups. To the extent that these groups crosscut important features of human experience, such as culture, language, or national origin, consistencies across these features provide statistical evidence toward the project of finding candidate universals of the relevant domain. Will this approach be fruitful in revealing either universals of human development or major dimensions of variation in human experience? Only time will tell. We hope, however, that we have created a potential template to enable such investigations.

Appendix A: Individual Datasets

This appendix gives the contributors and citations for each of the datasets used in this book.

American Sign Language

- **Form**: FormA
  Contributor: Diane Anderson, University of California, Berkeley
  Citation: Anderson, D., and Reilly, J. (2002). The MacArthur Communicative Development Inventory: Normative data for American Sign Language. *Journal of Deaf Studies and Deaf Education*, 7(2):83–106.
- **Form**: FormBOne
  Contributor: Diane Anderson, University of California, Berkeley
  Citation: Anderson, D., and Reilly, J. (2002). The MacArthur Communicative Development Inventory: Normative data for American Sign Language. *Journal of Deaf Studies and Deaf Education*, 7(2):83–106.
- **Form**: FormBTwo
  Contributor: Diane Anderson, University of California, Berkeley
  Citation: Anderson, D., and Reilly, J. (2002). The MacArthur Communicative Development Inventory: Normative data for American Sign Language. *Journal of Deaf Studies and Deaf Education*, 7(2):83–106.
- **Form**: FormC
  Contributor: Diane Anderson, University of California, Berkeley
  Citation: Anderson, D., & Reilly, J. (2002). The MacArthur Communicative Development Inventory: Normative data for American Sign Language. *Journal of Deaf Studies and Deaf Education*, 7(2):83–106.

British Sign Language

- **Form**: Words and Gestures (WG)
  Contributor: Bencie Woll, University College London
  Citation: Woolfe, T., Herman, R., Roy, P., and Woll, B. (2010). Early vocabulary development in deaf native signers: A British Sign Language adaptation of the communicative development inventories. *Journal of Child Psychology and Psychiatry*, 51(3):322–331.

Cantonese

- **Form**: Words and Sentences (WS)
  Contributor: Twila Tardif, University of Michigan
  Citation: Tardif, T., Fletcher, P., Liang, W., and Kaciroti, N. (2009). Early vocabulary development in Mandarin (Putonghua) and Cantonese. *Journal of Child Language*, 36(5):1115–1144.

Croatian

- **Form**: WG
  Contributor: Melita Kovacevic, University of Zagreb
  Citation: Kovacevic, M., Babic, Z., and Brozovic, B. (1996, July). *A Croatian language parent report study: Lexical and grammatical development*. [Paper presentation]. The Seventh International Congress for the Study of Child Language, Istanbul, Turkey.
- **Form**: WS
  Contributor: Melita Kovacevic, University of Zagreb
  Citation: Kovacevic, M., Babic, Z., and Brozovic, B. (1996, July). *A Croatian language parent report study: Lexical and grammatical development*. [Paper presentation]. The Seventh International Congress for the Study of Child Language, Istanbul, Turkey.

Czech

- **Form**: WS
  Contributor: Filip Smolik, Academy of Sciences of the Czech Republic
  Citation: Markova, G., and Smolík, F. (2014). What do you think? The relationship between person reference and communication about the mind in toddlers. *Social Development*, 23:61–79. https://doi.org/10.1111/sode.12044

Danish

- **Form**: WS
  Contributor: Dorthe Bleses, University of Southern Denmark
  Citation: Bleses, D., Vach, W., Slott, M., Wehberg, S., Thomsen, P., Madsen, T., and Basbøll, H. (2008). The Danish Communicative Development Inventories: Validity and main developmental trends. *Journal of Child Language,* 35:619–650
- **Form**: WG
  Contributor: Dorthe Bleses, University of Southern Denmark
  Citation: Bleses, D., Vach, W., Slott, M., Wehberg, S., Thomsen, P., Madsen, T., and Basbøll, H. (2008). The Danish Communicative Development Inventories: Validity and main developmental trends. *Journal of Child Language,* 35:619–650.

English (American)

- **Form**: WS
  Contributor: Larry Fenson, San Diego State University
  Citation: Fenson, L., Marchman, V. A., Thal, D., Dale, P., Reznick, J. S., and Bates, E. (2007). *MacArthur-Bates Communicative Development Inventories: User's guide and technical manual* (2nd ed.). Brookes Publishing.
- **Form**: WS
  Contributor: Virginia Marchman, Stanford University
- **Form**: WS
  Contributor: Virginia Marchman, Stanford University
- **Form**: WS
  Contributor: Linda Smith, Indiana University
- **Form**: WS
  Contributor: Linda Smith, Indiana University
- **Form**: WS
  Contributor: Krista Byers-Heinlein, Concordia University
- **Form**: WS
  Contributor: Donna Thal, San Diego State University
  Citation: Thal, D. J., Marchman, V. A., and Tomblin, J. B. (2013). Late talking toddlers: Characterization and prediction of continued delay. In L. Rescorla and P. Dale (Eds.)., *Late talkers: Language development, interventions, and outcomes.* Brookes Publishing.

- **Form**: WS
  Contributor: Donna Thal, San Diego State University
  Citation: Thal, D. J., Marchman, V. A., and Tomblin, J. B. (2013). Late talking toddlers: Characterization and prediction of continued delay. In L. Rescorla and P. Dale (Eds.), *Late talkers: Language development, interventions, and outcomes.* Brookes Publishing.
- **Form**: WG
  Contributor: Larry Fenson, San Diego State University
  Citation: Fenson, L., Marchman, V. A., Thal, D., Dale, P., Reznick, J. S., and Bates, E. (2007). *MacArthur-Bates Communicative Development Inventories: User's guide and technical manual* (2nd ed.). Brookes Publishing.
- **Form**: WG
  Contributor: Krista Byers-Heinlein, Concordia University
- **Form**: WG
  Contributor: Donna Thal, San Diego State University
  Citation: Thal, D. J., Marchman, V. A., and Tomblin, J. B. (2013). Late talking toddlers: Characterization and prediction of continued delay. In L. Rescorla & P. Dale (Eds.), *Late talkers: Language development, interventions, and outcomes.* Brookes Publishing.
- **Form**: WG
  Contributor: Donna Thal, San Diego State University
  Citation: Thal, D. J., Marchman, V. A., and Tomblin, J. B. (2013). Late talking toddlers: Characterization and prediction of continued delay. In L. Rescorla and P. Dale (Eds.), *Late talkers: Language development, interventions, and outcomes.* Brookes Publishing.
- **Form**: WG
  Contributor: Michael C. Frank, Stanford University
- **Form**: WS
  Contributor: Anne Fernald, Stanford University
  Citation: Fernald, A., Marchman, V. A., and Weisleder, A. (2013). SES differences in language processing skill and vocabulary are evident at 18 months. *Developmental Science,* 16:234–248. http://doi.org/10.1111/desc.12019

English (Australian)

- **Form**: WS
  Contributor: Marina Kalashnikova, MARCS Institute for Brain, Behaviour and Development

Citation: Kalashnikova, M., Schwarz, I.-C., and Burnham, D. (2016). OZI: Australian English communicative development. *First Language*, 36:407–427.

English (British)

- **Form**: Twins Early Development Study (TEDS) Twos
  Contributor: Philip Dale, University of New Mexico
  Citation: Dale, P. S., Price, T. S., Bishop, D. V. M., and Plomin, R. (2003). Outcomes of early language delay: I. Predicting persistent and transient difficulties at 3 and 4 years. *Journal of Speech-Language-Hearing Research*, 46:544–560.
- **Form**: TEDS Threes
  Contributor: Philip Dale, University of New Mexico
  Citation: Dale, P. S., Price, T. S., Bishop, D. V. M., and Plomin, R. (2003). Outcomes of early language delay: I. Predicting persistent and transient difficulties at 3 and 4 years. *Journal of Speech-Language-Hearing Research*, 46:544–560.
- **Form**: Oxford Communicative Development Inventory (CDI)
  Contributor: Caroline Floccia, Plymouth University
  Citation: Floccia, C. (2017). Data collected with the Oxford CDI over a course of 5 years in Plymouth Babylab, UK. With the permission of K. Plunkett and the Oxford CDI from Hamilton, A., Plunkett, K., and Schafer, G., (2000). Infant vocabulary development assessed with a British Communicative Development Inventory: Lower scores in the UK than the USA. *Journal of Child Language*, 27:689–705.
- **Form**: Oxford CDI
  Contributor: Caroline Floccia, Plymouth University
  Citation: Floccia, C. (2017). Data collected with the Oxford CDI over a course of 5 years in Plymouth Babylab, UK. With the permission of K. Plunkett and the Oxford CDI from Hamilton, A., Plunkett, K., and Schafer, G., (2000). Infant vocabulary development assessed with a British Communicative Development Inventory: Lower scores in the UK than the USA. *Journal of Child Language*, 27:689–705.

French (French)

- **Form**: WG
  Contributor: Christina Bergmann (Max Planck Institute for Psycholinguistics), Anne-Caroline Fievet (Laboratoire de Sciences Cognitives et Psycholinguistique [ENS, EHESS, CNRS], Département d'Etudes Cognitives, Ecole Normale Supérieure, PSL Research University)

- **Form**: WG
  Contributor: Katie Von Holzen, University of Maryland
  Citation: Von Holzen, K., Nishibayashi, L.-L., and Nazzi, T. (2018). Consonant and vowel processing in word form segmentation: An infant ERP study. *Brain Sciences*, 8(24):1–15. https://doi.org/10.3390/brainsci8020024
- **Form**: WS
  Contributor: Katie Von Holzen, University of Maryland
  Citation: Von Holzen, K., Nishibayashi, L.-L., and Nazzi, T. (2018). Consonant and vowel processing in word form segmentation: An infant ERP study. *Brain Sciences*, 8(24):1–15. https://doi.org/10.3390/brainsci8020024
- **Form**: WS
  Contributor: Sophie Kern, Centre national de la recherche scientifique (CNRS)

French (Quebecois)

- **Form**: WG
  Contributor: Natacha Trudeau, Université de Montréal
  Citation: Boudreault, M. C., Cabirol, E. A., Poulin-Dubois, D., Sutton, A., and Trudeau, N. (2007). MacArthur Communicative Development Inventories: Validity and preliminary normative data. *La Revue d'orthophonie et d'audiologie*, 31(1):27–37.
- **Form**: WS
  Contributor: Natacha Trudeau, Université de Montréal
  Citation: Trudeau, N., and Sutton, A. (2011). Expressive vocabulary and early grammar of 16-to 30-month-old children acquiring Quebec French. *First Language*, 31(4):480–507. https://doi.org/10.1177/0142723711410828

German

- **Form**: WS
  Contributor: Gisela Szagun, University College London
  Citation: Szagun, G., Stumper, B., and Schramm, A.S. (2009). *Fragebogen zur frühkindlichen Sprachentwicklung (FRAKIS) und FRAKIS-K (Kurzform).* Pearson Assessment.

Greek (Cypriot)

- **Form**: WS
  Contributor: Kleanthes K. Grohmann, University of Cyprus
  Citation: Taxitari, L., Kambanaros, M., and Grohmann, K. K. (2015). A Cypriot Greek adaptation of the CDI: Early production of translation equivalents in a bi(dia)lectal context. *Journal of Greek Linguistics*, 15:1–24.

Hebrew

- **Form**: WG
  Contributor: Hila Gendler Shalev, Tel-Aviv University
  Citation: Gendler-Shalev, H. (2005). התאמת שאלון הורים HCDI-WG לעברית [The adaptation of HCDI-WG parent questionnaire to Hebrew] [Master's Thesis, Tel Aviv University, Tev Aviv, Israel].
- **Form**: WS
  Contributor: Hila Gendler Shalev, Tel-Aviv University

Italian

- **Form**: WS
  Contributor: Christina Caselli, Institute of Cognitive Sciences and Technologies
  Citation: Caselli, M. C., Bates, E., Casadio, P., Fenson, J., Fenson, L., Sanderl, L., and Weir, J. (1995). A cross-linguistic study of early lexical development. *Cognitive Development*, 10(2):159–199.
- **Form**: WG
  Contributor: Christina Caselli, Institute of Cognitive Sciences and Technologies
  Citation: Caselli, M. C., Rinaldi, P., Stefanini, S., and Volterra, V. (2012). Early action and gesture "vocabulary" and its relation with word comprehension and production. *Child Development*, 83(2):526–542.

Kigiriama

- **Form**: WG
  Contributor: Katie Alcock, Lancaster University
  Citation: Alcock, K., Rimba, K., Holding, P., Kitsao-Wekulo, P., Abubakar, A., and Newton, C. R. J. C. (2015). Developmental inventories using illiterate parents as informants: Communicative Development Inventory (CDI) adaptation for two Kenyan languages. *Journal of Child Language*, 42:763–785.

- **Form**: WS
  Contributor: Katie Alcock, Lancaster University
  Citation: Alcock, K., Rimba, K., Holding, P., Kitsao-Wekulo, P., Abubakar, A., and Newton, C. R. J. C. (2015). Developmental inventories using illiterate parents as informants: Communicative Development Inventory (CDI) adaptation for two Kenyan languages. *Journal of Child Language*, 42:763–785.

Kiswahili

- **Form**: WG
  Contributor: Katie Alcock, Lancaster University
  Citation: Alcock, K., Rimba, K., Holding, P., Kitsao-Wekulo, P., Abubakar, A., and Newton, C. R. J. C. (2015). Developmental inventories using illiterate parents as informants: Communicative Development Inventory (CDI) adaptation for two Kenyan languages. *Journal of Child Language*, 42:763–785.
- **Form**: WS
  Contributor: Katie Alcock, Lancaster University
  Citation: Alcock, K., Rimba, K., Holding, P., Kitsao-Wekulo, P., Abubakar, A., and Newton, C. R. J. C. (2015). Developmental inventories using illiterate parents as informants: Communicative Development Inventory (CDI) adaptation for two Kenyan languages. *Journal of Child Language*, 42:763–785.
- **Form**: WS
  Contributor: Katie Alcock, Lancaster University
  Citation: Alcock, K., Rimba, K., Holding, P., Kitsao-Wekulo, P., Abubakar, A., and Newton, C. R. J. C. (2015). Developmental inventories using illiterate parents as informants: Communicative Development Inventory (CDI) adaptation for two Kenyan languages. *Journal of Child Language*, 42:763–785.

Korean

- **Form**: WG
  Contributor: Dongsun Yim, Ewha Womans University
- **Form**: WS
  Contributor: Dongsun Yim, Ewha Womans University
- **Form**: WS
  Contributor: Soyeong Pae, Hallym University
  Citation: Pae, S., and Kwak, K. (2011). *Korean MacArthur-Bates Communicative Development Inventories (K M-B CDI)*. Mindpress.

- **Form**: WG
  Contributor: Soyeong Pae, Hallym University
  Citation: Pae, S., and Kwak, K. (2011). *Korean MacArthur-Bates Communicative Development Inventories (K M-B CDI).* Mindpress.

Latvian

- **Form**: WG
  Contributor: Olga Urek, The Arctic University of Norway
  Citation: Urek, O., Vulāne, A., Darģis, R., Tauriņa, A., Zīriņa, T., and Simonsen, H. G. (2019). Latvian CDI: Methodology, developmental trends, and cross-linguistic comparison. *Journal of Baltic Studies,* 50(3):285–305. https://doi.org/10.1080/01629778.2019.1590433
- **Form**: WS
  Contributor: Olga Urek, The Arctic University of Norway
  Citation: Urek, O., Vulāne, A., Darģis, R., Tauriņa, A., Zīriņa, T., and Simonsen, H. G. (2019). Latvian CDI: Methodology, developmental trends, and cross-linguistic comparison. *Journal of Baltic Studies,* 50(3):285–305. https://doi.org/10.1080/01629778.2019.1590433

Mandarin (Beijing)

- **Form**: WS
  Contributor: Twila Tardif, University of Michigan
  Citation: Tardif, T., Fletcher, P., Liang, W., and Kaciroti, N. (2009). Early vocabulary development in Mandarin (Putonghua) and Cantonese. *Journal of Child Language,* 36(5):1115–1144.
- **Form**: Toddler Checklist (TC)
  Contributor: Ping Li, Pennsylvania State University
  Citation: Hao, M., Shu, H., Xing, A., and Li, P. (2008). Early vocabulary inventory for Mandarin Chinese. *Behavior Research Methods,* 40:728–733.
- **Form**: Infant Checklist (IC)
  Contributor: Ping Li, Pennsylvania State University
  Citation: Hao, M., Shu, H., Xing, A., and Li, P. (2008). Early vocabulary inventory for Mandarin Chinese. *Behavior Research Methods,* 40:728–733.

Mandarin (Taiwanese)

- **Form**: WG
  Contributor: Huei-Mei Liu, National Taiwan Normal University
  Citation: Liu, H. M., and Tsao, F. M. (2010). The standardization and application of Mandarin-Chinese communicative developmental inventory for infants and toddlers. *Formosa Journal of Mental Health*, 23(4):503–534.
  Liu, H. M., and Chen, Y. (2015). Developmental changes in the content and composition of early expressive vocabulary in Mandarin-speaking infants and toddlers. *Bulletin of Educational Psychology*, 24(7):217–242.
- **Form**: WS
  Contributor: Huei-Mei Liu, National Taiwan Normal University
  Citation: Liu, H. M., and Tsao, F. M. (2010). The standardization and application of Mandarin-Chinese communicative developmental inventory for infants and toddlers. *Formosa Journal of Mental Health*, 23(4):503–534.
  Liu, H. M., and Chen, Y. (2015). Developmental changes in the content and composition of early expressive vocabulary in Mandarin-speaking infants and toddlers. *Bulletin of Educational Psychology*, 24(7):217–242.

Norwegian

- **Form**: WS
  Contributor: Hanne Simonsen and Kristian Kristoffersen, University of Oslo
  Citation: Simonsen, H. G., Kristoffersen, K. E., Bleses, D., Wehberg, S., and Jørgensen, R. N. (2014). The Norwegian Communicative Development Inventories: Reliability, main developmental trends and gender differences. *First Language*, 34(1):3–23. https://doi.org/10.1177/0142723713510997
- **Form**: WS
  Contributor: Hanne Simonsen and Kristian Kristoffersen, University of Oslo
  Citation: Simonsen, H. G., Kristoffersen, K. E., Bleses, D., Wehberg, S., and Jørgensen, R. N. (2014). The Norwegian Communicative Development Inventories: Reliability, main developmental trends and gender differences. *First Language*, 34(1):3–23. https://doi.org/10.1177/0142723713510997
- **Form**: WG
  Contributor: Hanne Simonsen and Kristian Kristoffersen, University of Oslo
  Citation: Simonsen, H. G., Kristoffersen, K. E., Bleses, D., Wehberg, S., and Jørgensen, R. N. (2014). The Norwegian Communicative Development Inventories: Reliability, main developmental trends and gender differences. *First Language*, 34(1):3–23. https://doi.org/10.1177/0142723713510997

- **Form**: WG
  Contributor: Hanne Simonsen and Kristian Kristoffersen, University of Oslo
  Citation: Simonsen, H. G., Kristoffersen, K. E., Bleses, D., Wehberg, S., and Jørgensen, R. N. (2014). The Norwegian Communicative Development Inventories: Reliability, main developmental trends and gender differences. *First Language*, 34(1):3–23. https://doi.org/10.1177/0142723713510997

Portuguese (European)

- **Form**: WG
  Contributor: Irene Cadime, University of Minho
- **Form**: WS
  Contributor: Irene Cadime, University of Minho

Russian

- **Form**: WG
  Contributor: Stella Ceytlin, SPb Russian Pedagogical University
  Citation: Вершинина, Е. А., Елисеева, М. Б., Лаврова, Т. С., Рыскина, В. Л., и Цейтлин, С. Н. (2011). Некоторые нормативы речевого развития детей от 8 до 18 месяцев // Специальное образование: традиции и инновации: Сборник научно-методических трудов с международным участием. Издательство РГПУ им. А. И. Герцена. [Vershinina, E. A., Eliseeva, M. B., Lavrova, T. C., Ryskina, V. L., and Tseitlin, S. N. (2011). Some norms of language development of children from 8 to 18 months. In *Special education: Traditions and innovations: A collection of scientific and methodical works with international participation*. Publishing House of Herzen State Pedagogical University.]
- **Form**: WS
  Contributor: Stella Ceytlin, SPb Russian Pedagogical University
  Citation: Елисеева, М. Б., и Вершинина, Е. А. (2009). Некоторые нормативы речевого развития детей от 18 до 36 месяцев (по материалам МакАртуровского опросника) // Проблемы онтолингвистики—2009. Материалы международной конференции 17–19 июбня 2009 г. Санкт-Петербург, с. 22. [Eliseeva, M. B., and Vershinina, E. A. (2009, June 17–19). Some norms of language development of children aged 18 to 36 months (based on the MacArthur questionnaire). In *Problems in Developmental Linguistics—2009: Proceedings of the International Conference*, St. Petersburg, Russia.]

Slovak

- **Form**: WG
  Contributor: Svetlana Kapalková, Comenius University
- **Form**: WS
  Contributor: Svetlana Kapalková, Comenius University

Spanish (European)

- **Form**: WG
  Contributor: Alexandra Karousou, Democritus University of Thrace
  Citation: López Ornat, S., Gallego, C., Gallo, P., Karousou, A., Mariscal, S., and Martínez, M. (2005). *MacArthur: Inventario de desarrollo comunicativo. Manual y Cuadernillos.* Madrid, TEA Ediciones. ISBN: 84-7174- 820-7.
- **Form**: WS
  Contributor: Alexandra Karousou, Democritus University of Thrace
  Citation: López Ornat, S., Gallego, C., Gallo, P., Karousou, A., Mariscal, S., and Martínez, M. (2005). *MacArthur: Inventario de desarrollo comunicativo. Manual y Cuadernillos.* Madrid, TEA Ediciones. ISBN: 84-7174- 820-7.

Spanish (Mexican)

- **Form**: WS
  Contributor: Donna Jackson-Maldonado, Universidad Autónoma de Querétaro
  Citation: Jackson-Maldonado, D., Thal, D., Marchman, V., Newton, T., Fenson, L., and Conboy, B. (2003). *MacArthur Inventarios del Desarrollo de Habilidades Comunicativas: User's guide and technical manual.* Brookes.
- **Form**: WG
  Contributor: Donna Jackson-Maldonado, Universidad Autónoma de Querétaro
  Citation: Jackson-Maldonado, D., Thal, D., Marchman, V., Newton, T., Fenson, L., and Conboy, B. (2003). *MacArthur Inventarios del Desarrollo de Habilidades Comunicativas: User's guide and technical manual.* Brookes.
- **Form**: WG
  Contributor: Anne Fernald, Stanford University
  Citation: Weisleder, A., and Fernald, A. (2013). Talking to children matters: Early language experience strengthens processing and builds vocabulary. *Psychological Science,* 24(11):2143–2152. https://doi.org/10.1177/0956797613488145

- **Form**: WS
  Contributor: Anne Fernald, Stanford University
  Citation: Weisleder, A., and Fernald, A. (2013). Talking to children matters: Early language experience strengthens processing and builds vocabulary. *Psychological Science,* 24(11):2143–2152. https://doi.org/10.1177/0956797613488145

Swedish

- **Form**: WG
  Contributor: Mårten Eriksson, University of Gävle
  Citation: Eriksson, M., and Berglund, E. (2002). *Instruments, scoring manual and percentile levels of the Swedish Early Communicative Development Inventory, SECDI* (FoU-Rapport 17). Institutionen för pedagogik, didaktik och psykologi.
- **Form**: WS
  Contributor: Mårten Eriksson, University of Gävle
  Citation: Eriksson, M., and Berglund, E. (2002). *Instruments, scoring manual and percentile levels of the Swedish Early Communicative Development Inventory, SECDI* (FoU-Rapport 17). Institutionen för pedagogik, didaktik och psykologi.

Turkish

- **Form**: WG
  Contributor: Aylin Küntay, Koç University
  Citation: Acarlar, F., Aksu-Koç, A., Küntay, A. C., Maviş, İ., Sofu, H., Topbaş, S., and Turan, F. (2009). Adapting MB-CDI to Turkish: The first phase. In S. Ay, Ö. Aydın., İ. Ergenç, S. Gökmen, S. İşsever, and D. Peçenel (Eds.), *Essays on Turkish linguistics: Proceedings of the 14th International Conference on Turkish Linguistics,* August 6-8, 2008. Harrassowitz Verlag.
- **Form**: WS
  Contributor: Aylin Küntay, Koç University
  Citation: Acarlar, F., Aksu-Koç, A., Küntay, A. C., Maviş, İ., Sofu, H., Topbaş, S., and Turan, F. (2009). Adapting MB-CDI to Turkish: The first phase. In S. Ay, Ö. Aydın., İ. Ergenç, S. Gökmen, S. İşsever, and D. Peçenel (Eds.), *Essays on Turkish linguistics: Proceedings of the 14th International Conference on Turkish Linguistics,* August 6–8, 2008. Harrassowitz Verlag.

Appendix B: Measures of Variability

In chapter 5, we make use of nonparametric measures of variability, especially the mean absolute deviation from the median (MADM) rather than the more standard coefficient of variation (CV). In this brief appendix, we show that these measures are very highly correlated in the limit. We note however that the two measures (CV and MADM) produce quite different answers for individual data points, especially those that are at the floor or ceiling of a particular form.

Figure B.1 shows CV and MADM plotted together, with each point representing a single age group for a particular combination of form and language. The relationship between the two measures is strong ($r = 0.77$), and its slope is close to one, despite some variation. Overall, it appears that, for the majority of the data, CV is slightly lower than MADM but that it is dramatically higher for some individual datasets. We speculate that this relationship is due to floor/ceiling effects and small sample effects. This analysis suggests that MADM, the nonparametric estimate we use, is less subject to extreme fluctuations than CV.

Our second analysis, shown in figure B.2, is identical except that it plots Cohen's d by the median of the maximum absolute deviation (MMAD). Each of these is the reciprocal of the related measure plotted above. (For example, $d = \frac{\mu}{\sigma}$, whereas $CV = \frac{\sigma}{\mu}$). Thus, the same relation holds.

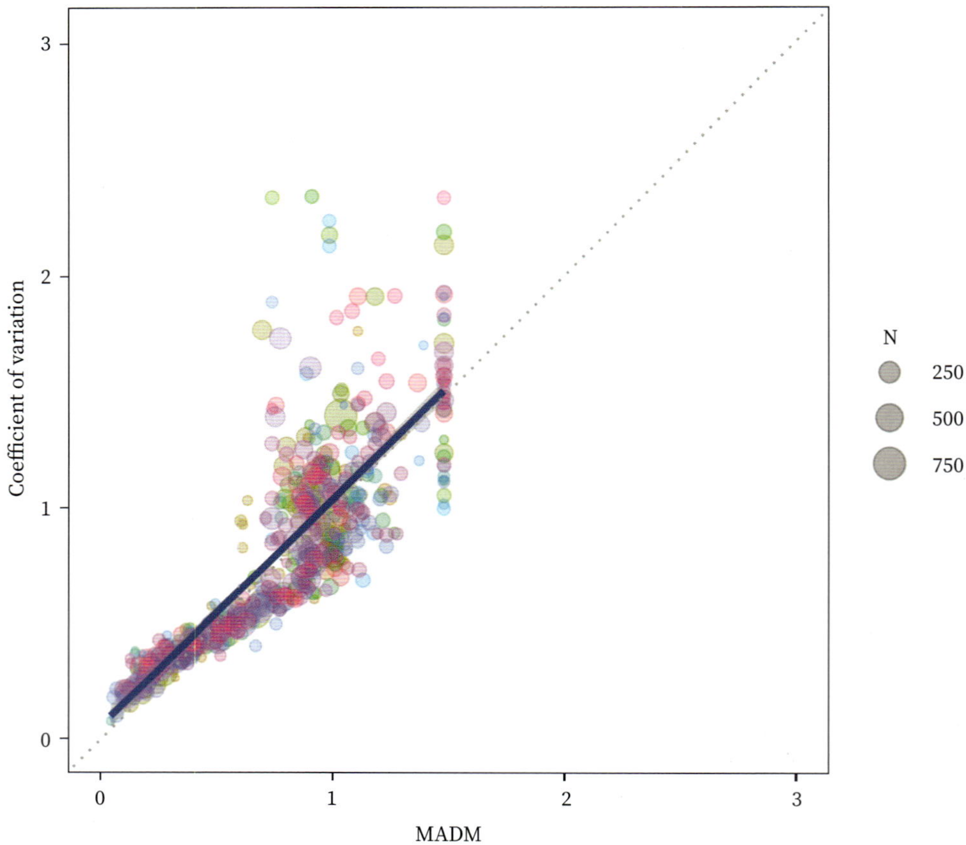

Figure B.1
Coefficient of variation and MADM, with each point showing a particular combination of language, form, and age and the line indicating a linear model fit.

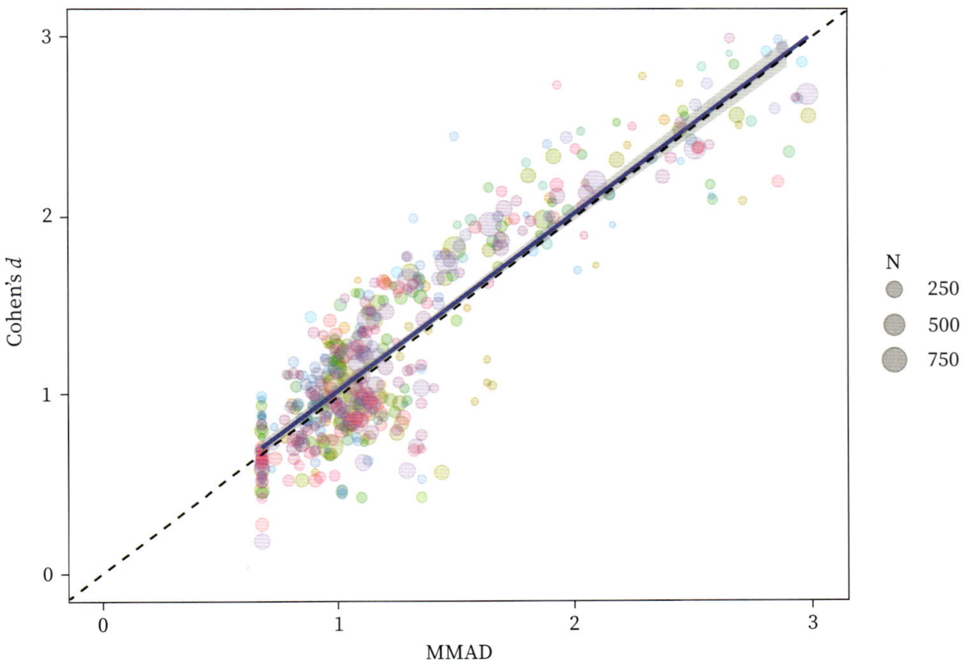

Figure B.2
Cohen's d and MMAD, with each point showing a particular combination of language, form, and age and the line indicating a linear model fit.

Appendix C: Stitching across Forms

Because we use different forms for different ages, there are sometimes good reasons to combine data across forms to get a broader range of ages in a particular analysis. We call this combination "stitching." This appendix provides some motivation for the practice.

The simplest stitching method is to use proportion measures derived each form. As shown in figure C.1, this naive method does not perform well. There is a clear gap between proportions derived from the two different instruments. This gap is presumably generated by the greater number of items on the Words and Sentences (WS) form. Because the WS form includes several hundred more items, many of these will likely be more difficult than those included on the Words and Gestures (WG) form; thus, the proportion for any individual will be lower.

A second method is to use absolute numbers, rather than relative proportions. This method is shown in figure C.2. While this method appears better, a gap is also visible between the smoothed means from the two instruments. Here, the vocabulary estimated from the WS instrument is *larger* presumably because the form affords more total items for parents to check, even if they are on average more difficult.

With these two negative examples in hand, we might be tempted to ask whether differences in the two forms allow data from them to be commensurate at all. For example, a longer form might lead parents to be make different choices (e.g., being more liberal in checking items so as to get through the form faster). To address this issue, we examine data from individual items in figure C.3, which shows twenty-five randomly sampled items from the American English data.

To a first approximation, production trajectories line up quite nicely with little or no visible gap between the two instruments. Thus, at least to the tolerances of visual inspection, we conclude that stitching is best accomplished at the level of individual items. Individual item reports seem robust to some of the details of form construction while percentiles and absolute scores are clearly not.

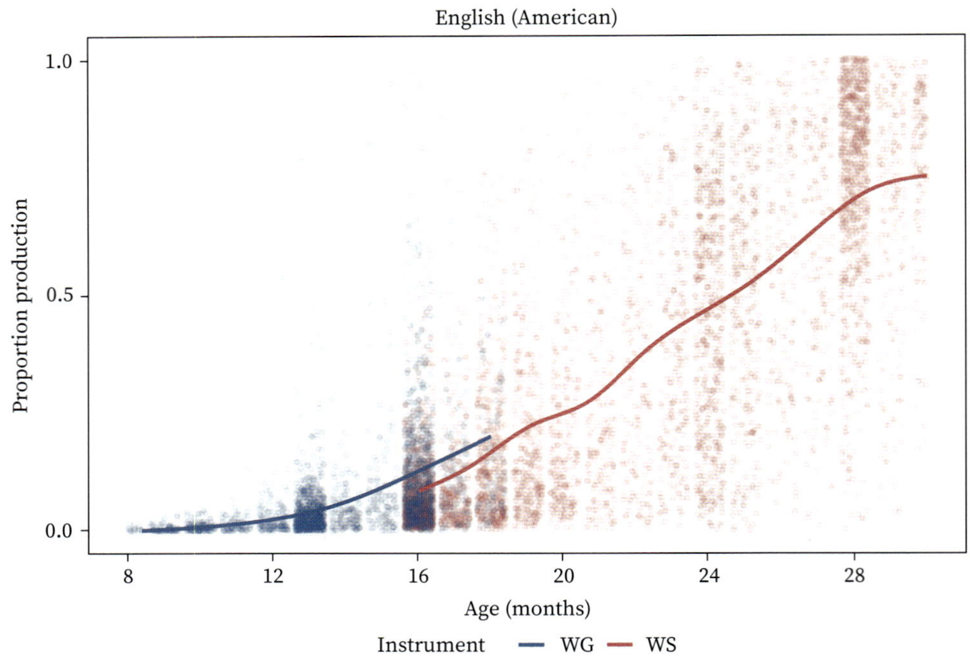

Figure C.1
Proportion WS and WG production scores plotted by form.

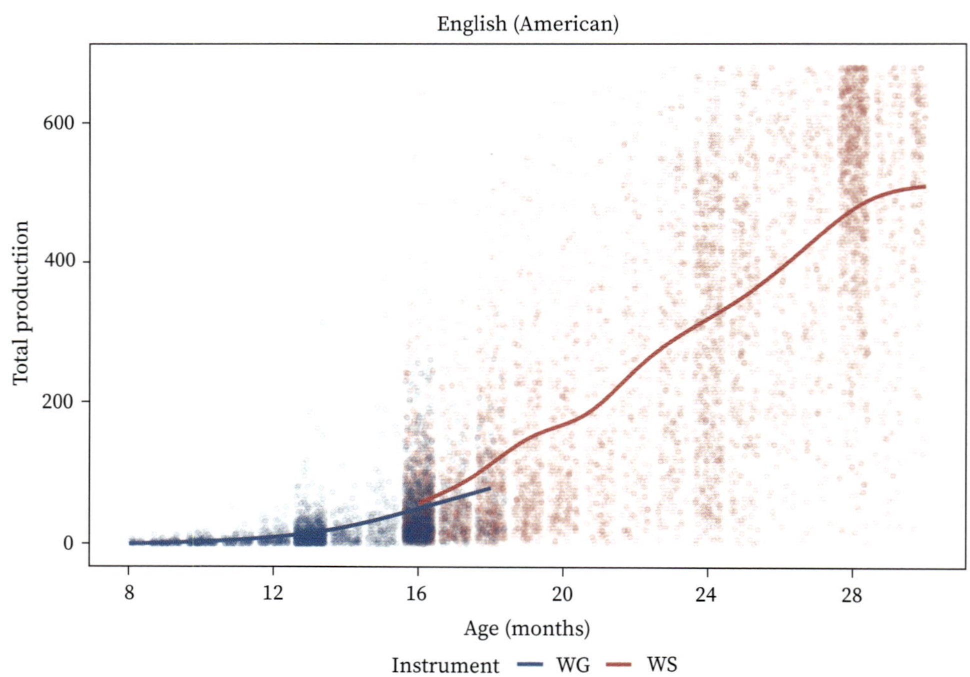

Figure C.2
Total WS and WG production scores plotted by form.

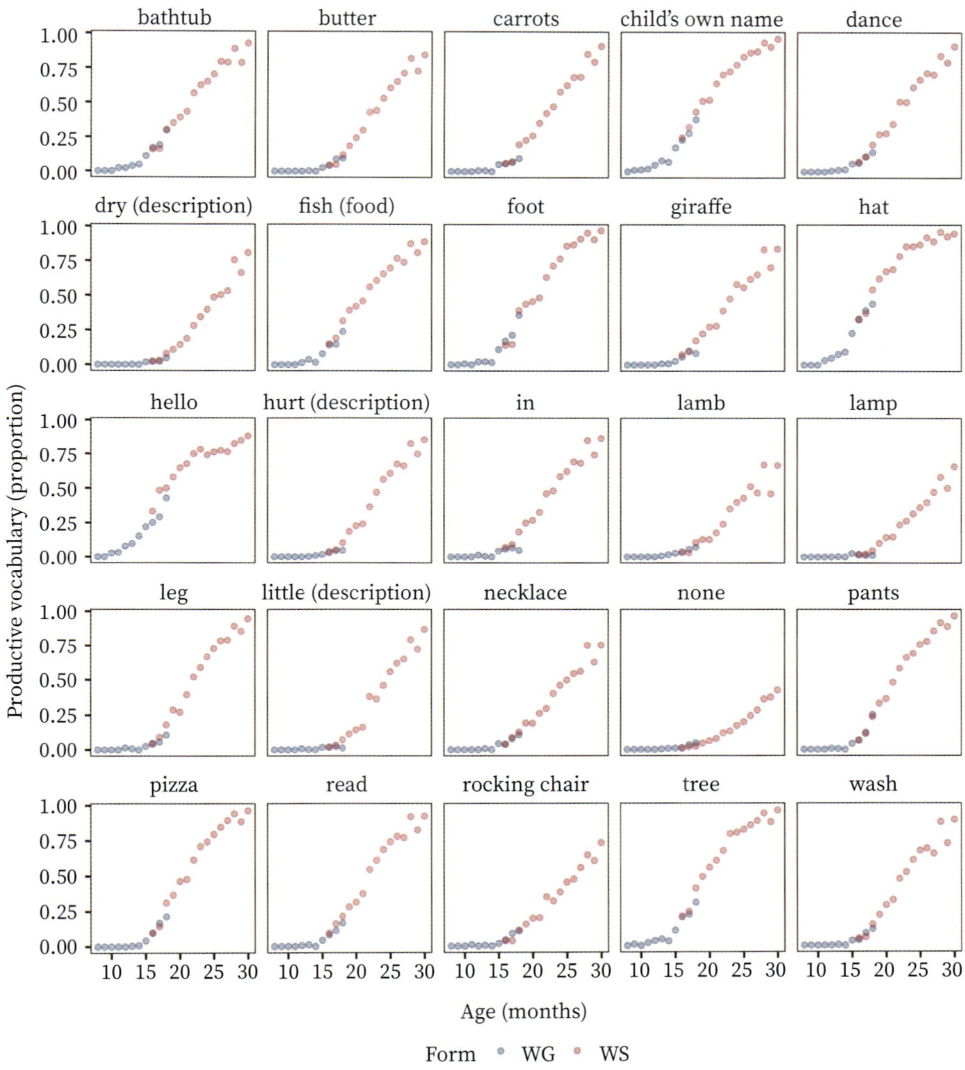

Figure C.3
WS and WG proportion production scores for a set of 25 randomly sampled examples.

Appendix D: Estimating Age of Acquisition

It is frequently useful to have an estimate of the age at which children produce a particular word with a probability greater than some threshold; these are commonly referred to as the word's age of acquisition (AoA) (Goodman et al., 2008). In this appendix, we compare methods for estimating age of acquisition, using the English Words and Sentences (WS) data as a case study.

The simplest and most obvious measure is to use the empirically determined first month at which the proportion producing a word exceeds the threshold. We will use 50 percent of children producing a word as our threshold in all subsequent discussion, following previous literature (Goodman et al., 2008). This approach is simple but results in the exclusion of a large number of words. Of the total, 11 percent do not reach 50 percent production by the ceiling of the form and must be discarded. Further, this method is very sensitive to sparse data. A dataset with highly clustered data will show clustered AoAs. For example, in the Swedish data, there are 307 22-month-olds and 0 21-month-olds. Thus, there will be no 21-month AoAs in this dataset. For these reasons, a model-based approach is likely to be more robust.

We initially investigated three model-based methods. Each of these models was individually fit to data from each word (proportion of children producing at each age), resulting in a continuous curve that can be used to predict AoA more precisely. The three models we examined were

1. standard generalized linear model (GLM) with a logistic link,
2. robust GLM, and
3. Bayesian GLM with hand-tuned prior parameters (Gelman et al., 2008).

For the Bayesian GLM, we took an ad hoc approach, experimenting substantially with prior setting to incorporate information about the slopes and intercepts we expected for words. We adopted default, long-tailed (Cauchy) priors over coefficients. We then used the empirical distribution of GLM slopes to set a strict prior over the age slopes we expected (guided in part by our investigations in chapter 4, which indicated that most items should show positive developmental change). We then set a much weaker prior over intercepts. While these choices are somewhat arbitrary, in practice, in larger datasets only the most extreme words (e.g. *mommy*, for which there is a ceiling effect for nearly every age) were affected by this choice.

In addition to these models, we investigated a hierarchical Bayesian model with shared distributional components across words. This model obviated the ad hoc prior determination that we performed for the individual Bayesian GLMs, but it appeared to perform very similarly (at least in the presence of sufficient data) and was quite expensive to fit in terms of computation, so we do not discuss it here.

Figure D.1 shows a comparison of these methods in the form of histograms of recovered 50 percent values for American English AoA data. The empirical AoAs are clearly clumpy in precisely the way we describe above, even with a substantial amount of data in the analysis ($N = 5{,}846$). In contrast, all three models smooth the AoA distribution substantially, which is likely beneficial to downstream analyses. Although there are some subtle differences in the shape of the main distribution between models, the main action is found on the tails. The different models treat floor and ceiling items differently.

Both the GLM and robust GLM recover two AoAs that are below zero, which is logically impossible (*mommy* and *daddy* are both presumed learned before birth). In contrast, the priors of the Bayesian GLM regularize these AoAs to be 8 and 9 months respectively. Further, the Bayesian GLM estimates 10 percent of AoAs above 30 months (the max value in the data) while the other two methods estimate slightly fewer: 8 percent and 7 percent respectively. These Bayesian GLM results strike us as more reasonable than those returned by the other methods, although they only affect a small minority of words.

To further validate the Bayesian GLM approach, we tested the accuracy of the method in recovering AoAs for much smaller datasets. We did this by taking a subsample of only 100 children from the full English (American) WS dataset. We then fit standard and Bayesian GLMs to this sparse subsample. The resulting AoA estimates are plotted in figure D.2. The Bayesian GLM shows the same minor bias to lower AoAs for hard words that the regular GLM does (a slightly below-diagonal slope), but the GLM shows noisier estimates for the earliest and (especially) the latest-learned words, suggesting that the regularization from the prior values in the Bayesian model is allowing it to deal with sparse data more effectively.

In sum, our analyses here suggest that a Bayesian approach is useful for estimating AoA values.

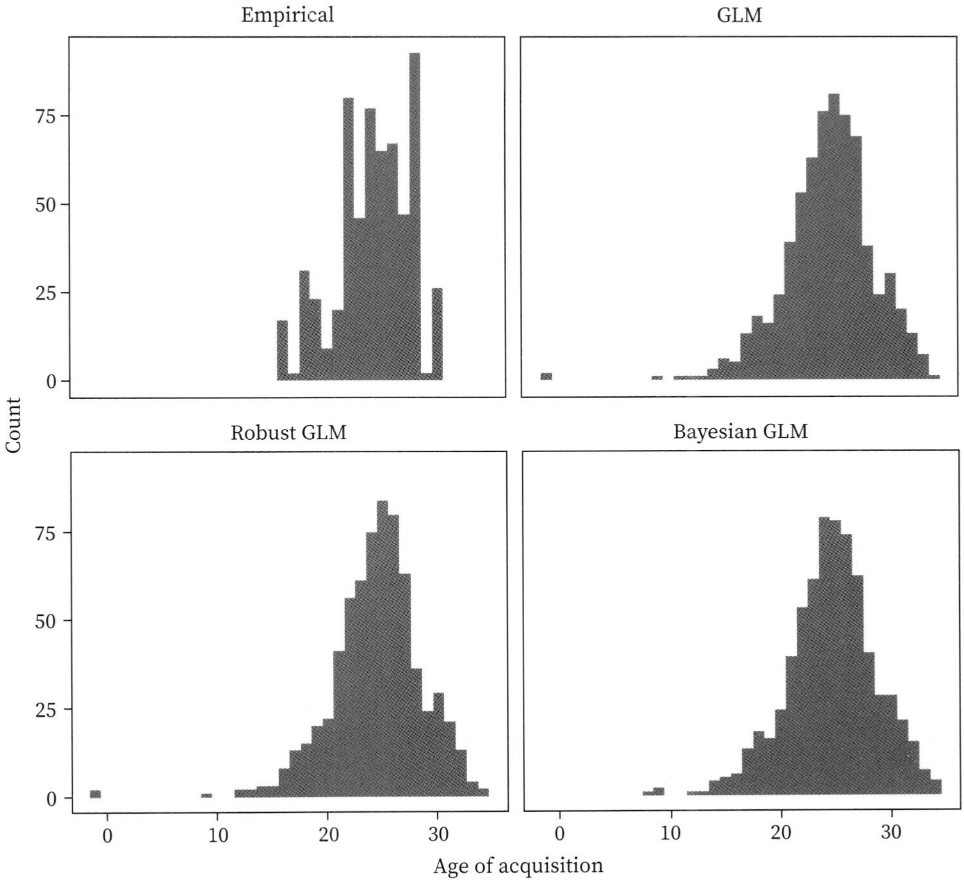

Figure D.1
Histogram of American English age of acquisition values as estimated via a variety of statistical methods (panels).

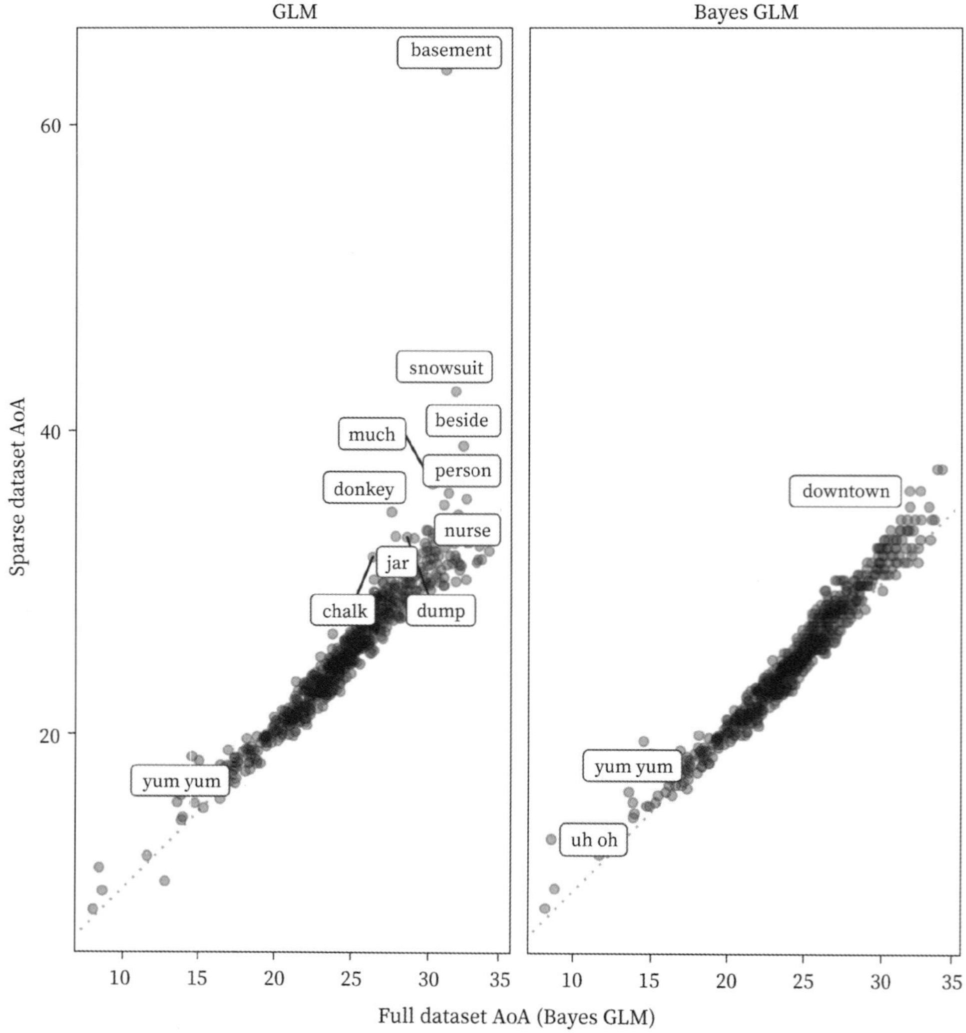

Figure D.2
Recovered AoAs from a sparse subsample (100 children), plotted by the Bayesian GLM AoAs from the full dataset. Left panel shows standard GLM, right panel shows Bayesian GLM. Differences of AoA > 4 months between methods are labeled.

References

Abend, O., Kwiatkowski, T., Smith, N. J., Goldwater, S., and Steedman, M. (2017). Bootstrapping language acquisition. *Cognition*, 164:116–143.

Acredolo, L. P., and Goodwyn, S. W. (1985). Symbolic gesturing in language development. *Human Development*, 28(1):40–49.

Ainsworth, M. D. S., Blehar, M. C., Waters, E., and Wall, S. N. (2015). *Patterns of attachment: A psychological study of the strange situation*. Psychology Press.

Akhtar, N. (2005). The robustness of learning through overhearing. *Developmental Science*, 8(2):199–209.

Akhtar, N., Callanan, M., Pullum, G. K., and Scholz, B. C. (2004). Learning antecedents for anaphoric one. *Cognition*, 93(2):141–145.

Akhtar, N., Jipson, J., and Callanan, M. A. (2001). Learning words through overhearing. *Child Development*, 72(2):416–430.

Alcock, K., and Alibhai, N. (2013). Language development in sub-Saharan Africa. In M. J. Boivin and B. Giordani (Eds.), *Neuropsychology of Children in Africa* (155–180). Springer.

Alcock, K., Rimba, K., Holding, P., Kitsao-Wekulo, P., Abubakar, A., and Newton, C. (2015). Developmental inventories using illiterate parents as informants: Communicative Development Inventory (CDI) adaptation for two Kenyan languages. *Journal of Child Language*, 42(4):763–785.

Alishahi, A., and Stevenson, S. (2008). A computational model of early argument structure acquisition. *Cognitive Science*, 32(5):789–834.

Ambridge, B., Kidd, E., Rowland, C. F., and Theakston, A. L. (2015). The ubiquity of frequency effects in first language acquisuisition. *Journal of Child Language*, 42(02):239–273.

Andersen, E. S. (1978). Lexical universals of body-part terminology. *Universals of Human Language*, 3:335–368.

Andonova, E. (2015). Parental report evidence for toddlers' grammar and vocabulary in Bulgarian. *First Language*, 35(2):126–136.

Andreassen, C., and Fletcher, P. (2007). *Early childhood longitudinal study, birth cohort (ECLS-B): Psychometric report for the 2-year data collection* (Metholodology Report NCES 2007-084). National Center for Education Statistics.

Anisfeld, M., Rosenberg, E. S., Hoberman, M. J., and Gasparini, D. (1998). Lexical acceleration coincides with the onset of combinatorial speech. *First Language*, 18(53):165–184.

Arriaga, R. I., Fenson, L., Cronan, T., and Pethick, S. J. (1998). Scores on the MacArthur Communicative Development Inventory of children from low and middle-income families. *Applied Psycholinguistics*, 19(2):209–223.

Aslin, R. N., and Newport, E. L. (2012). Statistical learning: From acquiring specific items to forming general rules. *Current Directions in Psychological Science*, 21(3):170–176.

Au, T. K.-f., Dapretto, M., and Song, Y.-K. (1994). Input vs. constraints: Early word acquisition in Korean and English. *Journal of Memory and Language*, 33(5):567–582.

Baker, F. B. (2001). *The basics of item response theory* (ED458219). ERIC.

Baker, M. C. (2005). Mapping the terrain of language learning. *Language Learning and Development,* 1(1):93–129.

Baldwin, D. (1993). Early referential understanding: Infants' ability to recognize referential acts for what they are. *Developmental Psychology,* 29(5):832–843.

Baldwin, D. A. (1995). Understanding the link between joint attention and language. In C. Moore and P. J. Dunham (Eds.), *Joint attention* (pp. 131–158). Lawrence Erlbaum Associates.

Bannard, C., Lieven, E., and Tomasello, M. (2009). Modeling children's early grammatical knowledge. *Proceedings of the National Academy of Sciences,* 106(41):17284.

Barner, D., Brooks, N., and Bale, A. (2011). Accessing the unsaid: The role of scalar alternatives in children's pragmatic inference. *Cognition,* 118(1):84–93.

Bartlett, E. J. (1977). *Semantic organization and reference: Acquisition of two aspects of the meaning of color terms.* [Paper presentation]. Biennial Meeting of the Society for Research in Child Development, New Orleans, LA, United States.

Bassano, D. (2000). Early development of nouns and verbs in French: Exploring the interface between lexicon and grammar. *Journal of Child Language,* 27(3):521–559.

Bates, D., Kelman, T., Kleinschmidt, D., Simon, A. B., Mogensen, P. K., Bouchet-Valat, M., Hatherly, M., Saba, E., Baldassari, A., and Noack, A. (2018). *dmbates/MixedModels.jl: Add adaptive Gauss-Hermite quadrature.* Zenodo. https://zenodo.org/record/1303403#.Xuu8AazQjcu

Bates, E. (1976). *Language and context: The acquisition of pragmatics* (Vol. 13). Academic Press.

Bates, E., Bretherton, I., and Snyder, L. (1988). *From first words to grammar: Individual differences and dissociable mechanisms.* Cambridge University Press.

Bates, E., Bretherton, I., and Snyder, L. (1991). *From first words to grammar: Individual differences and dissociable mechanisms,* (Vol. 20). Cambridge University Press. (Original work published 1988)

Bates, E., Bretherton, I., Snyder, L., Shore, C., and Volterra, V. (1980). Vocal and gestural symbols at 13 months. *Merrill-Palmer Quarterly of Behavior and Development,* 26(4):407–423.

Bates, E., Camaioni, L., and Volterra, V. (1975). The acquisition of performatives prior to speech. *Merrill-Palmer Quarterly of Behavior and Development,* 21(3):205–226.

Bates, E., Dale, P., and Thal, D. J. (1995). Individual differences and their implications for theories of language. In P. Fletcher and B. MacWhinney (Eds.), *The handbook of child language development* (pp. 96–151). Blackwell Publishing.

Bates, E., and Goodman, J. C. (1997). On the inseparability of grammar and the lexicon: Evidence from acquisition, aphasia and real-time processing. *Language and Cognitive Processes,* 12(5-6):507–584.

Bates, E., and Goodman, J. C. (1999). On the emergence of grammar from the lexicon. In B. MacWhinney, B. (Ed.) *The Emergence of Language.* Lawrence Erlbaum Associates.

Bates, E., and MacWhinney, B. (1987). Competition, variation, and language learning. In B. MacWhinney (Ed.), *Mechanisms of language acquisition* (pp. 157–193). Lawrence Erlbaum Associates.

Bates, E., MacWhinney, B., et al. (1989). Functionalism and the competition model. In B. MacWhinney and E. Bates (Eds.), *The crosslinguistic study of sentence processing* (pp. 73–112). Cambridge University Press.

Bates, E., Marchman, V., Thal, D., Fenson, L., Dale, P., Reznick, J. S., Reilly, J., and Hartung, J. (1994). Developmental and stylistic variation in the composition of early vocabulary. *Journal of Child Language,* 21(1):85–123.

Bauer, D. J., Goldfield, B. A., and Reznick, J. S. (2002). Alternative approaches to analyzing individual differences in the rate of early vocabulary development. *Applied Psycholinguistics,* 23(3):313–335.

Bellugi, U. (1967). *The acquisition of negation* [Unpublished doctoral dissertation]. Harvard University.

Benedict, H. (1979). Early lexical development: Comprehension and production. *Journal of Child Language,* 6(2):183–200.

Bergelson, E., and Aslin, R. N. (2017). Nature and origins of the lexicon in 6-mo-olds. *Proceedings of the National Academy of Sciences,* 114(49):12916–12921.

Bergelson, E., and Swingley, D. (2012). At 6–9 months, human infants know the meanings of many common nouns. *Proceedings of the National Academy of Sciences,* 109(9):3253–3258.

Bergelson, E., and Swingley, D. (2013). The acquisition of abstract words by young infants. *Cognition,* 127(3):391–397.

Bergelson, E., and Swingley, D. (2015). Early word comprehension in infants: Replication and extension. *Language Learning and Development,* 11(4):369–380.

Bergen, D. (2002). The role of pretend play in children's cognitive development. *Early Childhood Research & Practice,* 4(1):1–8.

Berglund, E., Eriksson, M., and Westerlund, M. (2005). Communicative skills in relation to gender, birth order, childcare and socioeconomic status in 18-month-old children. *Scandinavian Journal of Psychology,* 46(6):485–491.

Berko, J. (1958). The child's learning of English morphology. *Word,* 14:150–177.

Bilson, S., Yoshida, H., Tran, C. D., Woods, E. A., and Hills, T. T. (2015). Semantic facilitation in bilingual first language acquisition. *Cognition,* 140:122–134.

Bleses, D., Basbøll, H., and Vach, W. (2011). Is Danish difficult to acquire? Evidence from Nordic past-tense studies. *Language and Cognitive Processes,* 26(8):1193–1231.

Bleses, D., Vach, W., Slott, M., Wehberg, S., Thomsen, P., Madsen, T. O., and Basbøll, H. (2008). Early vocabulary development in Danish and other languages: A CDI-based comparison. *Journal of Child Language,* 35(3):619–650.

Bloom, L., Tinker, E., and Margulis, C. (1993). The words children learn: Evidence against a noun bias in early vocabularies. *Cognitive Development,* 8(4):431–450.

Bloom, L., Tinker, E., and Scholnick, E. (2001). The intentionality model and language acquisition: Engagement, effort, and the essential tension in development. *Monographs of the Society for Research in Child Development,* 66(4).

Bloom, P. (2000). *How children learn the meanings of words.* MIT Press.

Bohn, M., and Frank, M. C. (2019). The pervasive role of pragmatics in early language. *Annual Review of Developmental Psychology,* 1:223–249.

Bornstein, M. H. (1985). On the development of color naming in young children: Data and theory. *Brain and Language,* 26(1):72–93.

Bornstein, M. H. (2013). *Cultural approaches to parenting.* Psychology Press.

Bornstein, M. H., and Cote, L. R. (2005). Expressive vocabulary in language learners from two ecological settings in three language communities. *Infancy,* 7(3):299–316.

Bornstein, M. H., Cote, L. R., Maital, S., Painter, K., Park, S.-Y., Pascual, L., Pêcheux, M.-G., Ruel, J., Venuti, P., and Vyt, A. (2004a). Cross-linguistic analysis of vocabulary in young children: Spanish, Dutch, French, Hebrew, Italian, Korean, and American English. *Child Development,* 75(4):1115–1139.

Bornstein, M. H., Hahn, C.-S., and Haynes, O. M. (2004b). Specific and general language performance across early childhood: Stability and gender considerations. *First Language,* 24(3):267–304.

Bornstein, M. H., and Haynes, O. M. (1998). Vocabulary competence in early childhood: Measurement, latent construct, and predictive validity. *Child Development,* 69(3):654–671.

Bornstein, M. H., Hendricks, C., Hahn, C.-S., Haynes, O. M., Painter, K. M., and Tamis-LeMonda, C. S. (2003). Contributors to self-perceived competence, satisfaction, investment, and role balance in maternal parenting: A multivariate ecological analysis. *Parenting: Science and Practice,* 3(4):285–326.

Bornstein, M. H., and Putnick, D. L. (2012). Stability of language in childhood: A multiage, multidomain, multimeasure, and multisource study. *Developmental Psychology,* 48(2):477.

Borsboom, D. (2006). The attack of the psychometricians. *Psychometrika,* 71(3):425.

Bowerman, M. (1973). *Early syntactic development: A cross-linguistic study with special reference to Finnish* (Vol. 11). Cambridge University Press.

Bowerman, M. (1996). The origins of children's spatial semantic categories: Cognitive versus linguistic determinants. In J. J. Gumperz and S. C. Levinson (Eds.) *Rethinking Linguistic Relativity* (pp. 145–176). Cambridge University Press.

Bradley, R. H., and Corwyn, R. F. (2002). Socioeconomic status and child development. *Annual Review of Psychology,* 53(1):371–399.

Braginsky, M., Yurovsky, D., Marchman, V. A., and Frank, M. C. (2015). Developmental changes in the relationship between grammar and the lexicon. In D. C. Noelle, R. Dale, A. S. Warlaumont, T. Matlock, C. D. Jennings, and P. P. Maglio (Eds.), *Proceedings of the 37th annual meeting of the cognitive science society.* Cognitive Science Society.

Braginsky, M., Yurovsky, D., Marchman, V. A., and Frank, M. C. (2016). From uh-oh to tomorrow: Predicting age of acquisition for early words across languages. In A. Papafragou, D. Grodner, D. Mirman, and J. Trueswell (Eds.), *Proceedings of the 38th annual meeting of the cognitive science society.* Cognitive Science Society.

Braginsky, M., Yurovsky, D., Marchman, V. A., and Frank, M. C. (2019). Consistency and variability in children's word learning across languages. *OpenMind: Discoveries in Cognitive Science* 3:52–67.

Brent, M., and Siskind, J. (2001). The role of exposure to isolated words in early vocabulary development. *Cognition,* 81:33–44.

Brinchmann, E. I., Braeken, J., and Lyster, S.-A. H. (2018). Is there a direct relation between the development of vocabulary and grammar? *Developmental Science,* 22(1): e12709.

Brooks, R., and Meltzoff, A. (2008). Infant gaze following and pointing predict accelerated vocabulary growth through two years of age: A longitudinal, growth curve modeling study. *Journal of Child Language,* 35(01):207–220.

Brown, R. (1973). *A first language: The early stages.* Harvard Universirty Press.

Bruner, J. (1985). Child's talk: Learning to use language. *Child Language Teaching and Therapy,* 1(1):111–114.

Brysbaert, M., Warriner, A. B., and Kuperman, V. (2014). Concreteness ratings for 40 thousand generally known English word lemmas. *Behavior Research Methods,* 46(3):904–911.

Buhrmester, M., Kwang, T., and Gosling, S. D. (2011). Amazon's mechanical turk: A new source of inexpensive, yet high-quality, data? *Perspectives on Psychological Science,* 6(1):3–5.

Busby, J., and Suddendorf, T. (2005). Recalling yesterday and predicting tomorrow. *Cognitive Development,* 20(3):362–372.

Butterworth, G. (2003). Pointing is the royal road to language for babies. In S. Kita (Ed.), *Pointing* (pp. 17–42). Lawrence Erlbaum Associates.

Can, D. D., Ginsburg-Block, M., Golinkoff, R. M., and Hirsh-Pasek, K. (2013). A long-term predictive validity study: Can the CDI short form be used to predict language and early literacy skills four years later? *Journal of Child Language,* 40(4):821–835.

Carey, S. (1978). The child as word learner. In M. Halle, J. Bresnan, and G. A. Miller (Eds.), *Linguistic theory and psychological reality* (pp. 264–293). MIT Press.

Carey, S. (2009). *The origin of concepts.* Oxford University Press.

Carpenter, M., Nagell, K., Tomasello, M., Butterworth, G., and Moore, C. (1998). Social cognition, joint attention, and communicative competence from 9 to 15 months of age. *Monographs of the Society for Research in Child Development,* 63(4).

Cartmill, E. A., Armstrong, B. F., Gleitman, L. R., Goldin-Meadow, S., Medina, T. N., and Trueswell, J. C. (2013). Quality of early parent input predicts child vocabulary 3 years later. *Proceedings of the National Academy of Sciences,* 110(28):11278–11283.

Caselli, C., Casadio, P., and Bates, E. (1999). A comparison of the transition from first words to grammar in English and Italian. *Journal of Child Language,* 26(01):69–111.

Caselli, M. C., Bates, E., Casadio, P., Fenson, J., Fenson, L., Sanderl, L., and Weir, J. (1995). A cross-linguistic study of early lexical development. *Cognitive Development,* 10(2):159–199.

Cazden, C. B. (1968). The acquisition of noun and verb inflections. *Child Development,* 39(2):433–448.

Chalmers, R. P. (2012). mirt: A multidimensional item response theory package for the r environment. *Journal of Statistical Software,* 48(6):1–29.

Chalmers, R. P. (2016). Generating adaptive and non-adaptive test interfaces for multidimensional item response theory applications. *Journal of Statistical Software*, 71(5):1–39.

Charman, T., Drew, A., Baird, C., and Baird, G. (2003). Measuring early language development in preschool children with autism spectrum disorder using the MacArthur Communicative Development Inventory (infant form). *Journal of Child Language*, 30(1):213–236.

Chater, N., and Christiansen, M. H. (2018). Language acquisition as skill learning. *Current Opinion in Behavioral Sciences*, 21:205–208.

Choi, S., and Gopnik, A. (1995). Early acquisition of verbs in Korean: A cross-linguistic study. *Journal of Child Language*, 22(03):497–529.

Chomsky, N. (1957). *Syntactic structures*. Mouton.

Chomsky, N. (1981). Principles and parameters in syntactic theory. In N. Hornstein and D. Lightfoot (Eds.), *Explanation in linguistics: The logical problem of language acquisition* (pp. 32–75). Springer Longman.

Chomsky, N. (2014). *The Minimalist Program*. MIT Press.

Clahsen, H., Rothweiler, M., Woest, A., and Marcus, G. F. (1992). Regular and irregular inflection in the acquisition of German noun plurals. *Cognition*, 45(3):225–255.

Clark, E. V. (1971). On the acquisition of the meaning of before and after. *Journal of Verbal Learning and Verbal Behavior*, 10(3):266–275.

Clark, E. V. (1973). What's in a word? On the child's acquisition of semantics in his first language. In T. E. Moore (Ed.), *Cognitive development and acquisition of language* (pp. 65–110). Elsevier.

Clark, E. V. (1977). From gesture to word: On the natural history of deixis in language acquisition. In J. S. Bruner and A. Garton (Eds.), *Human growth and development: Wolfson College Lectures* (pp. 85–120). Oxford University Press.

Clark, E. V. (1988). On the logic of contrast. *Journal of Child Language*, 15:317–335.

Clark, E. V. (1993). *The lexicon in acquisition*. Cambridge University Press.

Clark, E. V. (2007). Young children's uptake of new words in conversation. *Language in Society*, 36(2):157–182.

Clark, E. V. (2016). *First language acquisition* (3rd ed.). Cambridge University Press.

Clark, E. V., and Estigarribia, B. (2011). Using speech and gesture to introduce new objects to young children. *Gesture*, 11(1):1–23.

Clark, E. V., and Hecht, B. F. (1983). Comprehension, production, and language acquisition. *Annual Review of Psychology*, 34(1):325–349.

Colombo, J. (2001). The development of visual attention in infancy. *Annual Review of Psychology*, 52(1):337–367.

Conwell, E., and Demuth, K. (2007). Early syntactic productivity: Evidence from dative shift. *Cognition*, 103(2):163–179.

Core, C., Hoff, E., Rumiche, R., and Señor, M. (2013). Total and conceptual vocabulary in Spanish–English bilinguals from 22 to 30 months: Implications for assessment. *Journal of Speech, Language, and Hearing Research*, 56:1637–1649.

Crain, S., and Thornton, R. (2000). *Investigations in universal grammar: A guide to experiments on the acquisition of syntax and semantics*. MIT Press.

Cristia, A., Seidl, A., Junge, C., Soderstrom, M., and Hagoort, P. (2014). Predicting individual variation in language from infant speech perception measures. *Child Development*, 85(4):1330–1345.

Cronbach, L. J., and Meehl, P. E. (1955). Construct validity in psychological tests. *Psychological Bulletin*, 52(4):281.

Curtiss, S. (1977). *Genie: A psychological study of a modern-day wild child*. Academic Press.

Dale, P. S. (2015). *Adaptations, not translations!* MacArthur-Bates CDI. https://mb-cdi.stanford.edu/documents/AdaptationsNotTranslations2015.pdf

Dale, P. S., Dionne, G., Eley, T. C., and Plomin, R. (2000). Lexical and grammatical development: A behavioural genetic perspective. *Journal of Child Language*, 27(3):619–642.

Dale, P. S., and Fenson, L. (1996). Lexical development norms for young children. *Behavior Research Methods, Instruments, & Computers*, 28(1):125–127.

Dale, P. S., Price, T. S., Bishop, D. V., and Plomin, R. (2003). Outcomes of early language delay: I. predicting persistent and transient language difficulties at 3 and 4 years. *Journal of Speech, Language, and Hearing Research*, 46(3):544–560.

Dale, P. S., Tosto, M. G., Hayiou-Thomas, M. E., and Plomin, R. (2015). Why does parental language input style predict child language development? A twin study of gene–environment correlation. *Journal of Communication Disorders*, 57:106–117.

Dapretto, M., and Bjork, E. L. (2000). The development of word retrieval abilities in the second year and its relation to early vocabulary growth. *Child Development*, 71(3):635–648.

Davidson, M. C., Amso, D., Anderson, L. C., and Diamond, A. (2006). Development of cognitive control and executive functions from 4 to 13 years: Evidence from manipulations of memory, inhibition, and task switching. *Neuropsychologia*, 44(11):2037–2078.

Decker, D. M., et al. (1999). *Handbook of the International Phonetic Association: A guide to the use of the International Phonetic Alphabet*. Cambridge University Press.

De Houwer, A., Bornstein, M. H., and Leach, D. B. (2005). Assessing early communicative ability: A cross-reporter cumulative score for the MacArthur CDI. *Journal of Child Language*, 32(4):735–758.

DeLoache, J. S., Simcock, G., and Macari, S. (2007). Planes, trains, automobiles—and tea sets: Extremely intense interests in very young children. *Developmental Psychology*, 43(6):1579.

Demuth, K., and Tremblay, A. (2008). Prosodically-conditioned variability in children's production of French determiners. *Journal of Child Language*, 35(1):99–127.

Devescovi, A., Caselli, M. C., Marchione, D., Pasqualetti, P., Reilly, J., and Bates, E. (2005). A crosslinguistic study of the relationship between grammar and lexical development. *Journal of Child Language*, 32(4):759–786.

Diamond, A. (2013). Executive functions. *Annual Review of Psychology*, 64:135–168.

Diamond, A., and Lee, K. (2011). Interventions shown to aid executive function development in children 4 to 12 years old. *Science*, 333(6045):959–964.

Dickinson, D. K., and Tabors, P. O. (2001). *Beginning literacy with language: Young children learning at home and school*. Paul H. Brookes Publishing.

Dionne, G., Dale, P. S., Boivin, M., and Plomin, R. (2003). Genetic evidence for bidirectional effects of early lexical and grammatical development. *Child Development*, 74(2):394–412.

Dixon, J. A., and Marchman, V. A. (2007). Grammar and the lexicon: Developmental ordering in language acquisition. *Child Development*, 78(1):190–212.

Dore, J. (1974). A pragmatic description of early language development. *Journal of Psycholinguistic Research*, 3(4):343–350.

Dromi, E. (1987). *Early lexical development*. University Press.

Duddington, J. (2012). Speak text to speech. http://espeak.sourceforge.net

Dunn, L. M., and Dunn, L. M. (2007). *Peabody Picture Vocabulary Test*. Pearson Assessments.

Dupoux, E. (2018). Cognitive science in the era of artificial intelligence: A roadmap for reverse-engineering the infant language-learner. *Cognition*, 173:43–59.

Eigsti, I.-M., de Marchena, A. B., Schuh, J. M., and Kelley, E. (2011). Language acquisition in autism spectrum disorders: A developmental review. *Research in Autism Spectrum Disorders*, 5(2):681–691.

Elman, J. L., Bates, E., Johnson, M. H., Karmiloff-Smith, A., Parisi, D., and Plunkett, K. (1996). *Rethinking innateness: A connectionist perspective on development*. MIT Press.

Eriksson, M., Marschik, P. B., Tulviste, T., Almgren, M., Pérez Pereira, M., Wehberg, S., Marjanovič-Umek, L., Gayraud, F., Kovacevic, M., and Gallego, C. (2012). Differences between girls and boys in emerging language skills: Evidence from 10 language communities. *British Journal of Developmental Psychology*, 30(2):326–343.

Evans, G. W. (2004). The environment of childhood poverty. *American Psychologist,* 59(2):77.

Evans, N., and Levinson, S. C. (2009). The myth of language universals: Language diversity and its importance for cognitive science. *Behavioral and Brain Sciences,* 32(5):429–448.

Farkas, G., and Beron, K. (2004). The detailed age trajectory of oral vocabulary knowledge: Differences by class and race. *Social Science Research,* 33(3):464–497.

Fazly, A., Alishahi, A., and Stevenson, S. (2010). A probabilistic computational model of cross-situational word learning. *Cognitive Science,* 34(6):1017–1063.

Feldman, H. M., Dale, P. S., Campbell, T. F., Colborn, D. K., Kurs-Lasky, M., Rockette, H. E., and Paradise, J. L. (2005). Concurrent and predictive validity of parent reports of child language at ages 2 and 3 years. *Child Development,* 76(4):856–868.

Feldman, H. M., Dollaghan, C. A., Campbell, T. F., Kurs-Lasky, M., Janosky, J. E., and Paradise, J. L. (2000). Measurement properties of the MacArthur Communicative Development Inventories at ages one and two years. *Child Development,* 71(2):310–322.

Feldman, N., Griffiths, T., and Morgan, J. (2009). The influence of categories on perception: Explaining the perceptual magnet effect as optimal statistical inference. *Psychological Review,* 116(4):752–782.

Fenson, L., Bates, E., Dale, P., Goodman, J. C., Reznick, J. S., and Thal, D. (2000a). Reply: Measuring variability in early child language: Don't shoot the messenger. *Child Development,* 71(2):323–328.

Fenson, L., Bates, E., Dale, P. S., Marchman, V. A., Reznick, J. S., and Thal, D. J. (2007). *MacArthur-Bates Communicative Development Inventories: User's Guide and Technical Manual,* (2nd Edition). Brookes Publishing.

Fenson, L., Dale, P., Reznick, J., Bates, E., Thal, D., Pethick, S., Tomasello, M., Mervis, C., and Stiles, J. (1994). Variability in early communicative development. *Monographs of the Society for Research in Child Development,* 59(5).

Fenson, L., Pethick, S., Renda, C., Cox, J. L., Dale, P. S., and Reznick, J. S. (2000b). Short-form versions of the MacArthur Communicative Development Inventories. *Applied Psycholinguistics,* 21(1): 95–116.

Fernald, A., and Marchman, V. A. (2012). Individual differences in lexical processing at 18 months predict vocabulary growth in typically developing and late-talking toddlers. *Child Development,* 83(1): 203–222.

Fernald, A., Marchman, V. A., and Weisleder, A. (2013). SES differences in language processing skill and vocabulary are evident at 18 months. *Developmental Science,* 16(2):234–248.

Fernald, A., and Morikawa, H. (1993). Common themes and cultural variations in Japanese and American mothers' speech to infants. *Child Development,* 64:637–656.

Fernald, A., Perfors, A., and Marchman, V. A. (2006). Picking up speed in understanding: Speech processing efficiency and vocabulary growth across the 2nd year. *Developmental Psychology,* 42(1):98.

Fernald, A., Pinto, J., Swingley, D., Weinbergy, A., and McRoberts, G. (1998). Rapid gains in speed of verbal processing by infants in the 2nd year. *Psychological Science,* 9(3):228.

Fernald, L. C., Kariger, P., Hidrobo, M., and Gertler, P. J. (2012). Socioeconomic gradients in child development in very young children: Evidence from India, Indonesia, Peru, and Senegal. *Proceedings of the National Academy of Sciences,* 109(Suppl 2):17273–17280.

Fiser, J., and Aslin, R. N. (2002). Statistical learning of new visual feature combinations by infants. *Proceedings of the National Academy of Sciences,* 99(24):15822–15826.

Fisher, C., Gertner, Y., Scott, R. M., and Yuan, S. (2010). Syntactic bootstrapping. *Wiley Interdisciplinary Reviews: Cognitive Science,* 1(2):143–149.

Floccia, C., Sambrook, T. D., Delle Luche, C., Kwok, R., Goslin, J., White, L., Cattani, A., Sullivan, E., Abbot-Smith, K., Krott, A., et al. (2018). Vocabulary of 2-year-olds learning English and an additional language: Norms and effects of linguistic distance. *Monographs of the Society for Research in Child Development,* 83(1):1–135.

Flournoy, J. C. (2017). *A better cross-lagged panel model, from Hamaker et al. (2015).* John C. Flournoy. jflournoy.github.io/2017/10/20/riclpm-lavaan-demo

Flynn, J. R. (1987). Massive IQ gains in 14 nations: What IQ tests really measure. *Psychological Bulletin*, 101(2):171.

Frank, M. C., Bergelson, E., Bergmann, C., Cristia, A., Floccia, C., Gervain, J., Hamlin, J. K., Hannon, E. E., Kline, M., Levelt, C., et al. (2017). A collaborative approach to infant research: Promoting reproducibility, best practices, and theory-building. *Infancy*, 22(4):421–435.

Frank, M. C., Braginsky, M., Yurovsky, D., and Marchman, V. A. (2016a). Wordbank: An open repository for developmental vocabulary data. *Journal of Child Language*, 44(3):677–694.

Frank, M. C., Goldwater, S., Griffiths, T., and Tenenbaum, J. (2010). Modeling human performance in statistical word segmentation. *Cognition*, 117(2):107–125.

Frank, M. C., Goodman, N. D., and Tenenbaum, J. B. (2009). Using speakers' referential intentions to model early cross-situational word learning. *Psychological Science*, 20:578–585.

Frank, M. C., Lewis, M., and MacDonald, K. (2016b). A performance model for early word learning. In A. Papafragou, D. Grodner, D. Mirman, and J. Trueswell (Eds.), *Proceedings of the 38th annual meeting of the cognitive science society*. Cognitive Science Society.

Frank, M. C., Tenenbaum, J. B., and Fernald, A. (2013). Social and discourse contributions to the determination of reference in cross-situational word learning. *Language, Learning, and Development*, 9(1):1–24.

Frank, S. L., Bod, R., and Christiansen, M. H. (2012). How hierarchical is language use? *Proceedings of the Royal Society of London B: Biological Sciences*, 279(1747).

Freudenthal, D., Pine, J., and Gobet, F. (2010). Explaining quantitative variation in the rate of optional infinitive errors across languages: A comparison of mosaic and the variational learning model. *Journal of Child Language*, 37(3):643–669.

Friend, M. and Keplinger, M. (2008). Reliability and validity of the Computerized Comprehension Task (CCT): Data from American English and Mexican Spanish infants. *Journal of Child Language*, 35(1):77–98.

Ganger, J., and Brent, M. R. (2004). Reexamining the vocabulary spurt. *Developmental Psychology*, 40(4):621.

Gathercole, S. E., Willis, C. S., Baddeley, A. D., and Emslie, H. (1994). The children's test of nonword repetition: A test of phonological working memory. *Memory*, 2(2):103–127.

Gathercole, V. C. M., Thomas, E. M., Roberts, E., Hughes, C., and Hughes, E. K. (2013). Why assessment needs to take exposure into account: Vocabulary and grammatical abilities in bilingual children. In V. C. M. Gathercole (Ed.), *Issues in the Assessment of Bilinguals* (pp. 20–55). Multilingual Matters.

Gelfand, M. J., Raver, J. L., Nishii, L., Leslie, L. M., Lun, J., Lim, B. C., Duan, L., Almaliach, A., Ang, S., Arnadottir, J., et al. (2011). Differences between tight and loose cultures: A 33-nation study. *Science*, 332(6033):1100–1104.

Gelman, A., Jakulin, A., Pittau, M. G., Su, Y.-S., et al. (2008). A weakly informative default prior distribution for logistic and other regression models. *The Annals of Applied Statistics*, 2(4):1360–1383.

Gentner, D. (1982). Why nouns are learned before verbs: Linguistic relativity versus natural partitioning. *Center for the Study of Reading Technical Report* (No. 257). Center for the Study of Reading, University of Illinois.

Gentner, D., and Boroditsky, L. (2001). Individuation, relativity, and early word learning. In M. Bowerman and S. Levinson (Eds.), *Language acquisition and conceptual development* (pp. 215–256). Cambridge University Press.

Gertner, Y., Fisher, C., and Eisengart, J. (2006). Learning words and rules. *Psychological Science*, 17(8):684.

Ghosh-Dastidar, B., Cohen, D., Hunter, G., Zenk, S. N., Huang, C., Beckman, R., and Dubowitz, T. (2014). Distance to store, food prices, and obesity in urban food deserts. *American Journal of Preventive Medicine*, 47(5):587–595.

Gillette, J., Gleitman, H., Gleitman, L., and Lederer, A. (1999). Human simulations of vocabulary learning. *Cognition*, 73(2):135–176.

Gleitman, L. (1990). The structural sources of verb meanings. *Language Acquisition*, 1(1):3–55.

Goldberg, A. E. (2006). *Constructions at work: The nature of generalization in language*. Oxford University Press.

Goldfield, B. A. (1993). Noun bias in maternal speech to one-year-olds. *Journal of Child Language*, 20(1):85–99.

Goldfield, B. A., and Reznick, J. S. (1990). Early lexical acquisition: Rate, content, and the vocabulary spurt. *Journal of Child Language*, 17(1):171–183.

Goldin-Meadow, S. (1998). The development of gesture and speech as an integrated system. *New Directions for Child and Adolescent Development*, 1998(79):29–42.

Goldin-Meadow, S., and Mylander, C. (1983). Gestural communication in deaf children: Noneffect of parental input on language development. *Science*, 221(4608):372–374.

Gómez, R., and Gerken, L. (1999). Artificial grammar learning by 1-year-olds leads to specific and abstract knowledge. *Cognition*, 70:109–135.

Goodman, J., Dale, P., and Li, P. (2008). Does frequency count? Parental input and the acquisition of vocabulary. *Journal of Child Language*, 35(3):515.

Gopnik, A., Choi, S., and Baumberger, T. (1996). Cross-linguistic differences in early semantic and cognitive development. *Cognitive Development*, 11(2):197–225.

Gopnik, A., Glymour, C., Sobel, D. M., Schulz, L. E., Kushnir, T., and Danks, D. (2004). A theory of causal learning in children: Causal maps and Bayes nets. *Psychological Review*, 111(1):3.

Greenberg, J. H. (Ed.). (1963). *Universals of language*. MIT Press.

Greenwald, A. G. (2012). There is nothing so theoretical as a good method. *Perspectives on Psychological Science*, 7(2):99–108.

Hamaker, E. L., Kuiper, R. M., and Grasman, R. P. (2015). A critique of the cross-lagged panel model. *Psychological Methods*, 20(1):102.

Hambleton, R. K., Swaminathan, H., and Rogers, H. J. (1991). *Fundamentals of item response theory* (Vol. 2). Sage.

Hammer, C. S., Farkas, G., and Maczuga, S. (2010). The language and literacy development of Head Start children: A study using the family and child experiences survey database. *Language, Speech, and Hearing Services in Schools*, 41(1):70–83.

Hao, M., Shu, H., Xing, A., and Li, P. (2008). Early vocabulary inventory for Mandarin Chinese. *Behavior Research Methods*, 40(3):728–733.

Hardwicke, T. E., Mathur, M., MacDonald, K., Nilsonne, G., Banks, G. C., Kidwell, M. C., Mohr, A. H., Clayton, E., Yoon, E. J., Tessler, M. H., et al. (2018). Data availability, reusability, and analytic reproducibility: Evaluating the impact of a mandatory open data policy at the journal *Cognition*. *Royal Society Open Science*, 5(8):180448.

Harner, L. (1975). Yesterday and tomorrow: Development of early understanding of the terms. *Developmental Psychology*, 11(6):864.

Hart, B., and Risley, T. (1995). *Meaningful differences in the everyday experience of young American children*. Brookes Publishing.

Hauser, M. D., Chomsky, N., and Fitch, W. T. (2002). The faculty of language: What is it, who has it, and how did it evolve? *Science*, 298(5598):1569–1579.

Hayiou-Thomas, M. E., Dale, P. S., and Plomin, R. (2012). The etiology of variation in language skills changes with development: A longitudinal twin study of language from 2 to 12 years. *Developmental Science*, 15(2):233–249.

Heilmann, J., Weismer, S. E., Evans, J., and Hollar, C. (2005). Utility of the MacArthur-Bates Communicative Development Inventory in identifying language abilities of late-talking and typically developing toddlers. *American Journal of Speech-Language Pathology*, 14(1):40–51.

Henrich, J., Heine, S. J., and Norenzayan, A. (2010). The weirdest people in the world? *Behavioral and Brain Sciences*, 33(2-3):101–102.

Hidaka, S. (2015). Estimating the latent number of types in growing corpora with reduced cost–accuracy trade-off. *Journal of Child Language*, 43(1):10–134.

Hills, T. T., Maouene, J., Riordan, B., and Smith, L. B. (2010). The associative structure of language: Contextual diversity in early word learning. *Journal of Memory and Language*, 63(3):259–273.

Hills, T. T., Maouene, M., Maouene, J., Sheya, A., and Smith, L. (2009). Longitudinal analysis of early semantic networks preferential attachment or preferential acquisition? *Psychological Science*, 20(6): 729–739.

Hirsh-Pasek, K., Adamson, L. B., Bakeman, R., Owen, M. T., Golinkoff, R. M., Pace, A., Yust, P. K., and Suma, K. (2015). The contribution of early communication quality to low-income children's language success. *Psychological Science*, 26(7):1071–1083.

Hoff, E. (2003). The specificity of environmental influence: Socioeconomic status affects early vocabulary development via maternal speech. *Child Development*, 74(5):1368–1378.

Hoff, E. (2006). How social contexts support and shape language development. *Developmental Review*, 26(1):55–88.

Hoff, E., Core, C., Place, S., Rumiche, R., Señor, M., and Parra, M. (2012). Dual language exposure and early bilingual development. *Journal of Child Language*, 39(1):1–27.

Hoff, E., and Naigles, L. (2002). How children use input to acquire a lexicon. *Child Development*, 73(2):418–433.

Hoff, E., Quinn, J. M., and Giguere, D. (2018). What explains the correlation between growth in vocabulary and grammar? new evidence from latent change score analyses of simultaneous bilingual development. *Developmental Science* 21(2):e12536.

Hoff-Ginsberg, E. (1998). The relation of birth order and socioeconomic status to children's language experience and language development. *Applied Psycholinguistics*, 19(4):603–629.

Hollich, G. J., Hirsh-Pasek, K., Golinkoff, R. M., Brand, R. J., Brown, E., Chung, H. L., Hennon, E., Rocroi, C., and Bloom, L. (2000). Breaking the language barrier: An emergentist coalition model for the origins of word learning. *Monographs of the Society for Research in Child Development*, 65(3):i–vi, 1–123.

Holman, E. W., Wichmann, S., Brown, C. H., Folia, V. V., Müller, A., and Bakker, D. (2008). Explorations in automated language classification. *Folia Linguistica*, 42:331–354.

Horowitz, A. C., Schneider, R. M., and Frank, M. C. (2017). The trouble with quantifiers: Exploring children's deficits in scalar implicature. *Child Development*, 89(6):e572–e593.

Huber, function" (Hambleton et al., 1991) and is S., Bookstein, F. L., and Fieder, M. (2010). Socioeconomic status, education, and reproduction in modern women: An evolutionary perspective. *American Journal of Human Biology*, 22(5):578–587.

Hurtado, N., Marchman, V., and Fernald, A. (2008). Does input influence uptake? Links between maternal talk, processing speed and vocabulary size in Spanish-learning children. *Developmental Science*, 11(6):F31–F39.

Huttenlocher, J., Haight, W., Bryk, A., Seltzer, M., and Lyons, T. (1991). Early vocabulary growth: Relation to language input and gender. *Developmental Psychology*, 27(2):236–248.

Huttenlocher, J., Vasilyeva, M., Cymerman, E., and Levine, S. (2002). Language input and child syntax. *Cognitive Psychology*, 45(3):337–374.

Hyde, J. S. and Linn, M. C. (1988). Gender differences in verbal ability: A meta-analysis. *Psychological Bulletin*, 104(1):53–69.

Inbar, Y. (2016). Association between contextual dependence and replicability in psychology may be spurious. *Proceedings of the National Academy of Sciences*, 113(34):E4933–E4934.

Iverson, J. M., Capirci, O., and Caselli, M. C. (1994). From communication to language in two modalities. *Cognitive Development*, 9(1):23–43.

Iverson, J. M., and Goldin-Meadow, S. (2005). Gesture paves the way for language development. *Psychological Science*, 16(5):367–371.

Jackson-Maldonado, D., Marchman, V. A., and Fernald, L. C. H. (2013). Short-form versions of the Spanish MacArthur–Bates Communicative Development Inventories. *Applied Psycholinguistics*, 34:837–868.

Jackson-Maldonado, D., Thal, D., Marchman, V., Bates, E., and Gutierrez-Clellen, V. (1993). Early lexical development in Spanish-speaking infants and toddlers. *Journal of Child Language*, 20(3):523–549.

Jakobson, R. (1960). Why "mama" and "papa"? In B. Kaplan and S. Wapner (Eds.), *Perspectives in psychological theory*. International Universities Press.

Jones, S., and Brandt, S. (2019). Do children really acquire dense neighbourhoods? *Journal of Child Language*, 46(6):1260–1273.

Jørgensen, R. N., Dale, P. S., Bleses, D., and Fenson, L. (2010). CLEX: A cross-linguistic lexical norms database. *Journal of Child Language*, 37(2):419–428.

Kachergis, G., Yu, C., and Shiffrin, R. M. (2012). An associative model of adaptive inference for learning word–referent mappings. *Psychonomic Bulletin & Review*, 19(2):317–324.

Kail, R. (1991). Developmental change in speed of processing during childhood and adolescence. *Psychological Bulletin*, 109(3):490.

Kamhi, A. G. (1986). The elusive first word: The importance of the naming insight for the development of referential speech. *Journal of Child Language*, 13(1):155–161.

Katsos, N., Cummins, C., Ezeizabarrena, M.-J., Gavarró, A., Kraljević, J. K., Hrzica, G., Grohmann, K. K., Skordi, A., De Lopez, K. J., Sundahl, L., et al. (2016). Cross-linguistic patterns in the acquisition of quantifiers. *Proceedings of the National Academy of Sciences*, 113(33):9244–9249.

Kauschke, C., and Hofmeister, C. (2002). Early lexical development in German: A study on vocabulary growth and vocabulary composition during the second and third year of life. *Journal of Child Language*, 29(4):735–757.

Kay, P., Berlin, B., Maffi, L., Merrifield, W. R., and Cook, R. (2009). *The world color survey*. CSLI Publications.

Kim, M., McGregor, K. K., and Thompson, C. K. (2000). Early lexical development in English-and Korean-speaking children: Language-general and language-specific patterns. *Journal of Child Language*, 27(2):225–254.

Kristoffersen, K. E., Simonsen, H. G., Bleses, D., Wehberg, S., Jørgensen, R. N., Eiesland, E. A., and Henriksen, L. Y. (2013). The use of the internet in collecting CDI data—An example from Norway. *Journal of Child Language*, 40(3):567–585.

Kuhl, P. (2004). Early language acquisition: Cracking the speech code. *Nature Reviews Neuroscience*, 5(11):831–843.

Kuhn, T. S. (1970). *The structure of scientific revolutions*. (2nd ed., pp. 84–85)) University ot Chicago Press.

Landau, B., Gleitman, L. R., and Landau, B. (2009). *Language and experience: Evidence from the blind child*. Harvard University Press.

Lany, J., and Saffran, J. R. (2010). From statistics to meaning: Infants acquisition of lexical categories. *Psychological Science*, 21(2):284–291.

Legate, J. A., and Yang, C. D. (2002). Empirical re-assessment of stimulus poverty arguments. *The Linguistic Review*, 18(1-2):151–162.

Levenshtein, V. I. (1966). Binary codes capable of correcting deletions, insertions, and reversals. *Cybernetics and Control Theory*, 10:707–710.

Libertus, M. E., Odic, D., Feigenson, L., and Halberda, J. (2015). A Developmental Vocabulary Assessment for Parents (DVAP): Validating parental report of vocabulary size in 2-to 7-year-old children. *Journal of Cognition and Development*, 16(3):442–454.

Lidz, J., Waxman, S., and Freedman, J. (2003). What infants know about syntax but couldn't have learned: Experimental evidence for syntactic structure at 18 months. *Cognition*, 89(3):295–303.

Lieven, E., Salomo, D., and Tomasello, M. (2009). Two-year-old children's production of multiword utterances: A usage-based analysis. *Cognitive Linguistics*, 20(3):481–507.

Lieven, E. V., Pine, J. M., and Baldwin, G. (1997). Lexically-based learning and early grammatical development. *Journal of Child Language*, 24(1):187–219.

Lieven, E. V., Pine, J. M., and Barnes, H. D. (1992). Individual differences in early vocabulary development: Redefining the referential-expressive distinction. *Journal of Child Language*, 19(2):287–310.

Liszkowski, U., Carpenter, M., and Tomasello, M. (2007). Pointing out new news, old news, and absent referents at 12 months of age. *Developmental Science*, 10(2):F1–F7.

Lutchmaya, S., Baron-Cohen, S., and Raggatt, P. (2001). Foetal testosterone and vocabulary size in 18-and 24-month-old infants. *Infant Behavior and Development,* 24(4):418–424.

Luyster, R., Lopez, K., and Lord, C. (2007). Characterizing communicative development in children referred for autism spectrum disorders using the MacArthur-Bates Communicative Development Inventory (CDI). *Journal of Child Language,* 34(3):623–654.

Maccoby, E. E. and Jacklin, C. N. (1974). *The psychology of sex differences.* Stanford University Press.

MacWhinney, B. (2000). *The CHILDES project: Tools for analyzing talk.* (3rd ed.). Lawrence Erlbaum.

Maital, S. L., Dromi, E., Sagi, A., and Bornstein, M. H. (2000). The Hebrew Communicative Development Inventory: Language specific properties and cross-linguistic generalizations. *Journal of Child Language,* 27(1):43–67.

Majid, A. (2010). Words for parts of the body. In B. Malt and P. Wolff (Eds.) *Words and the mind: How words capture human experience* (pp. 58–71). Oxford University Press.

Makransky, G., Dale, P. S., Havmose, P., and Bleses, D. (2016). An item response theory–based, computerized adaptive testing version of the MacArthur–Bates Communicative Development Inventory: Words & Sentences (CDI: WS). *Journal of Speech, Language, and Hearing Research,* 59(2):281–289.

Maratsos, M. (1993). Artifactual overregularizations? In E. Clark (Ed.), *Proceedings of the 24th Annual Child Language Research Forum* (pp. 139–148). Center for the Study of Language and Information.

Marchman, V. A. (1997). Children's productivity in the English past tense: The role of frequency, phonology, and neighborhood structure. *Cognitive Science,* 21(3):283–304.

Marchman, V. A., and Bates, E. (1994). Continuity in lexical and morphological development: A test of the critical mass hypothesis. *Journal of Child Language,* 21(2):339–366.

Marchman, V. A., and Dale, P. S. (2017). Assessing receptive and expressive vocabulary in child language. In A. M. B. de Groot and P. Hagoort (Eds.), *Research Methods in Psycholinguistics and the Neurobiology of Language: A Practical Guide* (p. 40). Wiley.

Marchman, V. A., and Fernald, A. (2008). Speed of word recognition and vocabulary knowledge in infancy predict cognitive and language outcomes in later childhood. *Developmental Science,* 11(3): F9–F16.

Marchman, V. A., and Martínez-Sussmann, C. (2002). Concurrent validity of caregiver/parent report measures of language for children who are learning both English and Spanish. *Journal of Speech, Language, and Hearing Research,* 45(5):983–997.

Marchman, V. A., Martínez-Sussmann, C., and Dale, P. S. (2004). The language-specific nature of grammatical development: Evidence from bilingual language learners. *Developmental Science,* 7(2):212–224.

Marchman, V. A., Wulfeck, B., and Weismer, S. E. (1999). Morphological productivity in children with normal language and SLI: A study of the English past tense. *Journal of Speech, Language, and Hearing Research,* 42(1):206–219.

Marcus, G. (1995). The acquisition of the English past tense in children and multilayered connectionist networks. *Cognition,* 56(3):271–279.

Marcus, G. F., Pinker, S., Ullman, M., Hollander, M., Rosen, T. J., Xu, F., and Clahsen, H. (1992). Overregularization in language acquisition. *Monographs of the Society for Research in Child Development,* 57(4):i–vi, 1–178.

Mariscal, S., and Gallego, C. (2012). The relationship between early lexical and grammatical development in Spanish: Evidence in children with different linguistic levels. *The Spanish Journal of Psychology,* 15(1):112–123.

Marjanovič-Umek, L., Fekonja-Peklaj, U., and Podlesek, A. (2013). Characteristics of early vocabulary and grammar development in Slovenian-speaking infants and toddlers: A CDI-adaptation study. *Journal of Child Language,* 40(4):779–798.

Markman, E. M. (1990). Constraints children place on word meanings. *Cognitive Science,* 14(1):57–77.

Markus, H. R. and Kitayama, S. (1991). Culture and the self: Implications for cognition, emotion, and motivation. *Psychological Review,* 98(2):224.

Masur, E. F. (1990). Gestural development, dual-directional signaling, and the transition to words. In V. Volterra and C. Erting (Eds.), *From gesture to language in hearing and deaf children* (pp. 18–30). Springer.

Mayor, J., and Mani, N. (2018). A short version of the MacArthur–Bates Communicative Development Inventories with high validity. *Behavior Research Methods*, 51(5):2248–2255.

Mayor, J., and Plunkett, K. (2011). A statistical estimate of infant and toddler vocabulary size from CDI analysis. *Developmental Science*, 14(4):769–785.

Mayor, J., and Plunkett, K. (2014). Shared understanding and idiosyncratic expression in early vocabularies. *Developmental Science*, 17(3):412–423.

McDonald, J. L., and Roussel, C. C. (2010). Past tense grammaticality judgment and production in non-native and stressed native English speakers. *Bilingualism: Language and Cognition*, 13(4):429–448.

McMurray, B. (2007). Defusing the childhood vocabulary explosion. *Science*, 317(5838):631–631.

McMurray, B., Horst, J. S., and Samuelson, L. K. (2012). Word learning emerges from the interaction of online referent selection and slow associative learning. *Psychological Review*, 119(4):831.

Medina, T., Snedeker, J., Trueswell, J., and Gleitman, L. (2011). How words can and cannot be learned by observation. *Proceedings of the National Academy of Sciences*, 108(22):9014.

Meylan, S. C., Braginsky, M., deMayo, B., Sanchez, A., Schonberg, C., Srinivasan, M., Vlach, H., Lupyan, G., Griffiths, T., and Frank, M. C. (2019). Wordful: Tracking early productive vocabulary growth with smartphones. [Poster session]. The 44th Annual Boston University Conference on Language Development, Boston, MA, United States.

Meylan, S. C., Frank, M. C., Roy, B. C., and Levy, R. (2017). The emergence of an abstract grammatical category in children's early speech. *Psychological Science*, 28(2):181–192.

Miller, D. I., and Halpern, D. F. (2014). The new science of cognitive sex differences. *Trends in Cognitive Sciences*, 18(1):37–45.

Mindell, J. A., Leichman, E. S., Composto, J., Lee, C., Bhullar, B., and Walters, R. M. (2016). Development of infant and toddler sleep patterns: Real-world data from a mobile application. *Journal of Sleep Research*, 25(5):508–516.

Mintz, T. H. (2003). Frequent frames as a cue for grammatical categories in child directed speech. *Cognition*, 90(1):91–117.

Mollica, F., and Piantadosi, S. T. (2017). How data drive early word learning: A cross-linguistic waiting time analysis. *Open Mind*, 1(2):67–77.

Moors, A., De Houwer, J., Hermans, D., Wanmaker, S., Van Schie, K., Van Harmelen, A.-L., De Schryver, M., De Winne, J., and Brysbaert, M. (2013). Norms of valence, arousal, dominance, and age of acquisuisition for 4,300 Dutch words. *Behavior Research Methods*, 45(1):169–177.

Moyle, M. J., Weismer, S. E., Evans, J. L., and Lindstrom, M. J. (2007). Longitudinal relationships between lexical and grammatical development in typical and late-talking children. *Journal of Speech, Language, and Hearing Research*, 50(2):508–528.

Munafò, M. R., Nosek, B. A., Bishop, D. V., Button, K. S., Chambers, C. D., du Sert, N. P., Simonsohn, U., Wagenmakers, E.-J., Ware, J. J., and Ioannidis, J. P. (2017). A manifesto for reproducible science. *Nature Human Behaviour*, 1:0021.

Naigles, L. (1990). Children use syntax to learn verb meanings. *Journal of Child Language*, 17(2):357–374.

Naroll, R. (1965). Galton's problem: The logic of cross-cultural analysis. *Social Research*, 32(4):428–451.

Nelson, C. A. (2007). A neurobiological perspective on early human deprivation. *Child Development Perspectives*, 1(1):13–18.

Nelson, K. (1973). Structure and strategy in learning to talk. *Monographs of the Society for Research in Child Development*, 38(1/2):1–135.

Nisbett, R. E., Peng, K., Choi, I., and Norenzayan, A. (2001). Culture and systems of thought: Holistic versus analytic cognition. *Psychological Review*, 108(2):291.

Nordmeyer, A. E., and Frank, M. C. (2014). The role of context in young children's comprehension of negation. *Journal of Memory and Language*, 77:25–39.

Nordmeyer, A. E. and Frank, M. C. (2018). *Uninformative negation is infelicitous to both adults and children.* Early understanding of pragmatic principles in children's judgments of negative sentences. *Language Learning and Development* 14(4): 262–278.

Nosek, B. A., Smyth, F. L., Sriram, N., Lindner, N. M., Devos, T., Ayala, A., Bar-Anan, Y., Bergh, R., Cai, H., Gonsalkorale, K., et al. (2009). National differences in gender–science stereotypes predict national sex differences in science and math achievement. *Proceedings of the National Academy of Sciences*, 106(26):10593–10597.

Open Science Collaboration. (2015). Estimating the reproducibility of psychological science. *Science*, 349(6251):aac4716.

Oshima-Takane, Y., Goodz, E., and Derevensky, J. L. (1996). Birth order effects on early language development: Do secondborn children learn from overheard speech? *Child Development*, 67(2):621–634.

Paul, R. (1996). Clinical implications of the natural history of slow expressive language development. *American Journal of Speech-Language Pathology*, 5(2):5–21.

Pea, R. D. (1982). Origins of verbal logic: Spontaneous denials by two-and three-year olds. *Journal of Child Language*, 9(3):597–626.

Pearson, B. Z., Fernandez, S. C., and Oller, D. K. (1993). Lexical development in bilingual infants and toddlers: Comparison to monolingual norms. *Language Learning*, 43(1):93–120.

Pejovic, V., Lathia, N., Mascolo, C., and Musolesi, M. (2016). Mobile-based experience sampling for behaviour research. In M. Tkalčič, B. De Carolis, M. de Gemmis, A. Odić, and A. Košir (Eds.), *Emotions and personality in personalized services* (pp. 141–161). Springer.

Pérez-Leroux, A. T., Castilla-Earls, A. P., and Brunner, J. (2012). General and specific effects of lexicon in grammar: Determiner and object pronoun omissions in child Spanish. *Journal of Speech, Language, and Hearing Research*, 55(2):313–327.

Perry, L. K., Perlman, M., and Lupyan, G. (2015). Iconicity in English and Spanish and its relation to lexical category and age of acquisition. *PLOS One*, 10(9):e0137147.

Pham-Gia, T. and Hung, T. (2001). The mean and median absolute deviations. *Mathematical and Computer Modelling*, 34(7–8):921–936.

Piaget, J. (1962). *Play, dreams and imitation in childhood*. Routledge.

Piantadosi, S. T., and Gibson, E. (2014). Quantitative standards for absolute linguistic universals. *Cognitive Science*, 38(4):736–756.

Pine, J. M., and Lieven, E. V. (1990). Referential style at thirteen months: Why age-defined cross-sectional measures are inappropriate for the study of strategy differences in early language development. *Journal of Child Language*, 17(3):625–631.

Pine, J. M., and Lieven, E. V. (1997). Slot and frame patterns and the development of the determiner category. *Applied Psycholinguistics*, 18(2):123–138.

Pinker, S. (1991). Rules of language. *Science*, 253(5019):530–535.

Pinker, S. (1998). Words and rules. *Lingua*, 106(1–4):219–242.

Pinker, S., and Jackendoff, R. (2005). The faculty of language: What's special about it? *Cognition*, 95(2):201–236.

Pinker, S., and Prince, A. (1988). On language and connectionism: Analysis of a parallel distributed processing model of language acquisition. *Cognition*, 28(1–2):73–193.

Plunkett, K., and Marchman, V. (1989). *Pattern association in a back propagation network: Implications for child language acquisition* (Technical report 8902). Center for Research in Language.

Plunkett, K., and Marchman, V. (1991). U-shaped learning and frequency effects in a multi-layered perception: Implications for child language acquisition. *Cognition*, 38(1):43–102.

Plunkett, K. and Marchman, V. (1993). From rote learning to system building: Acquiring verb morphology in children and connectionist nets. *Cognition*, 48(1):21–69.

Plunkett, K., and Marchman, V. (1996). Learning from a connectionist model of the acquisition of the English past tense. *Cognition*, 61(3):299–308.

Plunkett, K., Sinha, C., Møller, M. F., and Strandsby, O. (1992). Symbol grounding or the emergence of symbols? Vocabulary growth in children and a connectionist net. *Connection Science*, 4(3–4):293–312.

Poulin-Dubois, D., Graham, S., and Sippola, L. (1995). Early lexical development: The contribution of parental labelling and infants' categorization abilities. *Journal of Child Language*, 22(2):325–343.

Pullum, G. K. and Scholz, B. C. (2002). Empirical assessment of stimulus poverty arguments. *The Linguistic Review*, 18(1–2):9–50.

Quine, W. (1960). *Word and object*. MIT Press.

Redington, M., Crater, N., and Finch, S. (1998). Distributional information: A powerful cue for acquiring syntactic categories. *Cognitive Science*, 22(4):425–469.

Reese, E., and Read, S. (2000). Predictive validity of the New Zealand MacArthur Communicative Development Inventory: Words and sentences. *Journal of Child Language*, 27(2):255–266.

Regier, T., and Gahl, S. (2004). Learning the unlearnable: The role of missing evidence. *Cognition*, 93(2):147–155.

Rescorla, L. (1989). The language development surveya screening tool for delayed language in toddlers. *Journal of Speech and Hearing Disorders*, 54(4):587–599.

Rescorla, L. (2009). Age 17 language and reading outcomes in late-talking toddlers: Support for a dimensional perspective on language delay. *Journal of Speech, Language, and Hearing Research*, 52(1):16–30.

Rescorla, L., Dahlsgaard, K., and Roberts, J. (2000). Late-talking toddlers: MLU and IPSyn outcomes at 3;0 and 4;0. *Journal of Child Language*, 27(3):643–664.

Rescorla, L., Roberts, J., and Dahlsgaard, K. (1997). Late talkers at 2: Outcome at age 3. *Journal of Speech, Language, and Hearing Research*, 40(3):556–566.

Roberts, J. E., Burchinal, M., and Durham, M. (1999). Parents' report of vocabulary and grammatical development of African American preschoolers: Child and environmental associations. *Child Development*, 70(1):92–106.

Robinson, J. P., and Lubienski, S. T. (2011). The development of gender achievement gaps in mathematics and reading during elementary and middle school: Examining direct cognitive assessments and teacher ratings. *American Educational Research Journal*, 48(2):268–302.

Rogosa, D. (1980). A critique of cross-lagged correlation. *Psychological Bulletin*, 88(2):245.

Ross-Sheehy, S., Oakes, L. M., and Luck, S. J. (2003). The development of visual short-term memory capacity in infants. *Child Development*, 74(6):1807–1822.

Rovee-Collier, C. (1997). Dissociations in infant memory: Rethinking the development of implicit and explicit memory. *Psychological Review*, 104(3):467.

Rowe, M. L. (2012). A longitudinal investigation of the role of quantity and quality of child-directed speech in vocabulary development. *Child Development*, 83(5):1762–1774.

Rowe, M. L., and Goldin-Meadow, S. (2009). Early gesture selectively predicts later language learning. *Developmental Science*, 12(1):182–187.

Rowe, M. L., Suskind, D. L., and Hoff, E. (2013). *Early language gaps: Sources and solutions*. Bridge the Word Gap. https://bridgethewordgap.files.wordpress.com/2013/10/white-paper-rowe_suskind_hoff-9-26-2013.pdf

Roy, B. C., Frank, M. C., DeCamp, P., Miller, M., and Roy, D. (2015). Predicting the birth of a spoken word. *Proceedings of the National Academy of Sciences*, 112(41):12663–12668.

Roy, D., and Pentland, A. (2002). Learning words from sights and sounds: A computational model. *Cognitive Science*, 26:113–146.

Rumelhart, D. E., and McClelland, J. L. (1986). On learning the past tenses of English verbs. In *Parallel distributed processing: explorations in the microstructure of cognition, Vol. 2. Psychological and biological models* (pp. 216–271). MIT Press.

Saffran, J. R., Aslin, R., and Newport, E. (1996). Statistical learning by 8-month-old infants. *Science*, 274(5294):1926.

Saffran, J. R., and Kirkham, N. Z. (2018). Infant statistical learning. *Annual Review of Psychology*, 69 (1):1–23.

Schneider, R., Yurovsky, D., and Frank, M. C. (2015). *Large-scale investigations of variability in children's first words*. Proceedings of the Cognitive Science Society, Pasadena, CA, United States.

Schults, A., and Tulviste, T. (2016). Composition of Estonian infants expressive lexicon according to the adaptation of CDI/Words and Gestures. *First Language*, 36(5):485–504.

Schwartz, R. G., and Terrell, B. Y. (1983). The role of input frequency in lexical acquisuisition. *Journal of Child Language,* 10(01):57–64.

Shafto, P., Goodman, N. D., and Frank, M. C. (2012). Learning from others: The consequences of psychological reasoning for human learning. *Perspectives on Psychological Science,* 7(4):341–351.

Shneidman, L. A., and Goldin-Meadow, S. (2012). Language input and acquisition in a Mayan village: How important is directed speech? *Developmental Science,* 15(5):659–673.

Shukla, M., White, K. S., and Aslin, R. N. (2011). Prosody guides the rapid mapping of auditory word forms onto visual objects in 6-mo-old infants. *Proceedings of the National Academy of Sciences,* 108(15):6038–6043.

Singer Harris, N. G., Bellugi, U., Bates, E., Jones, W., and Rossen, M. (1997). Contrasting profiles of language development in children with Williams and Down Syndromes. *Developmental Neuropsychology,* 13(3):345–370.

Siskind, J. (1996). A computational study of cross-situational techniques for learning word-to-meaning mappings. *Cognition,* 61:39–91.

Slobin, D. I. (1970). *Suggested universals in the ontogenesis of grammar* (Working Paper 32). Language-Behavior Research Laboratory.

Slobin, D. I. (1973). Cognitive prerequisites for the development of grammar. In *Studies of child language development* (pp. 175–208). Holt, Rinehart, & Winston.

Slobin, D. I. (1985). *The crosslinguistic study of language acquisition: Theoretical issues* (Vol. 2). Psychology Press.

Slobin, D. I. (1996). From "thought and language" to "thinking for speaking." In J. J. Gumperz and S. C. Levinson (Eds.), *Rethinking Linguistic Relativity* (pp. 70–96). Cambridge University Press.

Smith, L., and Yu, C. (2008). Infants rapidly learn word-referent mappings via cross-situational statistics. *Cognition,* 106(3):1558–1568.

Smith, L. B., and Jones, S. S. (2011). Symbolic play connects to language through visual object recognition. *Developmental Science,* 14(5):1142–1149.

Snedeker, J. (2009). Word learning. In L. Squire (Ed.) *Encyclopedia of Neuroscience* (pp. 503–508). Elsevier.

Snedeker, J., Geren, J., and Shafto, C. L. (2007). Starting over: International adoption as a natural experiment in language development. *Psychological Science,* 18(1):79–87.

Snedeker, J., Geren, J., and Shafto, C. L. (2012). Disentangling the effects of cognitive development and linguistic expertise: A longitudinal study of the acquisition of English in internationally-adopted children. *Cognitive Psychology,* 65(1):39–76.

Spelke, E. S. and Kinzler, K. D. (2007). Core knowledge. *Developmental Science,* 10(1):89–96.

Stager, C. L. and Werker, J. F. (1997). Infants listen for more phonetic detail in speech perception than in word-learning tasks. *Nature,* 388(6640):381.

Steyvers, M., and Benjamin, A. S. (2019). The joint contribution of participation and performance to learning functions: Exploring the effects of age in large-scale data sets. *Behavior Research Methods,* 51(4):1531–1543.

Stivers, T., Enfield, N. J., Brown, P., Englert, C., Hayashi, M., Heinemann, T., Hoymann, G., Rossano, F., De Ruiter, J. P., Yoon, K.-E., et al. (2009). Universals and cultural variation in turn-taking in conversation. *Proceedings of the National Academy of Sciences,* 106(26):10587–10592.

Stoet, G., and Geary, D. C. (2013). Sex differences in mathematics and reading achievement are inversely related: Within- and across-nation assessment of 10 years of PISA data. *PLOS One,* 8(3):e57988.

Stokes, S. F. (2010). Neighborhood density and word frequency predict vocabulary size in toddlers. *Journal of Speech, Language, and Hearing Research,* 53(3):670–683.

Stolt, S., Haataja, L., Lapinleimu, H., and Lehtonen, L. (2009). Associations between lexicon and grammar at the end of the second year in Finnish children. *Journal of Child Language,* 36(4): 779–806.

Storkel, H. L. (2004). Do children acquire dense neighborhoods? An investigation of similarity neighborhoods in lexical acquisition. *Applied Psycholinguistics,* 25(2):201–221.

Størvold, G. V., Aarethun, K., and Bratberg, G. H. (2013). Age for onset of walking and prewalking strategies. *Early Human Development*, 89(9):655–659.

Swingley, D., and Aslin, R. N. (2007). Lexical competition in young children's word learning. *Cognitive Psychology*, 54(2):99–132.

Swingley, D., and Humphrey, C. (2018). Quantitative linguistic predictors of infants' learning of specific English Words. *Child Development*, 89(4):1247–1267.

Szagun, G., Steinbrink, C., Franik, M., and Stumper, B. (2006). Development of vocabulary and grammar in young German-speaking children assessed with a German language development inventory. *First Language*, 26(3):259–280.

Tager-Flusberg, H., Rogers, S., Cooper, J., Landa, R., Lord, C., Paul, R., Rice, M., Stoel-Gammon, C., Wetherby, A., and Yoder, P. (2009). Defining spoken language benchmarks and selecting measures of expressive language development for young children with autism spectrum disorders. *Journal of Speech, Language, and Hearing Research*, 52(3):643–652.

Tardif, T. (1996). Nouns are not always learned before verbs: Evidence from Mandarin speakers' early vocabularies. *Developmental Psychology*, 32(3):492.

Tardif, T., Fletcher, P., Liang, W., and Kaciroti, N. (2009). Early vocabulary development in Mandarin (Putonghua) and Cantonese. *Journal of Child Language*, 36(5):1115–1144.

Tardif, T., Fletcher, P., Liang, W., Zhang, Z., Kaciroti, N., and Marchman, V. A. (2008). Baby's first 10 words. *Developmental Psychology*, 44(4):929.

Tardif, T., Gelman, S. A., and Xu, F. (1999). Putting the noun bias in context: A comparison of English and Mandarin. *Child Development*, 70(3):620–635.

Tardif, T., Shatz, M., and Naigles, L. (1997). Caregiver speech and children's use of nouns versus verbs: A comparison of English, Italian, and Mandarin. *Journal of Child Language*, 24(3):535–565.

Thal, D., and Bates, E. (1988). Language and gesture in late talkers. *Journal of Speech, Language, and Hearing Research*, 31(1):115–123.

Thal, D., Jackson-Maldonado, D., and Acosta, D. (2000). Validity of a parent-report measure of vocabulary and grammar for spanish-speaking toddlers. *Journal of Speech, Language, and Hearing Research*, 43(5):1087–1100.

Thal, D., Tobias, S., and Morrison, D. (1991). Language and gesture in late talkers: A 1-year follow-up. *Journal of Speech, Language, and Hearing Research*, 34(3):604–612.

Thal, D. J., Bates, E., Goodman, J., and Jahn-Samilo, J. (1997). Continuity of language abilities: An exploratory study of late-and early-talking toddlers. *Developmental Neuropsychology*, 13(3):239–273.

Thal, D. J., Bates, E., Zappia, M. J., and Oroz, M. (1996). Ties between lexical and grammatical development: Evidence from early-talkers. *Journal of Child Language*, 23(2):349–368.

Thal, D. J., O'Hanlon, L., Clemmons, M., and Fralin, L. (1999). Validity of a parent report measure of vocabulary and syntax for preschool children with language impairment. *Journal of Speech, Language, and Hearing Research*, 42(2):482–496.

Thordardottir, E. T., Weismer, S. E., and Evans, J. L. (2002). Continuity in lexical and morphological development in Icelandic and English-speaking 2-year-olds. *First Language*, 22(1):3–28.

Tillman, K. A., and Barner, D. (2015). Learning the language of time: Children's acquisition of duration words. *Cognitive Psychology*, 78:57–77.

Tillman, K. A., Marghetis, T., Barner, D., and Srinivasan, M. (2017). Today is tomorrow's yesterday: Children's acquisition of deictic time words. *Cognitive Psychology*, 92:87–100.

Tincoff, R., and Jusczyk, P. W. (1999). Some beginnings of word comprehension in 6-month-olds. *Psychological Science*, 10(2):172–175.

Tincoff, R., and Jusczyk, P. W. (2012). Six-month-olds comprehend words that refer to parts of the body. *Infancy*, 17(4):432–444.

Tomasello, M. (2000). Do young children have adult syntactic competence? *Cognition*, 74(3):209–253.

Tomasello, M. (2003). *Constructing a language: A usage-based theory of language acquisition*. Harvard University Press.

Tomasello, M. (2010). *Origins of human communication*. MIT Press.

Tomasello, M., and Mannle, S. (1985). Pragmatics of sibling speech to one-year-olds. *Child Development*, 56(4):911–917.

Tomasello, M., and Mervis, C. B. (1994). The instrument is great, but measuring comprehension is still a problem. *Monographs of the Society for Research in Child Development*, 59(5):174–179.

Tomasello, M., and Todd, J. (1983). Joint attention and lexical acquisition style. *First Language*, 4(12): 197–211.

Trueswell, J. C., Medina, T. N., Hafri, A., and Gleitman, L. R. (2013). Propose but verify: Fast mapping meets cross-situational word learning. *Cognitive Psychology*, 66(1):126–156.

Trzaskowski, M., Harlaar, N., Arden, R., Krapohl, E., Rimfeld, K., McMillan, A., Dale, P. S., and Plomin, R. (2014). Genetic influence on family socioeconomic status and children's intelligence. *Intelligence*, 42:83–88.

Tsao, F.-M., Liu, H.-M., and Kuhl, P. K. (2004). Speech perception in infancy predicts language development in the second year of life: A longitudinal study. *Child Development*, 75(4):1067–1084.

Valian, V. (1986). Syntactic categories in the speech of young children. *Developmental Psychology*, 22(4):562.

Van Bavel, J. J., Mende-Siedlecki, P., Brady, W. J., and Reinero, D. A. (2016). Contextual sensitivity in scientific reproducibility. *Proceedings of the National Academy of Sciences*, 113(23):6454–6459.

Vihman, M. M. (1996). *Phonological development: The origins of language in the child*. Blackwell.

Vlach, H. A., and Johnson, S. P. (2013). Memory constraints on infants' cross-situational statistical learning. *Cognition*, 127(3):375–382.

Volterra, V., and Caselli, M. C. (1985). From gestures and vocalizations to signs and words. In W. Stokoe and V. Volterra (Eds.), *SLR '83: Proceedings of the III international symposium on sign language research* (pp. 1–9). Rome, Italy.

Vygotsky, L. S. (1980). *Mind in society: The development of higher psychological processes*. Harvard University Press.

Wagner, K., Dobkins, K., and Barner, D. (2013). Slow mapping: Color word learning as a gradual inductive process. *Cognition*, 127(3):307–317.

Warriner, A. B., Kuperman, V., and Brysbaert, M. (2013). Norms of valence, arousal, and dominance for 13,915 English lemmas. *Behavior Research Methods*, 45(4):1191–1207.

Weber, A. M., Marchman, V. A., Diop, Y., and Fernald, A. (2018). Validity of caregiver-report measures of language skill for Wolof-learning infants and toddlers living in rural African villages. *Journal of Child Language*, 45(4):939–958.

Weisleder, A., and Fernald, A. (2013). Talking to children matters: Early language experience strengthens processing and builds vocabulary. *Psychological Science*, 24(11):2143–2152.

Werner, H., and Kaplan, B. (1963). *Symbol formation: An organismic-developmental approach to language and the expression of thought*. Lawrence Erlbaum Associates.

Wexler, K. (1998). Very early parameter setting and the unique checking constraint: A new explanation of the optional infinitive stage. *Lingua*, 106(1–4):23–79.

Wichmann, S., Holman, E. W., Bakker, D., and Brown, C. H. (2010). Evaluating linguistic distance measures. *Physica A*, 389(17):3632–3639.

Xu, F., and Tenenbaum, J. (2007). Word learning as Bayesian inference. *Psychological Review*, 114:245.

Yang, C. D. (2002). *Knowledge and learning in natural language*. Oxford University Press.

Yang, C. D. (2004). Universal grammar, statistics or both? *Trends in Cognitive Sciences*, 8(10):451–456.

Yang, C. D. (2013). Ontogeny and phylogeny of language. *Proceedings of the National Academy of Sciences*, 110(16):6324–6327.

Yang, C. D. (2016). *The price of linguistic productivity: How children learn to break the rules of language*. MIT Press.

Yu, C., and Ballard, D. H. (2007). A unified model of early word learning: Integrating statistical and social cues. *Neurocomputing*, 70(13):2149–2165.

Yu, C., and Smith, L. (2007). Rapid word learning under uncertainty via cross-situational statistics. *Psychological Science*, 18(5):414–420.

Yurovsky, D., and Frank, M. C. (2015). An integrative account of constraints on cross-situational learning. *Cognition*, 145:53–62.

Yurovsky, D., Smith, L. B., and Yu, C. (2013). Statistical word learning at scale: The baby's view is better. *Developmental Science*, 16(6):959–966.

Yurovsky, D., Wagner, K., Barner, D., and Frank, M. C. (2015). Signatures of domain-general categorization mechanisms in color word learning. In D. C. Noelle, R. Dale, A. S. Warlaumont, T. Matlock, C. D. Jennings, and P. P. Maglio (Eds.), *Proceedings of the 37th annual meeting of the cognitive science society* (pp. 2775–2780). Cognitive Science Society.

Zajonc, R. B., and Markus, G. B. (1975). Birth order and intellectual development. *Psychological Review*, 82(1):74.

Zipf, G. K. (1935). The psycho-biology of language. Houghton, Mifflin.

Index

Page numbers in *italics* refer to figures and tables.